Experimental Fluid Mechanics

More information about this series at http://www.springer.com/series/3837

Michael Reader-Harris

Orifice Plates and Venturi Tubes

Springer

Michael Reader-Harris
NEL
Glasgow
UK

ISSN 1613-222X ISSN 2197-9510 (electronic)
Experimental Fluid Mechanics
ISBN 978-3-319-35943-4 ISBN 978-3-319-16880-7 (eBook)
DOI 10.1007/978-3-319-16880-7

Springer Cham Heidelberg New York Dordrecht London

Softcover reprint of the hardcover 1st edition 2015

Printed on acid-free paper

Springer International Publishing AG Switzerland is part of Springer Science+Business Media (www.springer.com)

Preface

Differential-pressure flowmeters are low-cost and robust and can achieve high accuracy. The majority of industrial flowmeters are based on the differential-pressure principle. The increasing use of diagnostics will help ensure that the use of differential-pressure flowmeters continues to grow.

For those who want to use an orifice plate or a Venturi tube, the ISO (International Organization for Standardization) standard (ISO 5167) and its associated Technical Reports give the instructions required. However, they rarely tell the users why they should follow certain instructions. The aim of this book is to help users of the ISO standards for orifice plates and Venturi tubes to understand the reasons why the standards are as they are, to apply them effectively and to understand the consequences of deviations from the standards. So this book gives the background to differential-pressure flow measurement and then in Chaps. 2–4 goes through the requirements explaining the reason for them.

The money involved in differential-pressure flow measurement is enormous. In the UK, most natural gas is measured at least once using an orifice plate: each year perhaps £25 billion of natural gas is so metered with an uncertainty of about £250 million. So the actual equations are of immense importance: these are covered in Chaps. 5 and 6 (and Chap. 7 covers the discharge coefficient for Venturi tubes at high Reynolds number).

High accuracy depends on correct installation: Chapter 8 covers installation requirements. Chapter 9 covers the performance of flow nozzles, particularly in order to help explain the performance of Venturi tubes. Dry-gas flows with occasional addition of liquid may be handled using orifice plates with drain holes: see Chap. 10. Wet-gas flow through Venturi tubes and orifice plates is covered in Chap. 11. Although proprietary designs of wet-gas and multiphase flowmeters are not covered in this book, most of them include a Venturi tube: the number of these meters has increased rapidly; this trend is expected to continue, and a fundamental understanding of Venturi tubes is essential for their continued development.

Sir Cyril Hinshelwood said that ‘in the 19th century fluid dynamicists were divided into hydraulic engineers who observed what could not be explained and mathematicians who explained things that could not be observed' (Lighthill, *Nature:* 178 No 4529: 343 18th Aug 1956). This book aims to provide physically well-founded equations that describe the actual experience.

It is expected that many of the users of this book will be from the oil and gas industries: accordingly, although equations require consistent units (most conveniently SI), nominal pipe sizes (and nominal nozzle diameters in Chap. 9) are in inches, static pressures are often given in bar and differential pressure in mbar.

This book will be particularly useful for members of standards committees: one of the difficulties in revising standards is to know where the data can be found on which a clause in the standard is based or whether the clause is based solely on engineering judgment. If there are data, a reviser must follow the original data (unless there are better data); if there are no data, modern engineering judgment may, perhaps, replace the older judgment. The author of this book has sought to ensure that any revisions to the ISO standards are adequately referenced; however, earlier versions of the standards had an absence of references: this book will be invaluable for members of standards committees in the future.

Following ISO 5167, this book only covers flow in circular pipes. Except where wet-gas flow is discussed, the pipes must run full.

The author is grateful to NEL (formerly the National Engineering Laboratory), for which he has worked for over 33 years and which provided some time for him to write this book, and to NMO (the UK National Measurement Office) and the Engineering and Flow Programme and their predecessors, which supported a significant part of the research described here, and many others, including Amoco, BHRA, BP, British Gas, CEESI, CNR, ConocoPhillips, Gasunie, NIST, Nova, Shell and SwRI, whose data and/or support have been particularly important to this work.

He is grateful to the following colleagues who read the book in draft form and took part in 'book clubs' at NEL: Tariq Boussouara, Neil Bowman, David Crawford, Christina Emans, Emmelyn Graham, Calum Hardie, Asaad Kenbar, Liz Lee, Terri Leonard, Gordon Lindsay, Marc MacDonald, Alick MacGillivray, Craig Marshall, Chris Mills, Ian Nicholson and Neil Ramsay.

He is particularly grateful to friends who read the book and sent detailed comments on it: Alistair Collins (Solartron ISA), David Irvine (previously of British Gas), David Michael (BSI) and Richard Paton (previously of NEL, now at Petrofac). Any remaining errors are of course his own.

He is very grateful to his wife, Susan, who has put up with piles of papers spread through the house and with a husband whose life has been increasingly dominated by writing this book.

He gives thanks to the Lord for a satisfying career exploring a fascinating part of His creation: 'Measuring lines have fallen unto me in pleasant places; indeed I have a beautiful inheritance' (Psalm 16 v 6).

Contents

Notations

A	$=(19000\beta/Re_D)^{0.8}$
A (*with a subscript*)	Area of pipe or throat
b	Bevel width ($=E - e$)
c	Pressure-tapping chamber width (see Sect. 7.3.2)
C	Discharge coefficient
C_h	Discharge coefficient for a drain hole
C_{notap}	Discharge coefficient without tappings
C_p	Heat capacity at constant pressure
c_s	Speed of sound in the fluid
C_v	Heat capacity at constant volume
C_{water}	Mean discharge coefficient in water
d	Diameter of orifice or throat of Venturi tube or nozzle at working conditions
D	Upstream internal pipe diameter (diameter of entrance cylinder for a Venturi tube) at working conditions
d'	Corrected orifice diameter to allow for the additional flowrate due to a drain hole
d_0	Diameter of orifice or throat of Venturi tube or nozzle at calibration conditions
d_c	Corrected orifice diameter to allow for the additional flowrate due to a drain hole following ISO/TR 15377:2007
d_h	Diameter of drain hole
D_r	Upstream internal pipe diameter at reference conditions
d_{tap}	Tapping diameter
e	Orifice (bore) thickness
E	Orifice plate thickness
E_i	Internal energy per unit mass
e_p	Increase in measured pressure
f	Pulsation frequency
$f_{n'}$	Acoustic resonance frequency
g	Acceleration due to gravity

h	Tapping depth (at constant diameter d_{tap})
H^*	Maximum percentage difference in u/u_{cl} from that in a fully developed flow over the range $r'/R \le 0.96$
h'	Dam height ($=D(1-\beta)/2$)
h_c	Contamination thickness
H_e	Total energy per unit mass
k	Uniform equivalent roughness
K	Pressure loss coefficient (the ratio of the pressure loss to the dynamic pressure, $\rho \bar{u}^2/2$)
l	Pressure tapping spacing
L	Pressure tapping spacing relative to the pipe diameter
L_1	Quotient of the distance of the upstream tapping from the *upstream* face of the orifice plate and the pipe diameter
L_2	Quotient of the distance of the downstream tapping from the *upstream* face of the orifice plate and the pipe diameter
L'_2	Quotient of the distance of the downstream tapping from the *downstream* face of the orifice plate and the pipe diameter
L_{p}	Length of impulse line
m	Area ratio: $m = \beta^2$
M	Pressure tapping spacing relative to the dam height, i.e. $D(1-\beta)/2$
M_2	Quotient of the distance of the downstream tapping from the *upstream* face of the orifice plate and the dam height, i.e. $D(1-\beta)/2$
M'_2	Quotient of the distance of the downstream tapping from the *downstream* face of the orifice plate and the dam height, i.e. $D(1-\beta)/2$
Ma	Mach number
n'	Mode number (1, 2, 3…)
p	Absolute static pressure of the fluid
P	Maximum permissible percentage shift in C
p_1	Absolute static pressure of the fluid at upstream pressure tapping
p_2	Absolute static pressure of the fluid at downstream pressure tapping
q_m	Mass flowrate
q_V	Volume flowrate
r	Radius of orifice sharp edge
R	Radius
r'	Distance from the centre line to a point (sometimes shown in a graph as negative in order to display a complete diameter)
R'	Temperature probe recovery factor
R_a	Arithmetical mean deviation of the (roughness) profile

r_c	Width of clean portion of orifice
Re	Reynolds number
$Re*$	Venturi-throat-tapping Reynolds number (see Eq. 7.11)
Re_D	Reynolds number referred to D
Re_d	Reynolds number referred to d
Re_{tap}	Tapping-hole Reynolds number (see Eq. 7.2)
s	Standard deviation of data about an equation
S	Percentage shift in discharge coefficient
t	Temperature of the fluid (°C)
T	Absolute (thermodynamic) (static) temperature of the fluid
T_0	Stagnation temperature of the fluid
$T_{\text{thermometer}}$	Temperature recorded by a thermometer
u	Axial velocity
U	Uncertainty
u_τ	Friction velocity, $\sqrt{(\tau/\rho)}$
u_{cl}	Axial velocity on the centre line of the pipe
$\bar{u}$	Mean axial velocity on a cross-section of the pipe or flowmeter
$\lvert\mathbf{u}\rvert$	Speed
x	Coordinate in the axial direction, e.g. the distance between a fitting and an orifice plate
z	Coordinate in the upward vertical direction
Z	Compressibility factor
β	Diameter ratio: $\beta = d/D$
γ	Ratio of specific heat capacities ($=C_p/C_v$)
δx	Absolute uncertainty in x
δ	Deflection
$\delta*$	Boundary-layer displacement thickness
Δp	Differential pressure
Δp_c	Pressure loss across a flow conditioner
$\overline{\Delta p}$	The time-mean value of the differential pressure
$\Delta p'$	The fluctuating component of the differential pressure
$\Delta p'_{\text{rms}}$	The r.m.s. value of $\Delta p'$
ΔT	Difference in temperature between the upstream tapping and the pressure recovery location (see Sect. 4.6.2)
$\Delta\varpi$	Pressure loss across a flowmeter (i.e. the permanent pressure loss, not the differential pressure)
ε	Expansibility [expansion] factor
θ	Angle between pressure tappings and drain hole
$\theta*$	Angular location for the pressure tappings at which the flowrate would be measured without error using Eq. 10.6
κ	Isentropic exponent
λ	Friction factor
λ_c	Linear expansion coefficient of a material

λ_{sp}	What λ would have been in a long straight pipe
μ	Dynamic viscosity of the fluid
μ_{JT}	Joule–Thomson coefficient
ν	Kinematic viscosity of the fluid: $\nu = \mu/\rho$
ρ	Density of the fluid
ρ_1	Density of the fluid at the upstream pressure tapping
τ	Pressure ratio: $\tau = p_2/p_1$
τ_w	Wall shear stress
ϕ	Over-reading (see Eq. 11.1)
ψ	The decay rate of swirl in a pipe
Ψ	Potential energy per unit mass

Chapter 1
Introduction and History

Abstract Differential-pressure flowmeters (called differential-pressure meters) remain the most common type of meter for large and medium flowrates: their demise has often been predicted, and yet large numbers of them are still both used and bought. The reasons for their continued popularity are considered: their advantages and disadvantages are listed. The theory (Bernoulli's Theorem) for incompressible and compressible flow is considered, and both the basic equation for flow through differential-pressure meters and the expansibility-factor equation for Venturi tubes are derived. An introduction to Reynolds number and velocity profile, pipe roughness and pressure loss is provided. The essential requirements for differential-pressure flow measurement are described and discussed, and the location of material in other chapters required to meet these requirements is given. The history of flow measurement by Venturi tubes and orifice plates is described, with particular reference to Clemens Herschel in the main text and an Appendix on Sextus Julius Frontinus.

1.1 Introduction

Differential-pressure flowmeters (called differential-pressure meters) remain the most common type of meter for large and medium flowrates: their demise has often been predicted, and yet large numbers of them are still both used and bought. The reasons for their continued popularity are considered below.

The chief types of differential-pressure meters are:

- orifice plates,
- Venturi tubes,
- nozzles,
- Venturi nozzles,
- cone meters,
- low-loss devices (e.g. Dall tubes), and
- inlet flowmeters (e.g. Borda inlets).

M. Reader-Harris, *Orifice Plates and Venturi Tubes*,
Experimental Fluid Mechanics, DOI 10.1007/978-3-319-16880-7_1

Variable-area meters are also differential-pressure meters, in which the differential pressure is balanced by the weight of a float or by a spring.

'Orifice plate' can refer just to the plate or to the whole meter; where it is important to be clear that the plate and pipework are meant, 'orifice meter' can be used.

It is estimated that at least 40 % of industrial flowmeters in use at present are differential-pressure-based devices, with the orifice plate being the most popular for accurate measurement of flow. Within this wide range of different devices this book restricts itself to those devices that are standardized in the ISO standard for differential-pressure meters, ISO 5167:2003, that is, the first four groups listed above (cone meters are likely to be standardized soon as ISO 5167-5); moreover, most of the material presented here is on orifice plates and Venturi tubes. These devices have been described in standards for many years and by different standards bodies. ISO 5167 now provides the most wide-ranging of these standards.

ISO 5167 was revised in 2003 into four parts: the first part is a general one ISO 5167-1:2003 (ISO 2003a); the other three parts cover orifice plates ISO 5167-2:2003 (ISO 2003b), nozzles ISO 5167-3:2003 (ISO 2003c) and Venturi tubes ISO 5167-4:2003 (ISO 2003d). ISO 5167 is used in the measurement of many different fluids across a wide range of applications; nevertheless, the measurement of natural gas is highest in terms of financial importance: about £25 billion (€31 billion) of natural gas each year in the UK is measured by devices conforming to this standard (see Reader-Harris 2007 and Chap. 12).

This chapter provides some simple theory and describes the history briefly. Subsequent chapters consider the design of orifice-plate and Venturi-tube systems, the equations that are used, installation effects, then nozzles, orifice plates with drain holes, wet gas and standards.

NOTE 1 '(ISO 19xx)' or '(ISO 20xx)' is not the number of an ISO standard: it is a reference to an ISO standard (published in 19xx or 20xx, respectively) listed in the References

NOTE 2 Sections or chapters of this book are normally referred to as 'Sect. x.x' or 'Chap. x'; sections of a standard are normally referred to as 'x.x of ISO x:x'

1.2 Theory

1.2.1 Bernoulli's Theorem

The theory required for differential-pressure meters is given by Bernoulli's Theorem (for Bernoulli see Sect. 1.10), which is that for steady motion of a frictionless non-conducting fluid the total energy per unit mass is constant along a streamline, provided that the total includes not only the kinetic and internal energies, but also the potential energies associated with the pressure field and the external body-force field, i.e., that the total energy per unit mass, H_e, given by

$$H_e = \frac{1}{2}|\mathbf{u}|^2 + E_i + \frac{p}{\rho} + \Psi \tag{1.1}$$

is constant along a streamline, where $|\mathbf{u}|$ is the speed, E_i the internal energy per unit mass, p the pressure and ρ the density, and the body force per unit mass is equal to $-\text{grad } \Psi$ (Batchelor 1967). Daniel Bernoulli established this in 1738 for the particular case of an incompressible fluid. Where the only external force acting on the fluid is that of gravity, $\Psi = gz$, where g is the acceleration due to gravity and the coordinate z is measured in the upward vertical direction. A very useful form of Bernoulli's Theorem for barotropic (i.e. the pressure is a function of the density only) or incompressible flow is that

$$\frac{1}{2}|\mathbf{u}|^2 + \int \frac{dp}{\rho} + gz \tag{1.2}$$

is constant on a streamline. This form, together with a simple derivation, is given by Curle and Davies (1968).

1.2.2 Method of Operation

1.2.2.1 General

Differential-pressure meters partially obstruct the flow and hence operate by creating a difference in static pressure between the upstream and the throat of the device. They are normally installed in line, i.e. with pipework both upstream and downstream of them, but they may be installed at the inlet to, or sometimes at the outlet from, a length of pipe.

1.2.2.2 Incompressible Flow

Figure 1.1 shows a converging section of pipe, typical of the entry cone of a Venturi tube (for a diagram of a Venturi tube see Fig. 3.1). If an incompressible flow through this section of pipe is considered and Bernoulli's Theorem is applied on a horizontal streamline crossing the upstream and the throat planes, then from Eq. 1.2 (since ρ and z are constant)

$$\frac{1}{2}\rho u_1^2 + p_1 = \frac{1}{2}\rho u_2^2 + p_2, \tag{1.3}$$

where p, ρ, and u are the pressure, density and axial velocity respectively and the subscripted 1 and 2 represent the upstream and downstream (throat) planes. If the velocity took a single value on each plane then u would be equal to $\bar{u}$, the mean axial velocity on that plane.

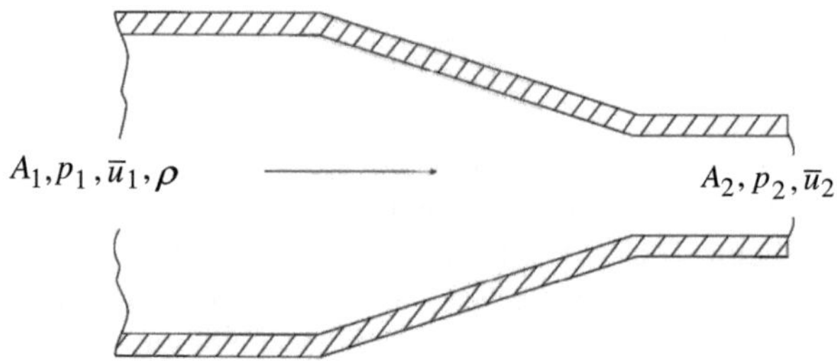

Fig. 1.1 Converging flow (labelled for incompressible flow)

Conservation of mass gives

$$q_m = \rho \bar{u}_1 A_1 = \rho \bar{u}_2 A_2 \tag{1.4}$$

where q_m is the mass flowrate and A the area of each plane. The area ratio, m, and the diameter ratio, β, are defined by

$$A_2/A_1 = m = (d/D)^2 = \beta^2, \tag{1.5}$$

where d and D are the throat and pipe diameters.

From Eqs. 1.3 (setting $u = \bar{u}$), 1.4 and 1.5 it follows that for incompressible flow

$$q_m = \frac{1}{\sqrt{1-\beta^4}} \frac{\pi d^2}{4} \sqrt{2\rho(p_1 - p_2)}. \tag{1.6}$$

$p_1 - p_2$ is the differential pressure, Δp.

Hence

$$q_m = \frac{1}{\sqrt{1-\beta^4}} \frac{\pi d^2}{4} \sqrt{2\rho \Delta p} \tag{1.7}$$

assuming no losses occur and the moving fluid completely fills the pipe.

1.2.2.3 Compressible Flow

If the fluid being metered is compressible, there will be a change in density when the pressure of the fluid falls from p_1 to p_2 on passing through the device. As the pressure changes quickly, it is assumed that no heat transfer occurs and, because no work is done by or on the fluid, the expansion is isentropic, i.e. p is proportional to ρ^κ, where κ is the isentropic exponent. So, in Venturi tubes (and nozzles), if Ψ also is constant, it follows from Eq. 1.2 that

$$\frac{1}{2}u_1^2 + \frac{\kappa}{\kappa - 1}\frac{p_1}{\rho_1} = \frac{1}{2}u_2^2 + \frac{\kappa}{\kappa - 1}\frac{p_2}{\rho_2} \tag{1.8}$$

Conservation of mass (Eq. 1.4 for incompressible flow) becomes

$$q_m = \rho_1 \bar{u}_1 A_1 = \rho_2 \bar{u}_2 A_2 = \rho_2 \bar{u}_2 \frac{\pi}{4} d^2 \tag{1.9}$$

From Eq. 1.9

$$\bar{u}_1 = \frac{\rho_2 A_2}{\rho_1 A_1}\bar{u}_2 = \tau^{\frac{1}{\kappa}}\beta^2 \bar{u}_2 \tag{1.10}$$

since $\frac{\rho_2}{\rho_1} = \left(\frac{p_2}{p_1}\right)^{\frac{1}{\kappa}} = \tau^{\frac{1}{\kappa}}$ where τ is the pressure ratio $\frac{p_2}{p_1}$ (p_2 is not measured directly but is equal to $p_1 - p$).

Setting $u = \bar{u}$ in Eq. 1.8, substituting from Eq. 1.10 into Eq. 1.8 and rearranging gives

$$\begin{aligned} \bar{u}_2^2\left(1 - \beta^4 \tau^{\frac{2}{\kappa}}\right) &= \frac{2\kappa}{\kappa - 1}\left(\frac{p_1}{\rho_1} - \frac{p_2}{\rho_2}\right) = \frac{2\kappa}{\kappa - 1}\frac{(p_1 - p_2)}{\rho_1}\left(\frac{p_1 - \tau^{-\frac{1}{\kappa}} p_2}{p_1 - p_2}\right) \\ &= \frac{2\kappa}{\kappa - 1}\frac{\Delta p}{\rho_1}\left(\frac{1 - \tau^{\frac{\kappa-1}{\kappa}}}{1 - \tau}\right) \end{aligned} \tag{1.11}$$

Substituting from Eq. 1.11 into Eq. 1.9 gives

$$\begin{aligned} q_m &= \rho_2 \frac{\pi}{4} d^2 \sqrt{\frac{2\kappa}{\kappa - 1}\left(\frac{\Delta p}{\rho_1}\right)\left(\frac{1}{1 - \beta^4 \tau^{\frac{2}{\kappa}}}\right)\left(\frac{1 - \tau^{\frac{\kappa-1}{\kappa}}}{1 - \tau}\right)} \\ &= \frac{1}{\sqrt{1 - \beta^4}}\frac{\pi}{4} d^2 \sqrt{2\rho_1 \Delta p}\sqrt{\left(\frac{\kappa \tau^{\frac{2}{\kappa}}}{\kappa - 1}\right)\left(\frac{1 - \beta^4}{1 - \beta^4 \tau^{\frac{2}{\kappa}}}\right)\left(\frac{1 - \tau^{\frac{\kappa-1}{\kappa}}}{1 - \tau}\right)} \end{aligned}$$

So

$$q_m = \frac{\varepsilon}{\sqrt{1 - \beta^4}}\frac{\pi}{4} d^2 \sqrt{2\rho_1 \Delta p}. \tag{1.12}$$

where the expansibility (or expansion) factor, ε, is given by:

$$\varepsilon = \sqrt{\left(\frac{\kappa \tau^{\frac{2}{\kappa}}}{\kappa - 1}\right)\left(\frac{1 - \beta^4}{1 - \beta^4 \tau^{\frac{2}{\kappa}}}\right)\left(\frac{1 - \tau^{\frac{\kappa-1}{\kappa}}}{1 - \tau}\right)}. \tag{1.13}$$

Because of the assumptions on which this equation is based, it has an uncertainty, given in 5.8 of ISO 5167-4:2003 as

$$\left(4 + 100\beta^8\right)\frac{\Delta p}{p_1}\%, \tag{1.14}$$

where Eq. 1.13 is used for Venturi tubes. As might have been expected this uncertainty increases with $\Delta p/p_1$.

This equation can only be used where the Mach number in the throat is not too close to 1: this is achieved by requiring that $p_2/p_1 \geq 0.75$. The same limit is also used with the orifice expansibility-factor equation in Chap. 6.

Equation 1.12 is identical to Eq. 1.7 except for the expansibility term.

1.2.2.4 Equation for Practical Use

In reality, there is a loss in pressure due to friction on the wall of the Venturi tube. Moreover, the calculation above is a simplification in that the velocity does not take a single value on each plane. In practice, the pressure is measured using a pressure tapping and so static-hole error (see Chap. 7) also needs to be taken into account. Therefore the flowrate in Eq. 1.12 is multiplied by the discharge coefficient C to take these effects into account. C is the ratio between the actual flowrate and the theoretical flowrate. Because the throat is immediately downstream of a contraction the velocity on the throat plane varies little with position except very close to the wall; so the error due to variation in velocity across that plane is small.

Thus

$$q_m = \frac{C\varepsilon}{\sqrt{1-\beta^4}}\frac{\pi d^2}{4}\sqrt{2\rho_1\Delta p}. \tag{1.15}$$

NOTE The expression $\frac{1}{\sqrt{1-\beta^4}}$ is known as the velocity of approach factor and is usually denoted by E. The term $\frac{C}{\sqrt{1-\beta^4}}$ is sometimes referred to as the flow coefficient of the meter and represented by α.

For an incompressible fluid $\varepsilon = 1$. It would be reasonable to ask whether the losses and the effect of variation in velocity across each plane are identical for incompressible and compressible flow; it appears that they are almost identical; any minor difference is included in the uncertainty of ε.

In Venturi tubes (and nozzles) the flow follows the boundary of the tube closely and the value of C is usually close to 1.

NOTE The pressure profile, particularly near the wall, is more complicated than a simple Bernoulli model might lead one to expect: see Fig. 3.5 (Lindley 1966; Sattary and Reader-Harris 1997).

In the case of an orifice plate (see Fig. 2.1) the flow continues to converge downstream of the plate with the location of maximum convergence called the *vena contracta*. The fluid then expands and re-attaches to the pipe wall; however, there is a relatively large net pressure loss across the plate which is not recovered. Bernoulli's Theorem can be applied between an upstream plane and the *vena contracta* (see Sect. 2.6), but the area of the *vena contracta* cannot practically be measured; therefore in Eq. 1.15 the diameter of the orifice bore (not that of the *vena contracta*) is used. This leads to a value of C of approximately 0.6 which, in effect, includes a coefficient of contraction.

The theoretical values of ε do not apply to orifice plates because the diameter of the *vena contracta* may be different in compressible and incompressible flow. This is covered in Chap. 6.

1.3 Essential Requirements

1.3.1 General

From Eq. 1.15 the essential requirements for a measurement of flowrate are to know the values of C, ε, d, D, ρ and Δp.

Because the upstream fittings and pipework affect the velocity profile (see Sect. 1.4) and the velocity profile affects C, there are minimum distances required between the differential-pressure meter and any upstream fittings (except those with which it was calibrated): these restrictions are covered in Chap. 8.

1.3.2 With a Calibration in a Flowing Fluid

If a differential-pressure meter is calibrated in a flowing fluid the discharge coefficient, C, is found by experiment: in this case accurate values of d and D are not required: the calibrated value of C (and the values of d and D used at the calibration), the differential pressure and the density would be needed. If the flowmeter is either calibrated or used or both calibrated and used in a compressible fluid then ε (usually from Eq. 1.13 for a Venturi tube or from Chap. 6 for an orifice plate) is also required. The density is determined either from measurement or from calculation from temperature and pressure. Change in dimensions due to temperature is considered in Sect. 4.6.4.

Calibration of an orifice plate (more precisely, an orifice meter) or a Venturi tube reduces the uncertainty of the measured flowrate. If the flowmeter is calibrated with the upstream fittings and pipework that will be used in the field then the minimum distances between the differential-pressure meter and its upstream fittings given in Chap. 8 will not be required.

1.3.3 Without a Calibration in a Flowing Fluid

One of the main advantages of a differential-pressure meter is that C can be calculated rather than measured, as long as the meter is manufactured to the specified design: using a differential-pressure meter may avoid the need for calibration in a flowing fluid. In this case the dimensions and an equation for C (and in compressible flow ε) are required in addition to the values of the differential pressure and the density.

Recent work to determine the values of C and ε is given in Chaps. 5–7. However, these values are only appropriate if other requirements are met:

- The differential-pressure meter in use (including its upstream piping) must be geometrically similar to those meters (and their piping) from which the original data for C (and ε if required) were determined. How similar they have to be to permit the equations in Chaps. 5–7 to be used has been a subject of research for many years.
- Therefore, there are restrictions on the design of the orifice plate or Venturi tube and its upstream and downstream pipes: these restrictions are covered in Chaps. 2 and 3 (and there are some general design requirements in Chap. 4). Some of the design restrictions in the standards have not been determined by experiment. Such restrictions tend to be sufficient, in that variations within these limits do not appear to have affected the discharge coefficient unduly; wider tolerances in some cases might be acceptable, but experiments would be required to confirm this.

If the orifice plate has a drain hole or the gas is wet there are additional requirements, which are covered in Chaps. 10 and 11, respectively.

1.4 Introduction to Reynolds Number and Velocity Profile

The discharge coefficient of an orifice plate or a Venturi tube, as of other differential-pressure meters, depends on the Reynolds number. It is this parameter that has a very significant effect on the behaviour of fluids flowing through pipes. This is defined as follows:

$$Re_D = \frac{\rho D \bar{u}_p}{\mu} \quad (1.16)$$

where $\bar{u}_p$ is the mean velocity in the pipe, D is the pipe diameter and ρ and μ are the density and the dynamic viscosity in the pipe.

Fig. 1.2 Osborne Reynolds (cropped photo of a painting painted in 1904 by John Collier)

The Reynolds number is named after Osborne Reynolds (see Fig. 1.2) (1842–1912): he was professor of engineering at Owens College in Manchester (now the University of Manchester). He is best known for his work on the conditions under which fluid flow in pipes changes from laminar to turbulent: his rig can still be seen in Manchester. From his experiments came the Reynolds number. Consider the numerator in Eq. 1.16: ρ is mass per unit volume. $\rho \bar{u}_p$ is therefore momentum per unit volume, and $\rho D \bar{u}_p$ is moment of momentum per unit volume. The numerator is therefore a measure of the flowing fluid's ability to generate dynamic forces, whilst the denominator, μ, is a measure of its ability to generate viscous forces.

This means that the Reynolds number indicates which kind of forces will predominate in the flowing fluid. When $\rho D \bar{u}_p$ is relatively large Re_D will be large and dynamic forces will prevail, but when μ is relatively large Re_D will be smaller and viscous forces will prevail.

Incidentally, Re_D is called Reynolds number because the dimensions of $\rho D \bar{u}_p$ are the same as those of μ, and Re_D is therefore a dimensionless ratio.

Sometimes differential-pressure meter performance is expressed in terms of throat Reynolds number (sometimes called orifice Reynolds number for an orifice plate): this is defined in 3.3.2.2 of ISO 5167-1:2003 as

$$Re_d = \frac{Re_D}{\beta}.$$

A fluid can flow along a pipe in one of two very different ways. Laminar flow occurs at Reynolds numbers below about 2000. This can be likened to the flow of traffic on a busy motorway, with the traffic in the various lanes travelling on parallel paths at different speeds; the slow lane is next to the pipe wall and the fast lane in the centre of the pipe. When studying laminar flow in pipes engineers usually assume that the 'traffic' never changes lanes, although in fact, gradual lane-changing does occur.

Turbulent flow occurs at Reynolds numbers above about 5000. It can be likened to the flight of a flock of starlings. The flock as a whole may be travelling in a straight line at a constant speed, but if you could watch the flight of any individual bird it will appear to be zig-zagging and gyrating wildly within the flock.

In much of industry, pipe Reynolds numbers are usually well above 5000 and laminar flow is rarely encountered, unless very viscous liquids are being piped. There is a transition region between turbulent and laminar flow at Reynolds numbers between about 2000 and 5000.

A graph showing how the velocity, u, varies across a diameter of a pipe is called a velocity profile. In laminar flow the velocity profile is a parabola, and the velocity in the centre of the pipe is twice the mean velocity, $\bar{u}$. See Fig. 1.3.

With turbulent flow, the velocity profile downstream of a very long length of straight pipe is much flatter, and the velocity at the centre is about 1.2 times the mean velocity. The turbulent velocity profile shown in Fig. 1.3 was obtained using laser Doppler velocimetry at a Reynolds number of 5×10^5 (Spearman et al. 1996). A turbulent velocity profile can often be approximated by a power-law profile:

$$\frac{u}{u_{\max}} = \left(1 - \frac{r'}{R}\right)^{\frac{1}{n}} \tag{1.17}$$

where r' is the distance from the axis of the pipe, R is the pipe radius and $u_{\max}$ is the maximum axial velocity in the pipe. Integrating Eq. 1.17 gives

$$\frac{\bar{u}}{u_{\max}} = \frac{2n^2}{(n+1)(2n+1)} \tag{1.18}$$

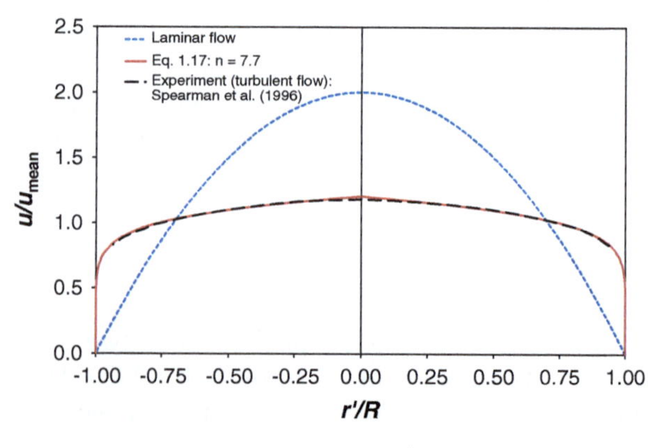

Fig. 1.3 Velocity profiles in *straight pipe*

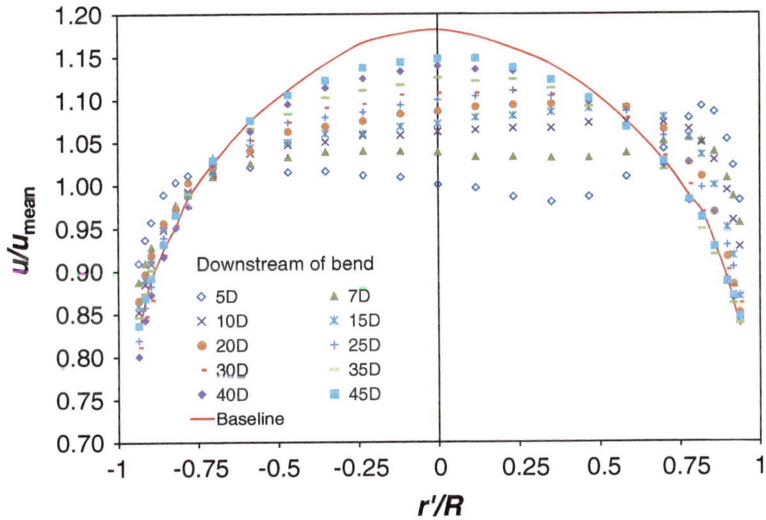

Fig. 1.4 Velocity profile at different locations downstream of a single bend in turbulent flow in a 4″ pipe (measurements made in the plane of the *bend*): note the *vertical scale*

The power-law equation with $n = 7.7$ gives very good agreement with the experimental measurements in Fig. 1.3.

This book (and ISO 5167) only covers the flow of Newtonian fluids: Newtonian fluids include water, air, natural gas and oil; non-Newtonian fluids include ketchup, custard, toothpaste and paint.

Downstream of a long length of straight pipe the profile is said to be fully developed. Figure 1.4 shows a velocity profile in a 4″ pipe that has been affected by a single upstream bend (Spearman et al. 1996). Although the velocity profile is gradually approaching the fully developed profile, even 45*D* downstream of the bend the profile is still not completely fully developed.

Changes in velocity profile due to Reynolds number (see Chaps. 5 and 7) or upstream fittings (see Chap. 8) affect the discharge coefficient. Flow conditioners may be used to improve the velocity profile (see Chap. 8).

Velocity profiles in wet gas have been studied in much less detail. The definition and measurement of wet gas is covered in Chap. 11.

1.5 Pipe Roughness

The velocity profile in a pipe is affected by both the Reynolds number and the pipe roughness, but for sufficiently high Reynolds number only depends on them through their effect on the friction factor. The friction factor, λ, can be measured directly, using

$$\Delta p = \frac{\lambda \rho \bar{u}^2 x}{2D} \tag{1.19}$$

where Δp is the difference in pressure between two tappings spaced a distance x apart in a pipe of diameter D and $\bar{u}$ is the mean velocity in the pipe.

This corresponds to

$$\tau_w = \frac{1}{8} \lambda \rho \bar{u}^2 \tag{1.20}$$

where τ_w is the shear stress at the wall.

It is quite rare to measure the friction factor of a pipe, but it may be estimated using the Colebrook-White Equation (Schlichting 1960):

$$\frac{1}{\sqrt{\lambda}} = 1.74 - 2lg\left(\frac{2k}{D} + \frac{18.7}{Re_D\sqrt{\lambda}}\right) \tag{1.21}$$

where k is the uniform equivalent roughness and lg is $\log_{10}$ (following ISO 80000-2:2009, $\log_{10}$ is written lg). Rather than measuring the pressure drop it is simpler to measure the arithmetic mean deviation of the roughness profile, R_a, and to deduce the uniform equivalent roughness, k, by taking it to be approximately equal to πR_a (based on the assumption that the wall has a sinusoidal profile) and then to calculate λ iteratively from Eq. 1.21. Although the errors inherent in using R_a to calculate λ are not insignificant, measuring pressure loss is not simple: any difference in pipe diameter at the pressure tappings causes an error, as do any defects in the pressure tappings. See also the NOTE in Sect. 2.3.3.1.

This topic is covered in detail in Chap. 20 of Schlichting (1960), which in its Fig. 20.25 gives a graph of the friction factor called the Moody Diagram. Modern data on friction factor in a pipe that at least for $Re_D \leq 1.36 \times 10^7$ is hydraulically smooth can be found in Zagararola and Smits (1998) and McKeon et al. (2004): their data are in good agreement with values obtained from Eq. 1.21 with $k = 0$: their values of λ lie between 0.0001 and 0.0003 below those given by Eq. 1.21.

1.6 Accuracy

In choosing a flowmeter, there are many factors to consider, and among them the question of accuracy is very important. While it is pointless to pay for higher accuracy than is necessary, a cheap meter that is not accurate may become expensive in use. Similarly, unless a meter is manufactured, calibrated and installed correctly (or designed, manufactured and installed in accordance with ISO 5167 or equivalent), it will not achieve its potential accuracy.

If an orifice plate (more precisely, an orifice meter) is designed and manufactured according to ISO 5167-2:2003 an uncertainty in the flowrate between 0.5 and

1 % will be achieved under good conditions. This uncertainty could be increased because of the effect of inadequate upstream length on the flow profile.

If a typical Venturi tube is designed and manufactured according to ISO 5167-4:2003, it is reasonable to expect that an uncertainty of a little more than 1 % will be obtained in water at maximum flowrate under ideal conditions. Because it has pressure tappings in the throat a Venturi tube is much more affected by burrs on the tappings than an orifice plate. It does, however, have the advantage of requiring a shorter upstream straight length than an orifice plate. There has been an increasing desire to use Venturi tubes in high-pressure gas flows. The accuracy obtained is discussed in Chap. 7.

By calibrating a differential-pressure meter an uncertainty of less than 0.5 % should be obtainable in a single-phase fluid. Reynolds numbers in ambient air and in water are very similar; therefore, one way of measuring ambient air flowrate is to calibrate an orifice plate (more precisely, an orifice meter) or a Venturi tube in water and to use it in air using the expansibility-factor equation in Chap. 6 (for an orifice plate) or Eq. 1.13 (for a Venturi tube) and to include the uncertainty of the expansibility-factor equation in the total uncertainty.

Because of the square-root relationship between differential pressure and flow-rate, the relative uncertainty of the flowrate measurement increases at low flowrates (see Sect. 4.3).

1.7 Pressure Loss

All differential-pressure meters cause a permanent pressure loss: values are given in the appropriate part of ISO 5167. Pressure loss is important because it is a waste of energy. The pressure in a differential-pressure meter, after decreasing to a minimum, recovers to give a smaller permanent pressure loss than the differential pressure. Because of the recovery of most of the pressure in the divergent outlet section, the Venturi tube (see Fig. 1.5) causes less overall pressure loss in a system than an orifice plate and thus saves energy. Further information on pressure loss is provided in

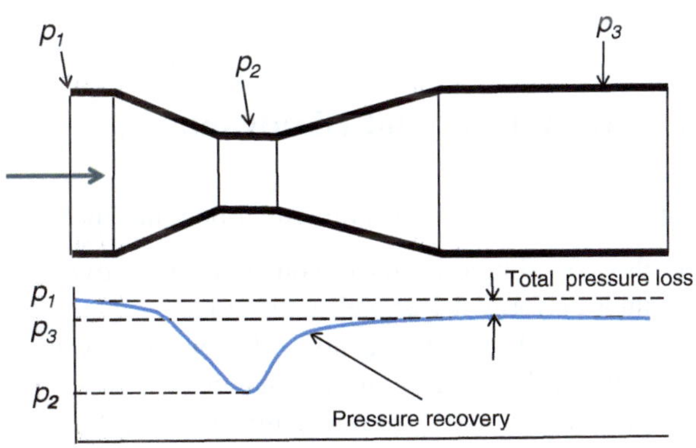

Fig. 1.5 Approximate pressure profile on the *centre line* of a Venturi tube

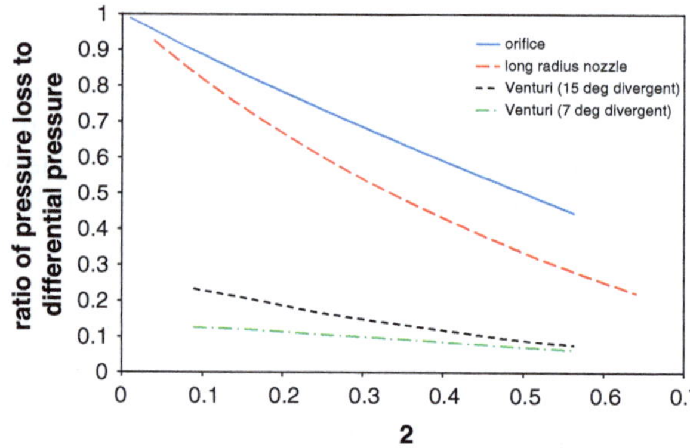

Fig. 1.6 Ratio of pressure loss to differential pressure

Chaps. 2 and 3. The approximate pressure profile through an orifice plate (more precisely, an orifice meter) is shown in Fig. 2.11. Approximate figures for the ratio of pressure loss to differential pressure based on ISO 5167 are given in Fig. 1.6.

When the outlet diameter of the divergent section is less than the pipe diameter a Venturi tube is called 'truncated'. More details are given in Sect. 3.3.

1.8 Standards

The advantage of an orifice plate or a Venturi tube is that, if designed and manufactured according to ISO 5167, the coefficient can be predicted by an internationally accepted equation, thus removing the need for calibration in a flowing fluid. Only differential-pressure meters at present have this capability.

Further guidance on the use of differential-pressure meters is given in other British and International Standards ISO/TR 3313:1998 (now withdrawn: see Sect. 12.3.2.6), ISO/TR 15377:2007, ISO/TR 12767:2007 and ISO/TR 9464:2008 (ISO 1998, 2007a, b, 2008).

Standards have enormous economic benefits: see Sect. 12.1. Most chapters of this book refer to standards; Chap. 12 in particular discusses standards.

1.9 Advantages and Disadvantages

The main advantages of differential-pressure meters are:

- they are simple to make, containing no moving parts,
- their performance is well understood,
- they are cheap—especially in larger pipes when compared with other meters,
- they can be used in any orientation,
- they can be used for most gases and liquids,

- they can be used in wet gas or as part of a multiphase meter,
- they can be calibrated in water and used in gas provided that the appropriate expansibility-factor equation is used,
- they generally have good standards, and
- the most common types do not generally require calibration.

Their main disadvantages are:

- their output signal is non-linear with flow,
- their rangeability (turndown) is less than for most other types of flowmeter,
- significant pressure loss may occur,
- the coefficient and accuracy may be affected by the pipe layout or the nature of the flow, and
- they may suffer from ageing effects, i.e. the build-up of deposits, or erosion of sharp edges.

This book explores these issues in depth. The continuing advantages of using orifice plates are considered by Lake (1999).

1.10 History

The special attraction of flowmeters used in accordance with ISO 5167 is that they can simply be made in accordance with a standard and the dimensions measured, and do not have to be calibrated using a flowing fluid. Compliance with the standardized design ensures a specified performance.

Perhaps the first person to undertake flow measurement in closed pipes and document a method whose descendant is that codified in modern differential-pressure standards was Sextus Julius Frontinus. He was Governor of Britain, wrote a book on stratagems in war, and came to flow measurement at the age of about 62, when he became Water Commissioner in Rome. He might have regarded this role as a sinecure, but instead produced a standard for flow measurement, took account of installation effects, and aimed for consistency and fairness. His work is described in Appendix 1.A.

A crucial event in making differential-pressure meters possible was Bernoulli's Theorem in the early eighteenth century. Daniel Bernoulli (1700–1782) (see Fig. 1.7) was born in Groningen. His father, Johann Bernoulli, encouraged him to study business. He then studied medicine. His father taught him mathematics privately. He became Professor of Mathematics in St Petersburg in his twenties. He successively held the chairs of medicine, metaphysics and natural philosophy at the University of Basel. He won ten prizes from the Paris Académie des Sciences: among them were ones for clepsydra for use at sea, tides (with Euler and Maclaurin), orbits (with his father), the best form of anchors (with his brother), and the nature & cause of ocean currents. He was the first person to derive Bernoulli's Theorem (Bernoulli 1738). However, his father Johann (1667–1748), who taught

Fig. 1.7 Daniel Bernoulli: engraved by Johann Jakob Haid (1704–1767), original artist Johann Rudolf Huber (1668–1748)

mathematics in Holland and then became Professor at Basel, subsequently published a better analysis, but dated it as if he had written it earlier (Bernoulli 1743). (Should the theorem be called Bernoullis' Theorem?)

After the Bernoullis came Giovanni Battista Venturi (1746–1822) (see Fig. 1.8). He was an abbé; he taught geometry and philosophy at the University of Modena; he was Ducal mathematician and State engineer, became Professor of Experimental Physics and wrote a history of Modena. He published the results of experimental work showing that fluids passing through a convergent section gain velocity but lose static pressure in return (Venturi 1797). His pioneering work made a practical differential-pressure meter possible.

Clemens Herschel (1842–1930) (see Fig. 1.9) graduated from Harvard in 1860, and studied in France and Germany; in 1879 he became Hydraulic Engineer of the Holyoke Water Power Company, and in 1881 he constructed the Holyoke Testing Flume. Kent (1912) records Herschel's words as follows: 'While Hydraulic Engineer

Fig. 1.8 Giovanni Battista Venturi: engraved by Francesco Rosaspina (1762–1841), original artist Mathias Pfenniger (1739–1813)

Fig. 1.9 Clemens Herschel [from Kent (1912)]

at Holyoke—I went there in 1879—I often had occasion to regret that I could not, in the pursuit of the duties of my office, meter the flow of large pipes, of 20″ or greater diameter. I finally resolved in 1887 that I would experiment on the subject. My first conception of such a pipe-meter was to place an orifice in line of the pipe, and to measure the head on it by noting the difference between the heads measured, both upstream and downstream from the orifice. Then to avoid an undue loss of head in the apparatus, I conceived the idea of having a gentle upstream adjutage, and a 'Venturi' expanding tube or nozzle downstream from the orifice. The experiments of 1887 were made under the influence of these ideas, the expanding adjutage being shaped in accordance with the results laid down in the book 'Lowell Hydraulic Experiments' of my former master, James B. Francis, of Lowell, Massachusetts. But so soon as it became necessary to compute and co-ordinate their results, the present accepted theory for explaining the action, and formulae for computing the discharges of the meter, were of necessity at once studied out, and the meter was named by me the 'Venturi' meter out of respect for Venturi's experiments of 1797.'

Herschel was a practical engineer, a man of delightful humility and of breadth of knowledge: he was the first translator of Frontinus' *de Aquis* (Frontinus c 97–98) into English: in 1899 he visited the monastery of Monte Cassino in Italy and secured a photographic copy of the original manuscript (it is an original manuscript in the sense that all other manuscripts of *de Aquis* in existence today are copies of the one at Monte Cassino). Herschel (1887, 1895) worked with Builders Iron Foundry in Providence, Rhode Island, to design and sell Venturi meters. He had an understanding of uncertainty: 'If this [discharge] co-efficient is taken without further measurement at 98 % [i.e. 0.98], we may be certain from experiments so far made that we shall rarely be over 2 % out of the way.' Venturi tubes were used in many water applications up to very large sizes (see Figs. 1.10 and 1.11). Herschel became Superintendent and Chief Engineer of the East Jersey Water Co.

When orifice plates were first used to measure fluids is unknown, but around 1890 Forrest M. Towl witnessed orifice meters being used to measure natural gas near Columbus, Ohio (AGA-ASME 1935). These were being used at the suggestion of Professor S.W. Robinson of Ohio State University. About 1904 Thomas R. Weymouth installed an orifice plate with flange tappings in series with a Pitot tube at Oil City, Pa., to make a comparison between orifice plates and Pitot tubes. This work was interrupted and not completed till 1911.

Hodgson (1917) commenced research on orifice flow measurement in 1909 and published his work in 1917; he records that the first orifice used for steam-flow measurement that was furnished to him came from Holbrook Gaskell of the United Alkali Company in 1910.

Some early work (second decade of 20th century) used tapping locations that are now rare: Metric Metal Work, which later became American Meter Company, used pipe tappings at 2½*D* upstream and 8*D* downstream, measuring the pressure loss; tests using pipe tappings were reported by Hickstein (1915); Bailey Meter Company located the downstream pressure tapping where the pressure is lowest,

Fig. 1.10 Cast bronze lining of throat, 93″ dia., for the 17′ 0″ Venturi meters, Catskill Aqueduct, New York water supply [photograph and caption from Kent (1912)]

Fig. 1.11 Installation of a mammoth "Venturi" meter for measuring 14,000,000 gallons per hour. Inserting the meter tube [photograph and caption from Kent (1912)]

which was thought to be at the *vena contracta*. Since the *vena contracta* is not a single location downstream of all orifice plates, this design is not very convenient.

NOTE There is doubt that the *vena contracta* occurs at the same axial location as the minimum pressure point on the wall: Wilson and Teyssandier (1975) present evidence that the *vena contracta* lies upstream of the minimum pressure point

The work undertaken in the USA between 1911 and 1935 is described in AGA-ASME (1935). Experiments in air at Edgewood Arsenal on 4″, 6″ and 8″ pipes are described in Bean et al. (1929) and those at Chicago on a 24″ pipe in Bean et al. (1931).

The equation for the orifice expansibility factor used in ANSI/API 2530-1985 (API 1985), ISO 5167:1980 (ISO 1980), API MPMS 14.3.1:1990 (API 1990) and ISO 5167-1:1991 (ISO 1991) was derived by Buckingham (1932) largely on the basis of data collected at tests in Los Angeles in 1929. Witte (1931) also undertook work on expansibility. The work of Howard Bean and Edgar Buckingham set a pattern for subsequent workers in this area. Murdock and Foltz (1953) tested the Buckingham Equation and found that it fitted their steam data very well.

The Buckingham orifice-plate discharge-coefficient Equation (see Chap. 5) was based on the data collected by Sam Beitler in the early 1930s at the Ohio State University Engineering Experimental Station (Beitler 1935). These data are of high quality, but were all collected in water; so any equations based on them require significant extrapolation in Reynolds number when used in high-pressure gas. They were taken in 1″ to 14″ pipes, and data from the smallest sizes are still especially valuable, since no data in pipes smaller than 2″ were taken during the European and API tests described in Sect. 5.3. The Refugio tests in 1952/3 measured the discharge coefficient in a 30″ orifice meter in natural gas by passing the flow through eight 10″ orifice meters downstream of a header and showed that the Buckingham Equation gave results in a 30″ pipe that were consistent with those in 10″ pipes (AGA 1954b).

Jean Stolz (1978) used 303 points from the Ohio State University Engineering Experimental Station data together with a table of flow coefficients from ISO/R 541:1967 (ISO 1967) to establish an equation that is dimensionally correct. His equation was an important advance because it could be used for flange, corner or D and $D/2$ pressure tappings. The ISO/R 541 table was established by the German VDI and is based on original data of Witte which are no longer available. The Stolz Equation was included in ISO 5167:1980 and ISO 5167-1:1991. Inconsistency between ISO standards and AGA (API) standards was identified as an important problem.

By 1980 improved flow measurement facilities were available: water facilities in a wide range of pipe sizes; high-pressure gas facilities, for example those in the UK (e.g. British Gas at Bishop Auckland) and the Netherlands (Gasunie), to give Reynolds numbers higher than those in water; and oil facilities, for example those in the UK (NEL) and in the US (CEESI), to give lower Reynolds numbers. Dr Tony Spencer received the OBE for his work at NEL.

Therefore, to resolve discrepancies between the ISO and AGA orifice-plate discharge-coefficient equations, data on orifice plate discharge coefficients were collected in Europe (supported by the EEC, now the EU) from 1980 to 1991 and in the United States around the same time in order to provide a new database from which an improved discharge-coefficient equation could be obtained which would receive international acceptance. Instead of just collecting data in water, data were also collected in air, natural gas and oil. Among those involved in the project were A. Aschenbrenner, R. Beaty, D. Bell, H. Bellinga, A. Benzoni, J. Bosio, J. Brennan, C. Britton, S. Caldwell, R. Ellington, W. Fling, J. Gallagher, D. Gould, B. Harbrink, M. Hobbs, P. Hoglund, E. Jones, Jr., P. Kervevan, F. Kinghorn, K. Kothari, P. LaNasa, G. Mattingly, R. Norman, M. Reader-Harris, J. Reid, J. Sattary, W. Seidl, C. Sindt, E. Spearman, E. Spencer, K. Starling, J. Stolz, J. Stuart, W. Studzinski, R. Teyssandier, P. van der Kam, J.-P. Vallet, K. West, J. Whetstone, P. Wilcox, and A. Willemse.

Details of how orifice-plate discharge-coefficient equations were derived and of those in use today are given in Chap. 5. Some history is also included in Chap. 12 on standards.

In parallel with collecting data on the discharge coefficient in good flow conditions, data were taken on the effect of upstream installation on the discharge coefficient. Early data were collected in 1926–1927 in the US at Columbus and Buffalo. Data were collected at Rockville in 1949–1953 (AGA 1951; 1954a). Spearman et al. (1995) summarized much of the data taken up to 1995. The BHRA data in Martin (1986) were taken in 8″ pipe and are the most comprehensive set up to 1995. The data collected in the EEC Orifice Plate project at the same time as the basic data described in Chap. 5 are summarized in Sattary (1991). Most of the best modern data have been collected in Studzinski et al. (2001), Morrow (2005) and Studzinski et al. (2000). For slightly more detail see Sect. 8.2.3.1.

During the 20th century much work on Venturi tubes was also carried out. With the development of electromagnetic meters for water, turbine meters and positive displacement meters for oils and orifice plates for gas there have been many fewer research papers on Venturi tubes than on orifice plates (perhaps a third of the number): from 1975 to 1995 very few papers were published on Venturi tubes: most work was being done on orifice plates.

Whereas the uncertainty of the discharge coefficient of uncalibrated orifice plates (more precisely orifice meters) is much smaller than that of uncalibrated Venturi tubes, Venturi tubes are more robust than orifice plates, and so are commonly used for subsea or wet-gas or multiphase applications. Chisholm (1967, 1977) at NEL carried out fundamental work on the wet-gas over-reading; this was followed by very important work by de Leeuw (1994, 1997), who produced the de Leeuw Equation for wet-gas flow through Venturi tubes. There has been a resurgence of interest in Venturi tubes in the last 20 years, and recent work on Venturi tubes is given in Chap. 7 and recent work on wet gas in Chap. 11.

This book aims to bring together much of the work undertaken on orifice plates and Venturi tubes up to 2014 and to present it so that those who come after may build upon it.

1.11 Conclusions

Differential-pressure meters remain the most common type of meter for large and medium flowrates: their demise has often been predicted, and yet large numbers of them are still both used and bought. The reasons for their continued popularity have been considered: their advantages and disadvantages have been listed. The theory (Bernoulli's Theorem) for incompressible and compressible flow has been considered, and both the basic equation for flow through differential-pressure meters and the expansibility-factor equation for Venturi tubes have been derived. An introduction to Reynolds number and velocity profile, pipe roughness and pressure loss has been provided. The essential requirements for differential-pressure flow measurement have been described and discussed, and the location of material in other chapters required to meet these requirements has been given. The history of flow measurement by Venturi tubes and orifice plates has been described, with particular reference to Clemens Herschel in the main text and an Appendix on Sextus Julius Frontinus.

NOTE 1 Designs of Orifice Plates and Venturi Tubes not Covered by this Book

This book only covers orifice plates and Venturi tubes as described in ISO 5167 in detail: designs of Venturi tube other than the classical are not covered with the exception of some research on those with modified convergent angles. Nozzles are described in Chap. 9 particularly for the light they shed on the performance of Venturi tubes at high Reynolds number.

- for other shapes of Venturi tube see Halmi (1973a, b);
- for inlet nozzles see Ito et al. (1985), Jaumotte and Onockx (1985), Kinghorn and McHugh (1978), Kinghorn et al. (1975) and Pearsall (1962);
- for slotted orifice plates see Morrison et al. (1993, 1994);
- for Dall tubes see Cousins (1975);
- for orifice plates of shapes other than sharp and square-edged see Ho and Leung (1985), McVeigh (1974), Stoll and Zientara (1974) and Turton (1975), which cover conical-entrance orifice plates: ISO/TR 15377:2007 covers conical-entrance and quarter-circle orifice plates;
- for eccentric and segment orifices see Witte (1951): ISO/TR 15377:2007 covers eccentric orifice plates;
- for orifice plates of different shapes at low Reynolds number see Eujen (1969), Kastner and McVeigh (1965–1966) and Vasy et al. (1968–1969).

This book only covers thin orifice plates. Small but relatively thick orifice plates of aspect ratio between 1 and 50 are considered by Ramamurthi and Nandakumar (1999)

NOTE 2 Background Work on Differential-Pressure Meters

Extensive work has been undertaken on the fluid mechanics of differential-pressure meters. Measurements of detailed velocity profiles in the neighbourhood of an orifice plate have been made by Bates (1981). Agarwal (1994) examined the reattachment point downstream of an orifice plate and obtained values in the range 9–12 dam heights, where the dam height is the distance from the upstream corner of the orifice plate to the nearest point of the pipe wall, i.e. $\frac{D(1-\beta)}{2}$. A very substantial amount of data has been collected by Morrison and co-workers from Texas A&M University. They measured velocity profiles and pressure profiles in orifice meters. The data are published by Morrison (1990, 1993). Data by Morrison and his co-workers are given in DeOtte et al. (1991) and Morrison et al. (1992b). Turbulence quantities including triple velocity correlations are given in Panak et al. (1992). The effect on orifice plates of measured flow profiles that are very far from fully developed is considered in Morrison et al. (1992a).

In addition to experimental work, computational work has been undertaken to compute the flow through differential-pressure meters: early studies included those of Mattingly and Davis (1977) and Mills (1968). Examples of theoretical work are Hall (1959) and Halmi (1973c). Theoretical work has helped to explain the experimental data, but experimental work has always been required to determine actual values for inclusion in standards.

Appendix 1.A: Sextus Julius Frontinus

Probably the first person to describe flow measurement in closed pipes using a method whose descendant is that codified in modern differential-pressure standards was Sextus Julius Frontinus. He was born about 35 AD and was Provincial Governor of Britain about 76 AD, where he subdued the Silures (a tribe in Wales) and constructed the Via Julia. After his return to Rome he wrote *Strategemata* (his work on the art of war) about 90 AD. He then became Water Commissioner in Rome: he might have regarded this as a sinecure, but instead he was an outstanding public servant and wrote a treatise *De Aquis* (Frontinus c 97–98) around 97–98 AD on the Aqueducts of Rome (quotations below from *De Aquis* are from the Loeb edition with a translation by Charles E Bennett, a revision of that by Clemens Herschel). He describes how the aqueduct system of ancient Rome (depicted in Figs. 1.A.1 and 1.A.2) operated, and places particular emphasis on the flow measurement aspects. His book is a remarkable achievement for one who came to flow measurement so late in life.

Standardized pipes

The system contained standardized pipes for offtakes and intakes from and to the main aqueducts. Frontinus described the permissible pipe sizes showing how they were constructed: the basic unit was the quinaria (or 5-pipe), a pipe with a diameter of 5 quarters of a digit, where a digit is 1/16 Roman foot (the Roman foot was equal to 0.967 imperial foot; the Roman foot could alternatively be divided into 12 inches). The senaria (or 6-pipe) had a diameter of 6 quarters of a digit and so up to the vicenaria (or 20-pipe) with a diameter of 20 quarters of a digit (i.e. 5 digits). Above

Fig. 1.A.1 Aqua Claudia in the Parco degli Acquedotti: August 2008 taken by Agricmarketing (talk) Wikipedia under Parco degli Acquedotti

Fig. 1.A.2 Porta Maggiore, Rome, formed by a supporting arch of the Aqua Claudia and Rome's highest elevated aqueduct Aqua Anio Novus, integrated into Aurelian Wall as a gate 271 AD. From Wikipedia under Roman aqueduct: it was taken by Diana

that size the system was different and based on area not diameter: the vicenum quinum (or 25-pipe) had an area of 25 square digits, the tricenaria (30-pipe) of 30 square digits and so up to the centenum vicenum (120-pipe). If the second definition based on area had been used for the vicenaria (or 20-pipe) then the difference from the definition based on diameter would have been very small (less than 2 per cent on area). So there was little problem at the changeover between systems.

Frontinus discovered that in four pipe sizes (duodenaria (12-pipe), vicenaria (20-pipe), centenaria (100-pipe) and centenum vicenum (120-pipe)) the common usage was different from the systematic method described above. Although for the first of these the error was small, for the second, which was commonly used for delivery, the common usage was 20 % smaller than it should have been, the third and fourth were 13 and 68 % larger, respectively, than they should have been. He sums this up as follows:

In all there are 25 ajutages.[1,2] They all conform to their computed and recorded capacities, barring these four which the water-men have altered. But everything embraced under the head of mensuration ought to be fixed, unchanged, and constant.[3] For only so will any special computation accord with general principles. Just as a *sextarius*,[4] for example, has a regular ratio to a *cyathus*,[5] and similarly a *modius*[6] to both a *cyathus* and *sextarius*, so also the multiplication of the *quinariae* in case of the larger ajutages must follow a regular progression. However, when less is found in the delivery ajutages and more in the receiving ajutages, it is obvious that there is not error, but fraud.

Frontinus was a faithful public servant, honest and conscientious, basing his conclusions on personal examination. For clarity he listed in detail the permissible 25 pipe sizes (those not actually in use are here in italics): 5, 6, *7*, 8, 10, *12*, 15, 20, *25*, 30, *35*, 40, *45*, 50, *55*, 60, *65*, 70, *75*, 80, *85*, 90, *95*, 100 and 120. The problem with his system was that he assumed the flowrate would be proportional to the area alone. He made no direct allowance for velocity.

Meter installation

Frontinus knew that meter installation (in terms of the angle of the meter to the main flow) does make a difference:

[1]In summa moduli sunt XX quinque. Omnes consentiunt et rationi et commentariis, exceptis his quattuor, quos aquarii novaverunt. Omnia autem quae mensura continentur, certa et immobilia congruere sibi debent; ita enim universitati ratio constabit. Et quemadmodum verbi gratia sextarii ratio ad cyathos, modii vero et ad sextarios et ad cyathos respondet; ita et quinariarum multiplicatio in amplioribus modulis servare consequentiae suae regulam debet. Alioqui cum in erogatorio modulo minus invenitur, in acceptorio plus, apparet non errorem esse sed fraudem (34).

[2]An offtake (or intake) tube, here effectively a water meter.

[3]'consistent' might be better.

[4]The Roman pint (about 0.55 l), close to an imperial pint.

[5]1/12th of a sextarius (about 45 ml), about 1.6 fl oz.

[6]The Roman peck: 16 sextarii (about 8.8 l), close to an imperial peck.

[7]But the position of the *calix* is also a factor. Placed at right angles and level, it maintains the normal quantity. Set against the current of the water, and sloping downward, it will take in more. If it slopes to one side, so that the water flows by, and if it is inclined with the current, that is, is less favourably placed for taking in water,[8] it will receive the water slowly and in scant quantity. The *calix*, now, is a bronze ajutage, inserted[9] into a conduit or reservoir, and to it the service pipes are attached. Its length ought not to be less than 12 digits, while its orifice[10] ought to have such capacity as is specified.

Calix is defined in the Oxford Latin Dictionary as ‘a short length of brass pipe of a specified diameter installed between a water main or reservoir and a private supply pipe to regulate the amount of water drawn off by the latter’.

Frontinus knew a little about the effect of pressure, but not enough to quantify the effect:

In setting ajutages[11,12] also, care must be taken to set them on the level, and not place the one higher and the other lower down. The lower one will take in more; the higher one will suck in less, because the current of water is drawn in by the lower one.

Frontinus knew about the effect of unmatched upstream pipework:

In some of the reservoirs, though their ajutages[13,14] were stamped in conformity with their lawful measurements, pipes of a greater diameter [than the ajutages] were at once attached to them. As a consequence, the water not being held together for the lawful distance,[15] and being on the contrary forced through the short restricted distance, easily filled the adjoining larger pipes. Care should therefore be taken, as

[7]Sed et calicis positio habet momentum. In rectum et ad libram conlocatus modum servat, ad cursum aquae oppositus et devexus amplius rapit, ad latus praetereuntis aquae conversus et supinus, id est ad haustum pronior, segniter et exiguum sumit. Est autem calix modulus aeneus, qui rivo vel castello induitur; huic fistulae applicantur. Longitudo eius habere debet digitos non minus duodecim, lumen capacitatem quanta imperata fuerit (36).

[8]This translation makes good sense but is not an obvious rendering of the Latin.

[9]The translation may be assuming that ‘induitur’ should be ‘inditur’.

[10]Interior area.

[11]Circa conlocandos quoque calices observari oportet ut ad lineam ordinentur nec alterius inferior calix, alterius superior ponatur. Inferior plus trahit; superior, quia cursus aquae ab inferiore rapitur, minus ducit (113).

[12]here *calices*.

[13]In quibusdam, cum calices legitimae mensurae signati essent, statim amplioris moduli fistulae subiectae fuerunt, unde acciderat ut aqua non per legitimum spatium coercita, sed per brevis angustias expressa facile laxiorem in proximo fistulam inpleret. Ideoque illud adhuc, quotiens signatur calix, diligentiae adiciendum est, ut fistulae quoque proximae per spatium, quod S. C. comprehensum diximus, signentur (112).

[14]here *calices*.

[15]i.e. 50 feet: see below.

often as an ajutage[16] is stamped, to stamp also the adjoining pipe over the length which we stated was prescribed by the resolution of the Senate.

Accreditation

Frontinus not only put forward a standard, but also an accreditation system based on standards and diligence:

The deputy must call in the levellers[17,18] and provide that the *calix* is stamped as conforming to the deeded quantity, and must study[19] the size of the ajutages[20] we have enumerated above, as well as have knowledge of their location, lest it rest with the caprice of the levellers to approve a *calix* of sometimes greater, or sometimes smaller, interior area, according as they interest themselves in the parties. Neither must the deputy permit the free option of connecting directly to the ajutages[21] any sort of lead pipe, but there must rather be attached for a length of fifty feet one of the same interior area as that which the ajutage[22] has been certified to have, as has been ordained by a vote of the Senate which follows.

For then and then only can the overseer be held to his full responsibility, when he understands that none but stamped pipes must be set in place.[23]

Frontinus's Legacy

So Frontinus produced a standard for flow measurement, took account of installation effects, and aimed for consistency and fairness. He even achieved pleasure in such worthwhile engineering as the aqueducts:

With such an array of indispensable structures carrying so many waters,[24] compare, if you will, the idle Pyramids or the useless, though famous, works of the Greeks!

[16]here *calix*.

[17]Procurator calicem eius moduli, qui fuerit impetratus, adhibitis libratoribus signari cogitet, diligenter intendat mensurarum quas supra diximus modum et positionis notitiam habeat, ne sit in arbitrio libratorum, interdum maioris luminis, interdum minoris pro gratia personarum calicem probare. Sed nec statim ab hoc liberum subiciendi qualemcumque plumbeam fistulam permittatur arbitrium, verum eiusdem luminis quo calix signatus est per pedes quinquaginta, sicut senatus consulto quod subiectum est cavetur (105).

[18]surveyors.

[19]diligently.

[20]here measures (i.e. meters).

[21]a pronoun that refers to the *calix*.

[22]here *calix*.

[23]Ita demum enim vilicus cum scierit non aliter quam signatas conlocari debere, omni carebit excusatione (112—immediately following previous extract from 112).

[24]Tot aquarum tam multis necessariis molibus pyramidas videlicet otiosas compares aut cetera inertia sed fama celebrata opera Graecorum (16).

As is stated in the Loeb edition of Frontinus's works 'it is the personality of the writer that one loves to contemplate, his sturdy honesty, his conscientious devotion to the duties of his office, his patient attention to details, his loyal attachment to the sovereign whom he delighted to serve, his willing labours in behalf of the people whose convenience, comfort and safety he aimed to promote'.

References

AGA (1951) Investigation of orifice meter installation requirements. American Gas Association, Washington

AGA (1954a) Investigation of orifice meter installation requirements. American Gas Association, Washington

AGA (1954b) Large diameter orifice meter tube tests. Final report of supervising committee. Research project NX-4. American Gas Association, Washington

AGA-ASME (1935) History of orifice meters and the calibration, construction and operation of orifices for metering. Report of the joint AGA—ASME committee on orifice coefficients

Agarwal NK (1994) Mean separation and reattachment in turbulent pipe flow due to an orifice plate. J Fluids Eng 115:373–376

API (1985) Manual of petroleum measurement standards, chapter 14—natural gas fluids measurement, section 3—orifice metering of natural gas and other related hydrocarbon fluids, 2nd edn. ANSI/API 2530

API (1990) Manual of petroleum measurement standards, chapter 14—natural gas fluids measurement, section 3—concentric, square-edged orifice meters, part 1—general equations and uncertainty guidelines. MPMS chapter 14, section 3, part 1, 3rd edn. American Petroleum Institute, Washington DC

Batchelor GK (1967) An introduction to fluid dynamics. Cambridge University Press, Cambridge

Bates CJ (1981) Laser Doppler anemometry measurements of the flow through an orifice plate. Flow 81 its measurement and control in science and industry 2. Instrument Society of America, Pittsburgh, pp 59–68

Bean HS, Benesh ME, Buckingham E (1931) Experiments on the metering of large volumes of air. Bur Stand J Res 7:93–145 (Research paper no 335)

Bean HS, Buckingham E, Murphy PS (1929) Discharge coefficients of square-edged orifices for measuring the flow of air. Bur Stand J Res 2:561–658 (Research paper no 49)

Beitler SR (1935) The flow of water through orifices. A study in 1-in., 1½-in., 2-in., 3-in., 6-in., 10-in, and 14-in. lines. Engineering experiment station bulletin no 89. Ohio State University

Bernoulli D (1738) Hydrodynamica, sive de viribus et motibus fluidorum commentarii. Strasbourg

Bernoulli J (1743) Hydraulica nunc primum detecta ac demonstrata directe ex fundamentis pure mechanicis, Anno 1732. Opera Omnia, vol 4, Lausanne and Geneva

Buckingham E (1932) Notes on the orifice meter: the expansion factor for gases. Bur Stand J Res 9 (Research paper no 459)

Chisholm D (1967) Flow of incompressible two-phase mixtures through sharp-edged orifices. J Mech Eng Sci 9(1):72–78

Chisholm D (1977) Research note: two-phase flow through sharp-edged orifices. J Mech Eng Sci 19(3):128–130

Cousins T (1975) An experimental investigation into the design parameters of the Dall tube. Fluid flow measurement in the mid-1970s, 2: paper J-2, National Engineering Laboratory

Curle N, Davies H J (1968) Modern fluid dynamics. Incompressible flow, vol 1. van Nostrand Reinhold, London de Leeuw H (1994) Wet-gas flow measurement by means of a Venturi meter and a tracer technique. In: Proceedings of the 12th North Sea Flow Meas Workshop, Peebles. National Engineering Laboratory Executive Agency, East Kilbride, Glasgow

de Leeuw H (1997) Liquid correction of Venturi meter readings in wet-gas flow. In: Proceedings of the 15th North Sea Flow Meas Workshop, paper 21, Norway

DeOtte RE, Morrison GL, Panak DL, Nail GH (1991) 3-D laser Doppler anemometry measurements of the axisymmetric flow field near an orifice plate. Flow Meas Instrum 2(2): 115–123

Eujen E (1969) Die Durchflußmessung mit Drosselgeräten bei laminarer Rohrströmung. Brennst-Wärme-Kraft 21(12):628–630

Frontinus SJ (c. 97–98) De aquis urbis Romae

Hall GW (1959) Application of boundary layer theory to explain some nozzle and Venturi flow peculiarities. Proc Inst Mech Eng 173(36):837–870

Halmi D (1973a) Metering performance investigation and substantiation of the "Universal Venturi Tube" (U.V.T.). Part 1—hydraulic shape and discharge coefficient. ASME paper 73-WA/FM-3

Halmi D (1973b) Metering performance investigation and substantiation of the "Universal Venturi Tube" (U.V.T.). Part 2—installation effect, compressible flow and head loss. ASME paper 73-WA/FM-4

Halmi D (1973c) Practical guide to the evaluation of the metering performance of differential producers. J Fluids Eng:127–141 (March 1973)

Herschel C (1887) The Venturi water meter: an instrument making use of a new method of gauging water; applicable to the cases of very large tubes, and of a small value only, of the liquid to be gauged. Trans Am Soc Civ Eng 17:228–258

Herschel C (1895) The Venturi meter patented by Clemens Herschel hydraulic engineer and by Builders Iron Foundry, made by Builders Iron Foundry, founders and machinists. Livermore and Knight Co., Providence

Hickstein EO (1915) The flow of air through thin-plate orifices. Trans ASME 37:765–793

Ho YS, Leung TP (1985) Performance of conical entrance orifice plates at low Reynolds numbers. Int J Heat Fluid Flow 6(2):122–125

Hodgson JL (1917) The commercial metering of air, gas and steam. Minutes Proc Inst Civ Eng CCIV 108–193

ISO (1967) Measurement of fluid flow by means of orifice plates and nozzles. International Organization for Standardization, Geneva. ISO/R 541:1967

ISO (1980) Measurement of fluid flow by means of orifice plates, nozzles and Venturi tubes inserted in circular cross-section conduits running full. International Organization for Standardization, Geneva. ISO 5167:1980

ISO (1991) Measurement of fluid flow by means of pressure differential devices—part 1: orifice plates, nozzles and Venturi tubes inserted in circular cross-section conduits running full. International Organization for Standardization, Geneva. ISO 5167-1:1991

ISO (1998) Measurement of fluid flow in closed conduits—guidelines on the effects of flow pulsations on flow measurement instruments. International Organization for Standardization, Geneva. ISO/TR 3313:1998

ISO (2003a) Measurement of fluid flow by means of pressure differential devices inserted in circular cross-section conduits running full—part 1: general principles and requirements. International Organization for Standardization, Geneva. ISO 5167-1:2003

ISO (2003b) Measurement of fluid flow by means of pressure differential devices inserted in circular cross-section conduits running full—part 2: orifice plates. International Organization for Standardization, Geneva. ISO 5167-2:2003

ISO (2003c) Measurement of fluid flow by means of pressure differential devices inserted in circular cross-section conduits running full—part 3: nozzles and Venturi nozzles. International Organization for Standardization, Geneva. ISO 5167-3:2003

ISO (2003d) Measurement of fluid flow by means of pressure differential devices inserted in circular cross-section conduits running full—part 4: Venturi tubes. International Organization for Standardization, Geneva. ISO 5167-4:2003

ISO (2007a) Measurement of fluid flow by means of pressure differential devices—guidelines for the specification of orifice plates, nozzles and Venturi tubes beyond the scope of ISO 5167. International Organization for Standardization, Geneva. ISO/TR 15377:2007

ISO (2007b) Measurement of fluid flow by means of pressure differential devices—guidelines on the effect of departure from the specifications and operating conditions given in ISO 5167. International Organization for Standardization, Geneva. ISO/TR 12767:2007

ISO (2008) Guidelines for the use of ISO 5167:2003. International Organization for Standardization, Geneva. ISO/TR 9464:2008

Ito H, Watanabe Y, Shoji Y (1985) A long-radius inlet nozzle for flow measurement. J Phys E Sci Instrum 18:88–91

Jaumotte AL, Onockx L (1985) Sur les valeurs du coefficient d'épaisseur d'un ajustage de Borda. Bulletin de la Classe des Sciences, 5e série, tome 71:138–146. Académie royale des Sciences, des Lettres et des Beaux-Arts, Bruxelles

Kastner LJ, McVeigh JC (1965–1966) A reassessment of metering orifices for low Reynolds numbers. Proc I Mech E 180(1): 331–345

Kent WG (1912) An appreciation of two great workers in hydraulics; Giovanni Battista Venturi… Clemens Herschel. Privately printed, London

Kinghorn FC, McHugh A (1978) The performance characteristics of Borda inlet air flowmeters. J Phys E: Sci Instrum 11:87–93

Kinghorn FC, McHugh A, Crawford DA (1975) Effect of cross-wind on three types of differential pressure inlet flowmeter. NEL report no 605. National Engineering Laboratory, East Kilbride

Lake WT (1999) Guidelines for choosing, installing, maintaining offshore metering: dealing with the "construction gorillas". Offshore 152, 153, 198. September 1999

Lindley D (1966) Venturimeters and boundary layer effects. Ph.D. thesis, Dept. of Mech. Eng, Univ. Coll. of South Wales and Monmouthshire, Cardiff

Martin CNB (1986) Effects of upstream bends and valves on orifice plate pressure distributions and discharge coefficients. NEL report no 702. National Engineering Laboratory, East Kilbride

Mattingly GE, Davis RW (1977) Numerical solutions for laminar orifice flow. ASME paper 77-WA/FE-13, American Society of Mechanical Engineers, New York

McKeon BJ, Li J, Jiang W, Morrison JF, Smits AJ (2004) Further observations on the mean velocity distribution in fully developed pipe flow. J Fluid Mech 501:135–147

McVeigh JC (1974) The effect of installation conditions on the discharge coefficient of the conical entrance orifice plate at low Reynolds numbers. In: Dowdell R (ed) Flow its measurement and control in science and industry, vol 1, no 2. Instrument Society of America, Pittsburgh, pp 533–537 (conference was in May 1971)

Mills RD (1968) Numerical solutions of viscous flow through a pipe orifice at low Reynolds numbers. J Mech Eng Sci 10(2):133–140

Morrison GL (1990) 3-D laser anemometer study of compressible flow through orifice plates. Part 1. GRI report no GRI-90/0036. Gas Research Institute, Chicago

Morrison GL (1993) 3-D laser anemometer study of compressible fluid flow through orifice plates. Final report to Gas Research Institute (September 1986–December 1992). GRI report no GRI-93/0314. Gas Research Institute, Chicago

Morrison GL, DeOtte RE, Beam EJ (1992a) Installation effects upon orifice flow meters. 71st Annual Gas Processors Association convention, Anaheim, California

Morrison GL, DeOtte RE, Nail GH, Panak DL (1992b) Mean velocity and turbulence fields inside a $\beta = 0.50$ orifice flow meter. AIChE 1992 spring national meeting, New Orleans, paper no 100G

Morrison GL, Hall KR, Holste JC, Macek M L, Ihfe LM, DeOtte RE, Terracina DP (1993) Slotted orifice flow meter/conditioner. Flow measurement for the utilities, Amsterdam, the Netherlands

Morrison GL, Hall KR, Holste JC, Macek ML, Ihfe LM, DeOtte RE, Terracina DP (1994) Comparison of orifice and slotted plate flowmeters. Flow Meas Instrum 5(2):71–77

Morrow TB (2005) Metering research facility program: additional studies of orifice meter installation effects and expansion factor. GRI report GRI-04/0246 on SwRI project no 18.10315. Gas Research Institute, Chicago

Murdock JW, Foltz CJ (1953) Experimental evaluation of expansion factors for steam. Trans ASME 75(5): 953–959

Panak DL, Morrison GL, DeOtte RE, Nail, GH (1992) Triple velocity correlations and other turbulence quantities from 3-D LDV measurements of the flow in an orifice meter. In: 6th International Symposium Application Laser Techniques to Fluid Mechanics. Lisbon, Portugal

Pearsall IS (1962) Calibration of three conical inlet nozzles. NEL report no 39. National Engineering Laboratory, East Kilbride

Ramamurthi K, Nandakumar K (1999) Characteristics of flow through small sharp-edged cylindrical orifices. Flow Meas Instrum 10(3):133–143

Reader-Harris MJ (2007) ISO flow measurement standards—report on the ISO/TC 30 meeting in November 2006. Flow Meas Instrum 18(3–4):114–120

Sattary JA (1991) EEC orifice plate programme—installation effects. Flow Meas Instrum 2(1): 21–33

Sattary JA, Reader-Harris MJ (1997) Computation of flow through Venturi meters. In: Proceedings of the 15th North Sea Flow Meas Workshop, Norway

Schlichting H (1960) Boundary layer theory. McGraw-Hill, New York

Spearman EP, Sattary JA, Reader Harris MJ (1996) Comparison of velocity and turbulence profiles downstream of perforated plate flow conditioners. Flow Meas Instrum 7 (3/4): 181–199

Spearman EP, Sattary JA, Reader-Harris MJ, Rhodes FS (1995) The effect of upstream installations on orifice meter discharge coefficients. Flow measurement memo FL/445. National Engineering Laboratory Executive Agency, East Kilbride

Stoll HW, Zientara D (1974) The conical entrance orifice plate, an investigation of its performance characteristics. In: Dowdell R (ed) Flow its measurement and control in science and industry, vol 1, no 2. Instrument Society of America, Pittsburgh, pp 517–522 (conference was in May 1971)

Stolz J (1978) A universal equation for the calculation of discharge coefficients of orifice plates. In: Dijstelbergen HH, Spencer EA (eds) Flow measurement of fluids. North Holland Publishing Company, Netherlands, pp 519–534

Studzinski W, Karnik U, LaNasa P, Morrow T, Goodson D, Husain Z, Gallagher J (2001) White paper on orifice meter installation configurations with and without flow conditioners. White paper prepared for API 14.3 Part 2. GRI report GRI 99/0262, Gas Research Institute, Chicago

Studzinski W, Weiss M, Attia J (2000) Effect of reducers, expanders and a gate valve on orifice meter performance. NRTC report # 01426, Nova Research and Technology Corporation, Calgary

Turton RK (1975) A note on flow through conical entrance orifice plates. In National Engineering Laboratory: Fluid Flow Measurement in the mid-1970s 1: paper A-3

Vasy GS, Kastner LJ, McVeigh JC (1968–1969) Profiles for low Reynolds number flows. Proc I Mech E 183(1):591–602

Venturi GB (1797) Recherches expérimentales sur le principe de la communication latérale du mouvement dans les fluides, appliqué à l'explication de différents phénomènes hydrauliques, Paris

Wilson MP, Teyssandier RG (1975) The paradox of the vena contracta. Trans ASME J Fluids Eng 97(3):366–371

Witte R (1931) Die Strömung durch Düsen und Blenden. Forschung auf dem Gebiet des Ingenieurwesens 2: 245 and 291 (two issues)

Witte R (1951) Neue Arbeiten zur Durchflußmessung. Brennst-Wärme-Kraft 3(11):378–380

Zagarola MV, Smits AJ (1998) Mean flow scaling in turbulent pipe flow. J Fluid Mech 373:33–79

Chapter 2
Orifice Design

Abstract This chapter describes the main design specifications for orifice plates (more precisely, orifice meters): it points the reader to important parts of ISO 5167 and gives reasons for the requirements in the standard. It covers the orifice plate (the circularity of the bore, the flatness, the parallelism of the two faces, the surface condition of the upstream face, and, above all, the edge sharpness), the pipe (the pressure tappings, the pipe roughness, the effect of upstream steps, the concentricity of the orifice plate in the pipe and the circularity and cylindricality of the pipe), the measurements of both the orifice plate and the pipe, and the pressure loss. A very significant incorrect installation of an orifice plate within the pipe, a reversed orifice plate, is also covered. Appendix 2.A considers the use of orifice plates of diameter smaller than that permitted in ISO 5167. The effect of upstream fittings is not covered here: it is Chap. 8. The basic instruction remains to follow ISO 5167. For some important deviations from ISO 5167 the errors in discharge coefficient can be calculated using what is described in this chapter.

2.1 Introduction

An orifice plate is fundamentally a plate with a hole machined through it which is inserted into a pipe. As flow passes through the hole it produces a pressure difference across the hole (some of which is recovered). The plate must be sufficiently thin that over the range of permitted thickness its thickness does not affect the discharge coefficient but thick enough not to be distorted by the forces imposed by the pressure difference. The pressure difference is proportional to the square of the flowrate (mass or volume).

The hole may be of any shape if the discharge coefficient is determined by calibration. A few shapes and designs have been produced for which the discharge coefficient can be predicted. Of these only for a round sharp-edged centrally located hole are there sufficient data to allow prediction of flowrate with an uncertainty which is low enough to use for trade in the most valuable fluids: only this shape is covered in this book. Other shapes are listed in NOTE 1 of Chap. 1.

M. Reader-Harris, *Orifice Plates and Venturi Tubes*,
Experimental Fluid Mechanics, DOI 10.1007/978-3-319-16880-7_2

The main design specifications are in ISO 5167-1:2003 and ISO 5167-2:2003 (ISO 2003a, b). This chapter points the reader to important parts of ISO 5167 and gives reasons for the requirements in the standard. It covers the orifice plate (the circularity of the bore, the flatness, the parallelism of the two faces, the surface condition of the upstream face, and, above all, the edge sharpness), the pipe (the pressure tappings, the pipe roughness, the effect of upstream steps, the concentricity of the orifice plate in the pipe and the circularity and cylindricality of the pipe), the measurements of both the orifice plate and the pipe, and the pressure loss. A very significant incorrect installation of an orifice plate within the pipe, a reversed orifice plate, is also covered. Appendix 2.A considers the use of orifice plates of diameter smaller than that permitted in ISO 5167-2:2003. The installation of the orifice meter (the orifice plate and pipe) in terms of upstream straight lengths, flow conditioners and pulsations is covered in Chap. 8.

To design an orifice plate, given the range of mass flowrates, the density, the desired range of differential pressures (discussed in Sect. 4.3), the discharge coefficient (from Chap. 5), the expansibility factor (from Chap. 6) and the desired pipe diameter (D), the throat diameter (d) can be determined from Eq. 1.15:

$$q_m = \frac{C\varepsilon}{\sqrt{1-\beta^4}}\frac{\pi d^2}{4}\sqrt{2\rho_1\Delta p}. \tag{1.15}$$

If this gives too large a value of diameter ratio (β) ($=d/D$) then a larger value of D is required (or a second orifice meter in parallel could be used). For best accuracy $0.2 \le \beta \le 0.6$ is recommended. The uncertainty increases if β is outside this range. If there is no preferred value of D then it would be good to choose a value of D so that $0.5 \le \beta \le 0.6$. As β increases the effects of pipe roughness (see Sect. 2.3.3), installation (see Chap. 8) and eccentricity (see Sect. 2.3.5) increase: this is reflected in an increase in the uncertainty of the discharge-coefficient equation (see 5.3.3.1 of ISO 5167-2:2003 and Fig. 5.12). As β is reduced the pressure loss increases. If $\beta < 0.2$ there is a scarcity of data (see Chap. 5) and unless D is large there are increasing problems with edge sharpness (see Sect. 2.2.4); there is an increase in the uncertainty of the discharge coefficient.

2.2 Orifice Plate

2.2.1 General

The orifice plate is as in Fig. 2.1 (see also Fig. 2.2: the plate pictured is for insertion between flanges and has a handle for ease of removal). Good quality of manufacture is very important, especially the following:

- The circularity of the bore (5.1.8.3 of ISO 5167-2:2003) (see Sect. 2.4)
- the flatness (5.1.3.1 of ISO 5167-2:2003; see Sect. 2.2.2),

Fig. 2.1 Orifice plate nomenclature

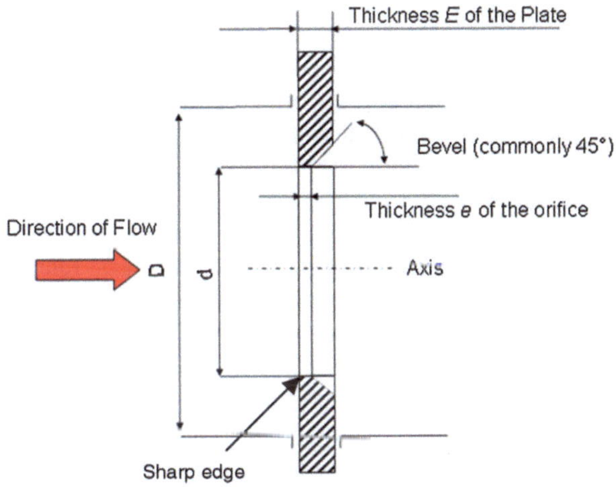

Fig. 2.2 An orifice plate

- the parallelism of the two faces (5.1.5.4 of ISO 5167-2:2003),
- the surface condition of the upstream face (5.1.3.2 of ISO 5167-2:2003; see Sect. 2.2.3),
- and, above all, the edge sharpness (5.1.7.2 of ISO 5167-2:2003; see Sect. 2.2.4).

The requirement for parallelism is not, as far as the author knows, based on experiment, but simply on what can reasonably be manufactured; that is why there are different requirements for $D < 200$ mm. This lack of similarity suggests that the requirement for larger orifice plates may be sufficient but not necessary. The required values for the thicknesses of the orifice plate (E) and of the orifice bore (e) are based on experiment (see Sect. 2.2.5).

2.2.2 Flatness

The effect of buckling an orifice plate was measured by Jepson and Chipchase (1973, 1975) who measured the effect of bending $\beta = 0.2$ and $\beta = 0.7$ plates in an 8″ line. The deflections of the orifice bore were in the range $0.0039D$ to $0.0625D$ [0.79 to 12.7 mm (1/32″ to ½″)]. An under-estimation of flow of up to 14 % was measured. The experimental work was compared with theoretical work in which the effects of changing the bore position relative to the pressure tappings, of changing the inward radial momentum of the fluid and of changing the bore size were considered. The first was shown to be smaller than the other two and was neglected. Quite good agreement between experiment and theory was obtained. Work was also undertaken by Gorter (1978). Ting (1993) reported results with 4″ and 6″ orifice plates that had been bent by a press, and his results were compared with the theory of Mason et al. (1975). The data are presented in Fig. 2.3. With a slope of 1 % the theoretical model of Jepson and Chipchase shown gives a shift in discharge coefficient of a little less than 0.2 % in absolute value at maximum.

The fundamental need is that the orifice plate not be bent in such a way as to cause a significant shift in discharge coefficient. It can be bent both by poor manufacture and by the effect of the differential pressure across it. The former can be measured in the laboratory; the latter (except in the case of permanent distortion) cannot. It must be calculated. The requirement on which ISO 5167-2:2003 is based is that the discharge coefficient should not change by more than 0.2 % from the value given by a flat plate perpendicular to the pipe axis (Norman et al. 1983). This corresponds to a requirement that the plate should not have a slope ($=2\delta/(D-d)$) greater than 1 % (this assumes that the experimental data for $\beta = 0.2$ in Fig. 2.3 have quite high

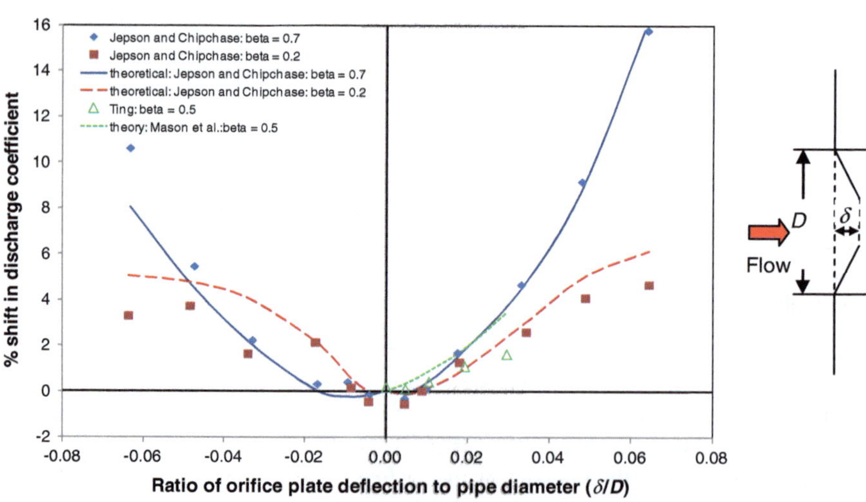

Fig. 2.3 The effect of bending an orifice plate: positive deflection (δ) is in the downstream direction

uncertainty, but that the theoretical model of Jepson and Chipchase can be used). To ensure this, the slope of the plate should not exceed 0.5 % when measured in the laboratory and the differential pressure should not cause a slope of more than 0.5 %. The slope in the laboratory can be measured using feeler gauges. Advice on how to calculate the maximum differential pressure so that the orifice plate is neither elastically deformed beyond the permitted limit nor permanently buckled is given in 5.2.5.1.2.3 of ISO/TR 9464:2008 (ISO 2008). To carry out the calculation the material of the plate, the thickness and how it is supported at the outside edge must be known. More accurate predictions can be calculated using finite element analysis; however, ISO/TR 9464 will normally provide adequate guidance.

2.2.3 Surface Condition of the Upstream Face of the Plate

The requirement (in 5.1.3.2 of ISO 5167-2:2003) on the roughness of the upstream face of the orifice plate that $R_a < 10^{-4}d$ is normally fairly easy to meet at manufacture (definitions of R_a and k are given in Sect. 2.3.3.1). If it were exceeded it would make it difficult to meet the edge-sharpness requirement. It appears to be based on work of McVeigh (1962) and Dall (1958): McVeigh recommends that $k/d \leq 3 \times 10^{-4}$, which is equivalent to the ISO limit: some of McVeigh's data were taken in a 1″ pipe. Data were taken by Studzinski and Berg (1988), but these are surprising: the effect of R_a = 2 μm is greater in an 8″ pipe than in a 4″ pipe. Computational work by Reader-Harris (1991, 1993) concluded that the ISO limit is reasonable for $Re_d \approx 10^7$ and conservative for smaller Re_d. 2.4.1 of API MPMS 14.3.2 (API 2000) requires a maximum of 1.27 μm (50 μin.); this is sufficient (but not non-dimensional); the plates used by API to provide data used to develop the orifice discharge-coefficient equation had R_a in the range 0.10–0.53 μm (4–21 μin.).

One important cause of poor surface condition in practice is the presence of contamination due to deposition. This is covered in 9.3 of ISO/TR 12767:2007 (ISO 2007c), which contains some examples of the effect of contamination (Pritchard et al. 2003, 2004), but does not give a general formula for its effect. The experiments by Pritchard et al. used grease on an orifice plate; this grease moved during the test. Experiments on the effect of grease deposits on an orifice plate have also been performed by Botros et al. (1992) and by Burgin (1971).

The importance of the final portion of the orifice plate approaching the sharp edge can be seen from Morrison et al. (1990): the pressure on the front face of the orifice changes very slowly near the corner between the orifice plate and the pipe, but very rapidly over the portion of the face between a circle whose diameter is about 1.1d and the sharp edge.

There are two patterns of contamination that may occur: one is contamination up to the orifice edge, the other contamination in which a clean ring (kept clean by the flow) is left. The first gives an error that is impossible to predict (it can be directly measured), because the contamination changes the edge sharpness and has a potentially very large effect; some examples of shifts in discharge coefficient are

Fig. 2.4 A contaminated orifice plate

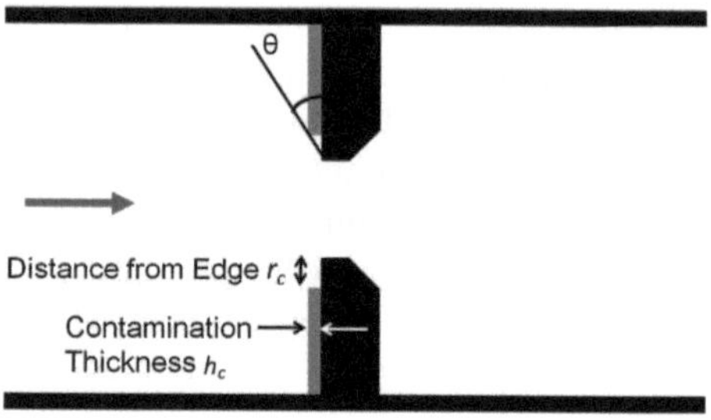

given in Tables 5 and 6 of ISO/TR 12767:2007. The second has been analysed by Reader-Harris et al. (2010): both computational and experimental work were carried out with a uniform layer of contamination of thickness h_c on the front face of an orifice plate, starting a distance r_c from the orifice edge, as shown in Fig. 2.4. The key idea of using θ, the largest angle between the sharp edge and the contamination, was due to Neil Barton at NEL.

The percentage shift, S, in discharge coefficient due to contamination of this pattern was shown to be

$$S = 22.8\left(\frac{h_c}{r_c}\right)^{1.4}\left(1 - \frac{2r_c}{D(1-\beta)}\right)^4 \quad (2.1)$$

This has an uncertainty of 0.28 % based on 2 standard deviations (Reader-Harris et al. 2010).

The effect of very small amounts of liquid in a gas flow can be similar to that of contamination (Ting and Corpron 1995).

The effects of oil coating on orifice plates and pipes have been considered by Johansen et al. (1966) and Johansen (1966). Generally coating the pipe has more effect than coating the plate. Moreover, as would be expected, the effect of coating the pipe increases rapidly with β (see also Morrow 1998). The discharge coefficient increases, but the effect generally reduces with time as the flow removes the oil. Botros et al. (1992) found that very thin liquid films reduce surface roughness and thus the discharge coefficient, an effect not seen in Johansen's or Morrow's work.

The condition of the downstream face of the orifice plate is much less important than that of the upstream (Hobbs and Humphreys 1990): when a plate was roughened by sticking sandpaper of particle size about 1 mm to the downstream face of orifice plates of diameter ratio 0.4, 0.6 and 0.75 in a 12″ line there was no significant change in discharge coefficient; it did not matter whether the roughness was over the whole of the downstream face or simply on the bevel.

In conclusion, contamination is the main cause of problems with surface condition: the main aim must be to avoid contamination. Where this is not possible, there is one pattern of contamination where the error can be predicted from the contamination adhering to the plate; in other situations the error needs to be directly measured.

2.2.4 Edge Sharpness

A rounded orifice upstream edge leads to error because the flow remains attached to the initial part of the rounded edge: therefore there is an increase in the area of the *vena contracta* and thus in the discharge coefficient.

The edge sharpness of an orifice plate may be destroyed by erosion, cavitation or poor handling.

This is a particularly critical area for small orifice plates. To achieve the required sharpness in large plate sizes, i.e. $d > 100$ mm, is straightforward at manufacture, although subsequent damage is possible. When $d < 50$ mm it is difficult, and for $d < 25$ mm very difficult.

The effect of edge radius, i.e. the radius of the sharp edge, r, is a well-worked area. It is discussed in Hobbs and Humphreys (1990). The effect is given in Fig. 2.5, which shows their data, taken in 12″ pipe. The data points with the smallest value of r/d for each β are taken as having no shift in discharge coefficient. The slope of the fitted line is approximately 550; so the effect of a change of edge radius of $0.0002d$ on C is 0.11 %.

Data have also been taken in 2″ and 4″ pipe (Herning 1962), 6″ pipe (Herning and Wolowski 1963), 3″ pipe (Crocket and Upp 1973) and 4″ pipe (Benedict et al. 1974). These are shown together with the NEL data in Fig. 1 of ISO/TR 12767:2007 (ISO 2007c). Data with r/d up to 0.025 were taken by Burgin (1971).

The effect of edge sharpness is also discussed in Spencer et al. (1969) (see below).

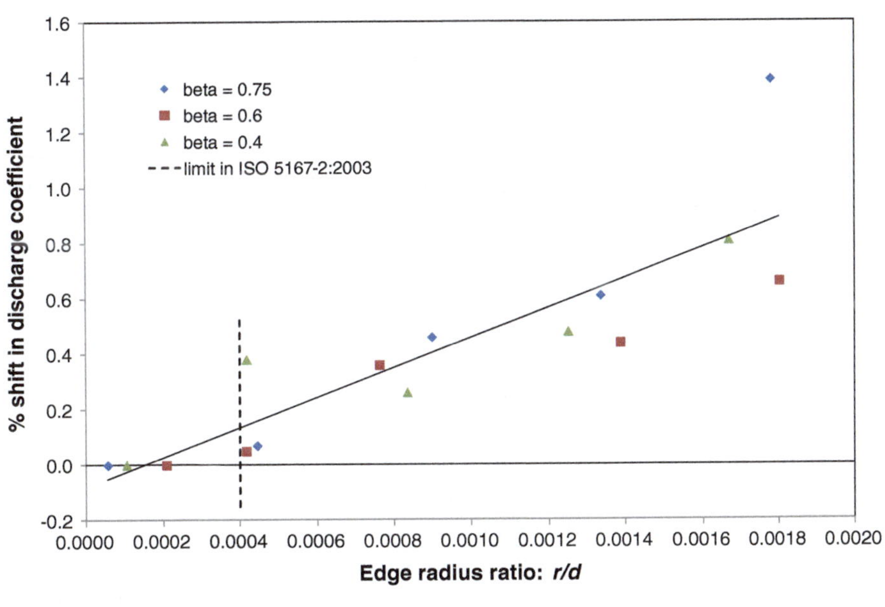

Fig. 2.5 Effect of orifice edge radius (r) on discharge coefficient: NEL tests

Measurement techniques for edge radius are available including casting (Gallagher 1968), lead foil impression (Crocket and Upp 1973), and an optical method in which the condition of the orifice edge is determined by observing the distortion caused by a fine beam of light falling onto the edge (Benedict et al. 1974): the results are compared in Brain and Reid (1973).

Measurements of the edge radii of the plates used for the EEC data for the discharge-coefficient equation are given in Spencer (1987): techniques for the same plate included lead foil, casting and profilometry, in which a Talysurf or equivalent machine is used: the typical spread of measurements of the same plate in different laboratories was around ±5 to ±20 μm, based on the average value of edge radius taken in a laboratory; the spread of measurements at different points of the circumference for a single plate in one laboratory could be as large as ±20 μm.

In almost all cases it is better to manufacture the edge sharpness at the required value rather than to attempt to measure it and make appropriate corrections.

From this work comes the requirement that $r \leq 0.0004d$, where r is the radius of the edge. In practice the uncertainty of the measurement tends to be about 10 μm at best. So for small d it may not be clear whether or not a plate has passed. Difficulties in measurement of edge radius led to difficulty in determining how the plates with small diameters should be used in determining the orifice-plate discharge-coefficient equation. Certainly for d less than about 38 mm edge radius is a cause of additional uncertainty. The fact that the plates in 3″ pipe gave data in agreement with the data for larger sizes is more remarkable than that the plates in 2″ pipe did not. It is common to look at an orifice upstream edge and to decide that it is unacceptable if it reflects a beam of light when viewed without magnification. This method enables some poor plates to be identified but takes no account of the fact that the required edge sharpness is proportional to orifice diameter.

One possible solution for small orifices is to spark-erode them. The consequences of using this technique require more research, but it appears that one manufacturer gave a consistent edge radius of about 9 μm. This gives sharp edges for $d \geq 25$ mm. This area is investigated in Appendix 2.A.

Another source of error is damage to an edge by a chip, nick or groove in the sharp edge. It is helpful to distinguish between a notch that extends completely through the orifice bore (increasing the orifice area) and a groove that cuts the upstream face of the plate and the sharp edge but does not increase the orifice area. The tests of Humphreys and Hobbs (1990) were of the latter type and gave remarkably small effects even where a large amount of damage was inflicted on the plate: for example, a cold chisel was held at an angle of 45° against the sharp edge and struck with a hammer: the metal was not removed but displaced. After one strike the dimensions of the damage were of the order of 2 mm for $\beta = 0.34$ and 3 mm for $\beta = 0.75$ in a 12″ line: the area of the bore was not increased; the increase in discharge coefficient in the case of both plates was in the range 0.02 or 0.03 to 0.18 % depending on the location of the damage relative to the pressure tappings. A further strike increased the damage to the order of 3 mm for $\beta = 0.34$ and 4 mm for $\beta = 0.75$: then there was a *decrease* in discharge coefficient from that of the undamaged plate of 0.33 % to 0.51 % for $\beta = 0.34$ and 0.10 % to 0.20 % for

$\beta = 0.75$ depending on the location of the damage relative to the pressure tappings. Botros et al. (1992) undertook a study of the effect of both a groove and a notch. The effect of a notch that extends through the plate is approximately $\frac{4}{3}$ times the effect of the area increase (and at maximum twice the area increase). Botros et al. state that a typical small nick encountered in a pipeline system causes a measurement error of the order of −0.002 %.

In conclusion, the sharpness of the upstream edge of the orifice plate is of the utmost importance to the accuracy of the flow measurement. In fiscal measurement systems the orifice plate is checked regularly mainly because of the risk that its edge might be damaged (the check will also find any contamination adhering to the surface). A large defect on a small part of the sharp edge has a small effect on the discharge coefficient, whereas a small defect over the whole of the sharp edge has a large effect.

2.2.5 *Plate Thickness* E *and Orifice (Bore) Thickness* e

2.2.5.1 General

Essentially the plate thickness E has to be sufficiently large that the plate does not bend significantly (see Sect. 2.2.2) but sufficiently small that the installation is geometrically similar to those used for determining the orifice discharge-coefficient equation (Chap. 5). The orifice (bore) thickness e has to be sufficiently small that the orifice is essentially a thin plate but sufficiently large that a square edge can be obtained. Where $e < E$ the plate is bevelled.

The plate thickness is easy to measure with a micrometer caliper. Measuring the orifice (bore) thickness using a depth micrometer is less accurate.

2.2.5.2 Plate Thickness *E*

Husain and Teyssandier (1986b) measured the shift in discharge coefficient in a 6″ flange-tapped orifice meter between a baseline with E = 3.2 mm (1/8″) and tests with values of E up to 19 mm (3/4″). Throughout e was 2.8 mm (0.11″). Bevel angles of 45° and 30° (to the pipe axis) were used. To do this the plate was initially 19 mm (¾″) thick and was successively machined thinner. Part of the effect of changing the plate thickness E is to move downstream corner or flange tappings relative to the upstream face of the plate: the difference in discharge coefficient between an orifice plate of standard thickness with the downstream tapping (E − 3.2 mm) downstream of the flange tapping and the baseline (which is an orifice plate of standard thickness with flange tappings) was calculated from the Reader-Harris/Gallagher (1998) Equation (Eq. 5.22) [5.3.2.1 of ISO 5167-2:2003 (ISO 2003b)]; this is described in Fig. 2.6 as the simple model.

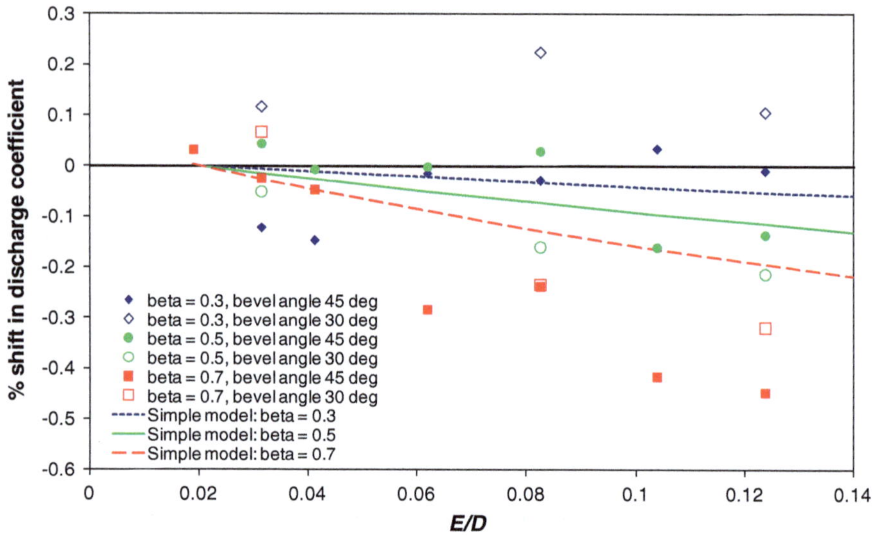

Fig. 2.6 Effect of changing *E/D*. In the simple model the shift is based on the location of the downstream tapping relative to the downstream face of the baseline plate

In Fig. 2.6 the experimental data for $\beta = 0.3$ are surprisingly scattered; those for $\beta = 0.5$ are in good agreement with the simple model; those for $\beta = 0.7$ show a larger shift than the simple model, presumably because in the simple model there is a flow in the part of the orifice meter which in the experiment is the downstream part of the plate.

2.2.5.3 Orifice (Bore) Thickness *e*

The effect of increasing the orifice (bore) thickness *e* is to change the orifice from a thin orifice, in which it is as if *e* were as close as possible to 0 given that a square edge is required on the orifice, to a thick orifice, whose discharge coefficient is around 0.8, because the flow has now reattached to the orifice bore. The discharge coefficient changes slowly where e/d is small but more rapidly around $e/d = 0.5$. This is well exhibited in a set of data from NBS (now NIST) (see Lansverk 1990) given in Fig. 2.7. Flow is unstable where the flow reattaches intermittently to the orifice bore.

In addition to the tests described in Sect. 2.2.5.2, Husain and Teyssandier (1986) measured the shift in discharge coefficient using an unbevelled orifice plate in a 6″ flange-tapped orifice meter from a baseline with $E = e = 3.2$ mm (1/8″) to tests with values of E (=e) up to 19 mm (3/4″). To do this the plate was initially 19 mm (¾″) thick and was successively machined thinner. The data are shown in Fig. 2.8.

The famous OSU data (Beitler 1935) were collected by Ohio State University in the 1930s on many plates, some used in more than one pipe. In some cases the plate

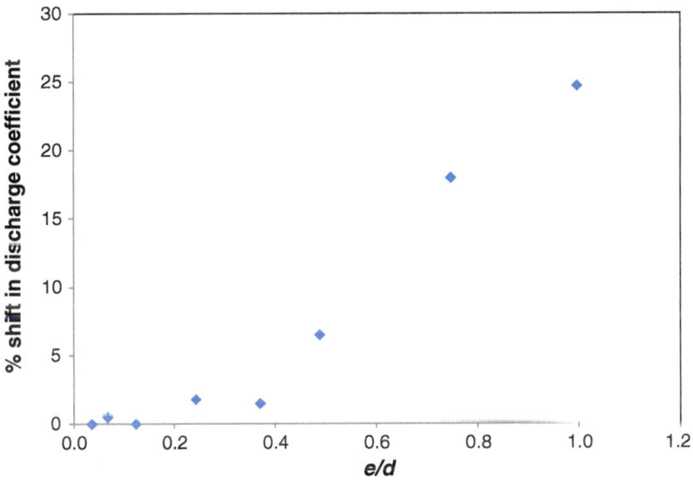

Fig. 2.7 Shift in discharge coefficient as a function of *e*/*d*: 4″ $\beta = 0.25$

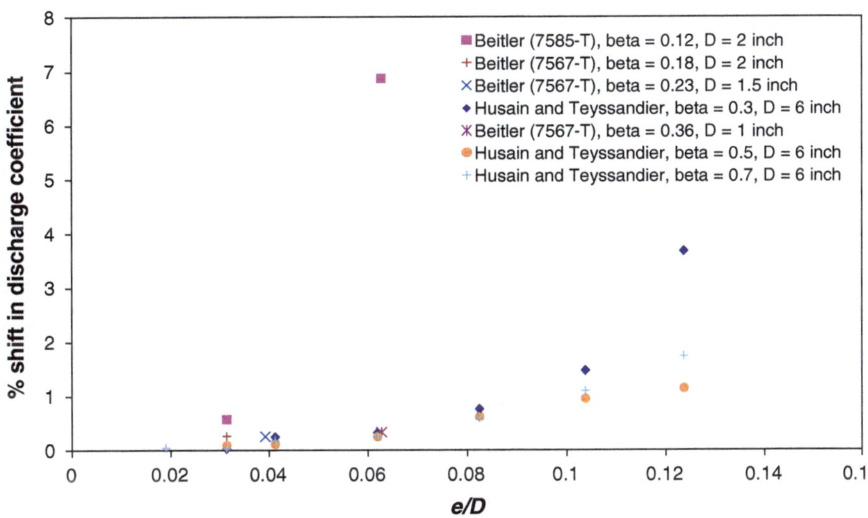

Fig. 2.8 Effect of changing *e*/*D*

was re-bevelled to change the value of *e*. Those for small pipe diameters are presented in Reader-Harris et al. (2008). It is then possible using the points with $e/d < 0.09$ as the baseline to plot the effect of *e*/*D* (see Fig. 2.8). A point from Beitler (7585-T, $\beta = 0.12$, $D = 2''$, $e/d = 0.125$) was omitted because the spread of the deviations from the Reader-Harris/Gallagher (RG) (API) Equation (Eq. 5.23) for different Reynolds numbers was 1.39 %, more than three times the spread of the

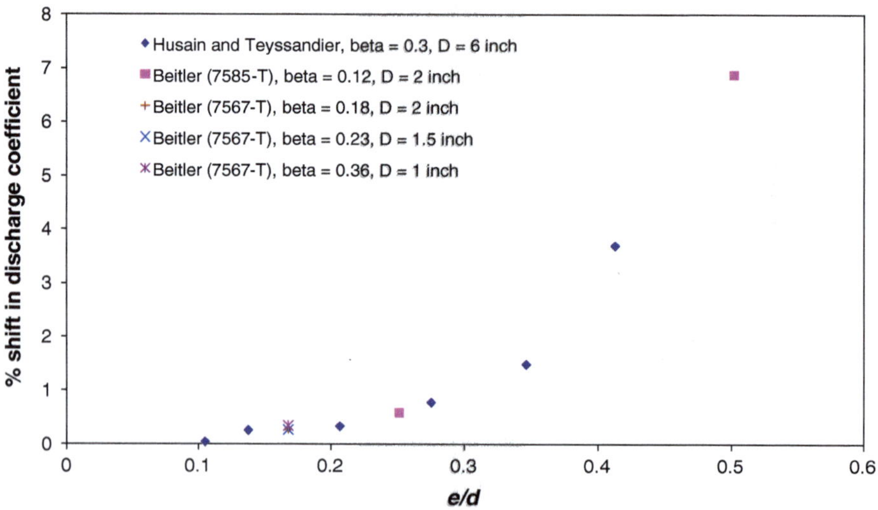

Fig. 2.9 Effect of orifice (bore) thickness (e) for $\beta < 0.4$: shift in discharge coefficient from that obtained when $e/d < 0.09$

deviations for any plotted points (except for the point from Beitler (7585-T), $\beta = 0.12$, $D = 2''$, $e/D = 0.0625$). Where no points are available with $e/d < 0.09$ the data are not plotted. For $\beta \geq 0.2$ the ISO limit of $e/D \leq 0.02$ is adequate, but for smaller β it is inadequate. The data for β up to 0.4 are plotted in Fig. 2.9 against e/d: a common curve for all data as a function of e/d is obtained. A new analysis of the OSU data for larger β and for larger D could be undertaken: the conclusions of Beitler (1935) are that there are two requirements: that $e/d \leq 0.125$ and that $e/h' \leq 0.25$, where h' is the dam height, the distance from the sharp edge of the orifice plate to the nearest point of the pipe wall (i.e. $h' = D(1 - \beta)/2$). The conclusions as regards requirements in Sect. 2.2.5.4 are more demanding than Beitler's and on the basis of the work here should prove adequate.

Husain and Goodson (1986) took data in 2″ pipes with plates that are on the maximum limit in ISO 5167 for plate thickness or exceed it; these data show that an unbevelled plate with $E = 3.57$ mm (5/32″) did not significantly change its discharge coefficient when it was machined to be an unbevelled plate with $E = 3.18$ mm (1/8″) and then when it was bevelled with $e = 0.79$ mm. Larger shifts would have been expected from Figs. 2.8 and 2.9.

2.2.5.4 Requirements

On the basis of the work in Sects. 2.2.5.2 and 2.2.5.3 the requirements in ISO 5167-2:2003 are reasonable:

$$0.005D \leq e \leq 0.02D.$$
$$e \leq E \leq 0.05D.$$

However, there is an additional requirement that should be included in the next revision:

$$e \leq 0.1d.$$

This last requirement is only additional to the previous requirements when $\beta < 0.2$.

The angle of bevel makes no difference to the discharge coefficient, provided that it is 45° ± 15°. Husain and Goodson (1986) showed that having a 30° angle to the bore gave the same discharge coefficient as having a 45° angle; they also state that Ohio State data show that having a 60° angle to the bore makes a negligible difference from having a 45° angle [only data with a 45° bevel are given in Beitler (1935)].

If the plate is unbevelled ($e = E$), then provided that both faces meet the specification for the front face and that flange or corner tappings are used the plate can be used bi-directionally. It will be fairly thin ($E/D \leq 0.02$); so there may be issues with plate bending.

2.2.6 Circularity

The circularity of the orifice plate (and that of the pipe) is considered in Sect. 2.4.

2.3 The Pipe

2.3.1 General

The orifice plate can only be used as a flowmeter in accordance with ISO 5167 when it has upstream and downstream pipe (for an example see Fig. 2.10).

In addition to the effect of upstream (and downstream) fittings, considered in Chap. 8, it is necessary to consider

- the location of the pressure tappings (Sect. 2.3.2),
- the pipe roughness (Sect. 2.3.3),
- the effect of upstream steps (Sect. 2.3.4),
- the concentricity of the orifice plate in the pipe (Sect. 2.3.5)
- and the circularity and cylindricality of the pipe (see Sect. 2.4).

Fig. 2.10 Orifice plate held between flanges

2.3.2 Pressure Tappings

2.3.2.1 General

In ISO 5167 there are three choices for the tappings: flange tappings (Sect. 2.3.2.2), *D* and *D*/2 tappings (Sect. 2.3.2.2) and corner tappings (Sect. 2.3.2.3): see Fig. 2.11.

2.3.2.2 Flange and *D* and *D*/2 Tappings

General

For fiscal oil and gas measurement flange tappings are normally specified: they are located in the flanges 25.4 mm (1″) upstream of the upstream face of the plate and 25.4 mm (1″) downstream of the downstream face of the plate. This distance will normally place them within the thickness of a flange holding the plate (see Fig. 2.12 for an example).

Flange tappings are generally the simplest in terms of manufacturing. Because of the large difference between the inside and outside diameters of the flanges no significant pressure expansion of the pipe will occur at the tappings and no correction in pipe diameter due to static pressure is required; moreover, the smaller the flow velocity is at the tapping, the less important are the roundness, sharpness and depth of the tappings (the main issues with tapping quality occur with Venturi tubes or throat-tapped nozzles: see Chaps. 7 and 9). The disadvantage in the use of orifice

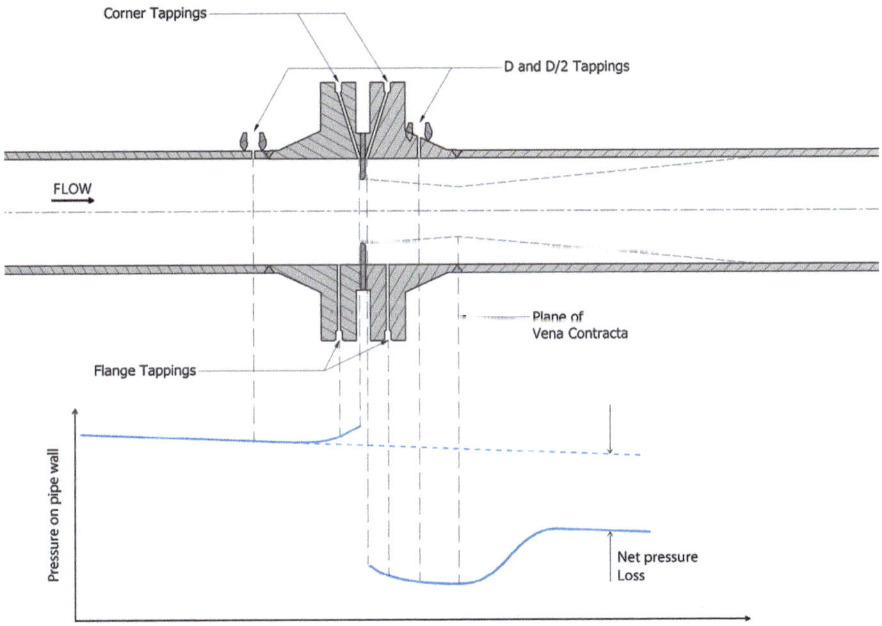

Fig. 2.11 Flow through an orifice plate, showing the pressure tapping positions and approximate pressure profile

Fig. 2.12 8″ orifice meter with flange tappings

plates with flange tappings is that because they are not geometrically similar they introduce additional complexity to the discharge-coefficient equation; moreover, they may be unsuitable for pipes smaller than those covered by ISO 5167-2:2003 (i.e. $D < 50$ mm) because the downstream tapping might be in the pressure-recovery zone, which starts a little to the right of the *vena contracta* in Fig. 2.11.

For D and $D/2$ tappings the tappings are located $1D$ upstream and $D/2$ downstream, both measured from the upstream face of the orifice plate: the upstream tapping is upstream of any disturbance to the pressure from the plate; the downstream tapping is near the pressure minimum for large β. Great care should be taken if tappings are formed by welding a boss to a pipe wall that the welding heat does not create an internal bulge which alters the local velocity at the tapping as well as altering the pipe diameter at the tapping. Welding of bosses should be carried out prior to machining a bore, or an alternative method of preventing distortion employed.

Tapping Diameter

For both flange and D and $D/2$ tappings the tapping diameter is not very critical as long as the centre of the tapping is correctly located: in 5.2.2.7 of ISO 5167-2:2003 the diameter must be less than both $0.13D$ and 13 mm. In the equivalent API standard, API MPMS 14.3.2 (API 2000), the tapping diameters must be 9.5 mm (3/8″) for 2″ and 3″ pipes and 12.7 mm (1/2″) for 4″ pipes and larger. When the data on which the discharge-coefficient equation is based were collected, all the European data were collected in tubes to ISO specification; the American data were collected in tubes to API specification: for example, for 2″ tubes the European tappings were generally 2 mm in diameter, i.e. $0.04D$, and the American tappings were 9.5 mm (3/8″) in diameter, i.e. $0.19D$. There was no noticeable difference between European data and American data: the mean deviation for European water data from the Reader-Harris/Gallagher (1998) Equation was 0.005 %, for American water data it was 0.008 %. All the 3″ data were American (with tappings 9.5 mm (3/8″) in diameter, i.e. $0.122D$) and 95 % of the 3″ data are within about 0.26 % of the Reader-Harris/Gallagher (1998) Equation in ISO 5167-2:2003, an exceptionally good result. So there was no sign that tapping diameter had much effect on the discharge coefficient.

Tapping Location

The tolerances on location are given in 5.2.2.2 and 5.2.2.3 of ISO 5167-2:2003. The permitted upstream and downstream spacings for flange tappings are:

25.4 mm ± 0.5 mm when $\beta > 0.6$ and $D < 150$ mm
25.4 mm ± 1 mm in all other cases, i.e. $\beta \leq 0.6$, or $\beta > 0.6$, but 150 mm $\leq D \leq$ 1000 mm

The Reader-Harris/Gallagher (1998) Equation (Eq. 5.22a, 5.22b) can be used to calculate the effect of moving the tappings: the effect is shown in Figs. 2.13 and 2.14. In Eq. 5.22a, 5.22b A was taken as 0; this gives the maximum shift.

Work has also been carried out by Zedan and Teyssandier (1990).

From Figs. 2.13 and 2.14 the largest errors can occur for large β at the upstream flange-tapping location.

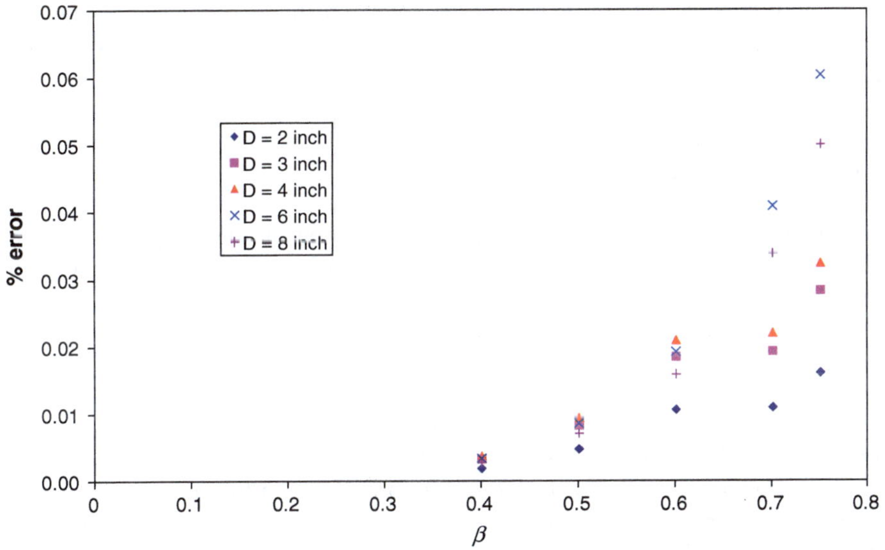

Fig. 2.13 Maximum absolute error due to permissible variation of ISO upstream flange-tapping location

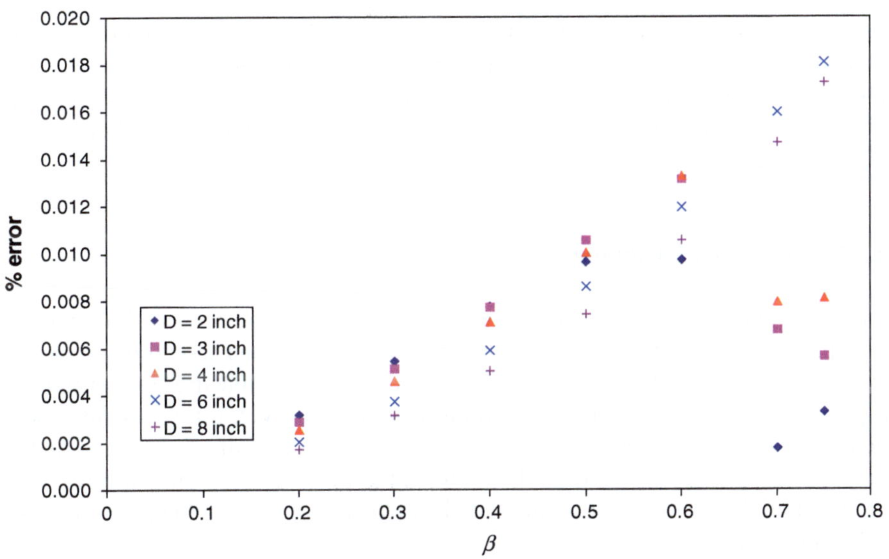

Fig. 2.14 Maximum absolute error due to permissible variation of ISO downstream flange-tapping location

Because the pressure profile is very flat both D upstream of the orifice plate and near the vena contracta the effect of small errors in the location of D and $D/2$ tappings is very small.

2.3.2.3 Corner Tappings

Corner tappings are located in the corners between the plate and the pipe wall. They may be either single tappings or annular slots. In 5.2.3.3 of ISO 5167-2:2003 the three lines headed 'For clean fluids and vapours' give requirements for sizes of corner tappings based on geometrical similarity to the original orifice runs on which the discharge coefficient is based. As β increases the discharge coefficient changes more rapidly with tapping position (as can be seen from the Reader-Harris/Gallagher (1998) Equation) and so the orifice meters need to be closer to the original ones. However, it was accepted that to require a tapping size of a maximum of 1 mm was not sensible; so for $D < 100$ mm values up to 2 mm are acceptable even if not permitted by instructions 1 or 2 in 5.2.3.3 of ISO 5167-2:2003. The three lines headed 'For any values of β' give requirements based on what is sensible for particular fluids. For certain fluids in certain sizes it is not possible to design a system using single corner tappings that is compliant with ISO 5167.

Except for pipes smaller than those covered by ISO 5167-2:2003 flange tappings are much more commonly used than corner tappings.

NOTE Corner tappings may be installed in a carrier ring: this is shown in Fig. 4 of ISO 5167-2:2003. The permitted diameter of the carrier ring is given in 5.2.3.6 of ISO 5167-2:2003. If the tappings are installed in a carrier ring then differences between the ring diameter and that of the pipe become important: see 7.3 of ISO/TR 12767:2007.

2.3.2.4 Number of Tappings

Single tappings or multiple tappings in each tapping plane are permitted. If there are 4 tappings in each tapping plane, those in each tapping plane should be connected in a triple-T arrangement as in Fig. 4.8. If there is an asymmetric flow profile the triple-Ts reduce the effect of upstream installation. However, if there is any risk of liquid in a gas flow tappings should only be in the top half of the pipe, and in practice a single tapping on the side or on the top is common. If there is any risk of gas in a liquid flow tappings should only be in the bottom half of the pipe. See Sect. 4.2.2 for more details.

See ISO 2186:2007 (ISO 2007a) for more information on steam measurement with orifice plates.

2.3.3 *Pipe Roughness*

2.3.3.1 Uniform Roughness

New roughness limits for pipework upstream of orifice plates were included in ISO 5167-2:2003. These are derived from a physical understanding of the effect of pipe roughness on orifice plate discharge coefficients and the knowledge of the roughness of the pipes on which the discharge-coefficient equation in ISO 5167-2:2003 is based. The physical understanding was obtained from computational work at NEL (Reader-Harris and Keegans 1986; Reader-Harris 1990). Pipe roughness limits have been calculated to ensure that roughness will not shift the discharge coefficient from that given by the discharge-coefficient equation by more than an appropriate fraction of its uncertainty.

The friction factor, λ, can be measured directly (see Sect. 1.5), using:

$$\Delta p = \frac{\lambda \rho \bar{u}^2 x}{2D} \tag{1.19}$$

where Δp is the difference in pressure between two tappings spaced a distance x apart in a pipe of diameter D and $\bar{u}$ is the mean velocity in the pipe. It is simpler to measure the arithmetic mean deviation of the roughness profile, R_a, (the average of the absolute values of the deviation from the mean), to deduce the uniform equivalent roughness, k, by taking it to be approximately equal to πR_a (based on the assumption that the wall has a sinusoidal profile) and to calculate λ using the Colebrook-White Equation (Schlichting 1960):

$$\frac{1}{\sqrt{\lambda}} = 1.74 - 2lg\left(\frac{2k}{D} + \frac{18.7}{Re_D\sqrt{\lambda}}\right) \tag{1.21}$$

(following ISO 80000-2:2009, $\log_{10}$ is written lg).

The arithmetic mean deviation of the roughness profile, R_a, or the friction factor, λ, (or both) was measured for each of the pipes used to collect the data to which the discharge-coefficient equation was fitted. Values of roughness are given in the reports listed in Chap. 5. The API (US) pipes had rms (R_q) approximately 1–6 μm (R_a values would be similar, probably slightly smaller) (Whetstone et al. 1989).

NOTE When the 24″ pipework (D = 585.95 mm) for the orifice equation tests was made British Gas (Stewart 1989) measured both its R_a value, obtaining 4.2, 6.1, 5.5 and 3.3 μm at four locations along it, an average of 4.8 μm, and the pressure loss along it, which gave a friction factor of 0.0095 at a Reynolds number of 2.8×10^7, which corresponded to k = 13.8 μm. On this occasion $k/R_a = 2.9$, in good agreement with expectation.

Using the measurements of pipe roughness it was possible to fit a discharge-coefficient equation containing an explicit friction factor term; this was done at the time when the PR14 Equation (see Chap. 5) was developed and, using the PR14 tapping terms, the following equation for the C_∞ and slope terms was obtained for $Re_D > 3700$ (see Eq. 5.C.4):

$$C_\infty + C_s = 0.5945 + 0.0157\beta^{1.3} - 0.2417\beta^8 + 0.000514(10^6\beta/Re_D)^{0.7} + (3.134 + 4.726A')\beta^{3.5}\lambda$$

where (2.2)

$$A' = \left(\frac{2100\beta}{Re_D}\right)^{0.9}.$$

This gives the change in discharge coefficient due to roughness, $\Delta C_{\mathrm{rough}}$, as

$$\Delta C_{\mathrm{rough}} = 3.134\beta^{3.5}\Delta\lambda \tag{2.3}$$

provided that Re_d is sufficiently large that A' is negligible. This equation was evaluated by Morrow and Morrison (1999): 'they compared the CFD results with the Reader-Harris correlation. The changes in pipe roughness and C_d values were in agreement with the equation predictions of Reader-Harris. They also found that the new Reader-Harris equation compensates for systematic variations of C_d with the β ratio that have been included in the past equations' (Morrow et al. 2002).

Figure 2.15 gives measured and computed (using CFD) values of ΔC as a function of $\beta^{3.5}\Delta\lambda$. The computed values (Reader-Harris 1990) and the European experimental data (Clark and Stephens 1957; Herning and Lugt 1958; Spencer et al. 1969; Thibessard 1960; Witte 1953) were obtained using corner tappings. The North American experimental data (Bean and Murdock 1959; Brennan et al. 1989; Studzinski et al. 1990) were obtained using flange tappings. In computational work the effect of roughness on discharge coefficients obtained with different pairs of pressure tappings was considered and it was shown that the effect of pipe roughness on the discharge coefficient using *D* and *D*/2 tappings is about 25 % less than on that using corner tappings. Since all the computational and most of the experimental data in Fig. 2.15 were collected using corner tappings this may explain why Eq. 2.3 lies below the majority of the plotted data.

Nevertheless there is a large scatter in the plotted data, and so a single equation is used to describe the effect of pipe roughness for all tappings. The equation used to determine limits of pipe roughness is again taken from Eq. 2.2, but the A' term is included:

$$\Delta C_{\mathrm{rough}} = (3.134 + 4.726A')\beta^{3.5}\Delta\lambda \tag{2.4}$$

It is not known whether the effect of change in friction factor increases for small Reynolds number, but it is safer to include the term in A' in calculating the limits of pipe roughness. Moreover, there is little disadvantage in its inclusion since it causes a slight reduction in the limits of pipe roughness in a range of Reynolds number where they are already wide.

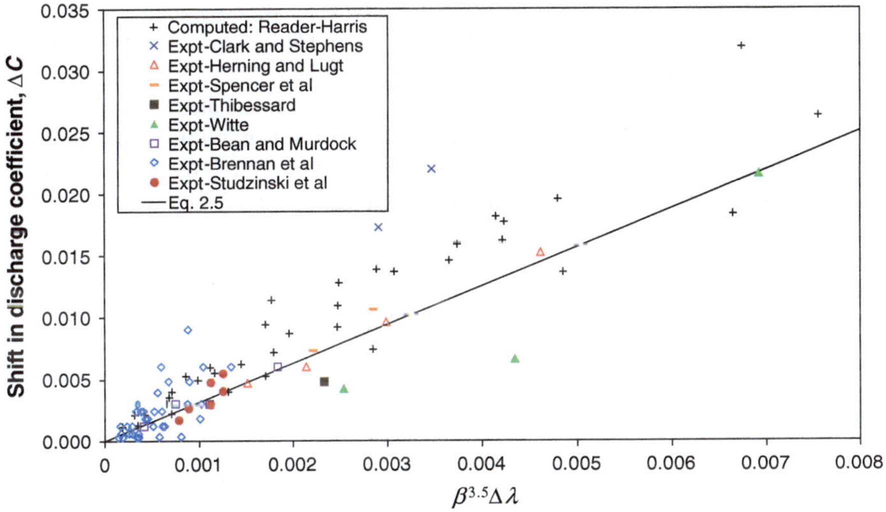

Fig. 2.15 The effect of rough pipe on the orifice plate discharge coefficient

In order to calculate the limits of pipe roughness for the discharge-coefficient equation in ISO 5167-2:2003, it is necessary to use the measured values of relative roughness for each pipe used to collect the data to obtain an approximate mean pipe relative roughness for the data (and thus for the equation fitted to it). The mean pipe relative roughness is a function of Re_D: the estimates are given in Table 2.1. k/D reduces with Re_D because the higher Reynolds numbers generally occurred in the larger pipes, which were generally relatively smoother. The values of friction factor are consistent with the values of relative roughness if the Colebrook-White Equation is used.

The maximum permissible shift in C depends on U, the stated uncertainty of C. It is assumed here that the percentage shift, P, should not exceed the values in Table 2.2.

Table 2.1 Values of k/D and λ associated with the Reader-Harris/Gallagher (1998) Equation

Re_D	10^4	3×10^4	10^5	3×10^5	10^6	3×10^6	10^7	3×10^7	10^8
$10^4 k/D$	1.75	1.45	1.15	0.9	0.7	0.55	0.45	0.35	0.25
λ	0.031	0.024	0.0185	0.0155	0.013	0.0115	0.0105	0.010	0.0095

Table 2.2 Maximum permissible percentage shift, P, in C due to pipe roughness

P	β
0.5β	$\beta \leq 0.5$
0.25	$0.5 < \beta \leq 0.6$
$0.5(1.667\beta - 0.5)$	$0.6 < \beta \leq 0.71$
$1.13\beta^{3.5}$	$0.71 < \beta$

This restriction ensures that for $\beta \leq 0.5$, where other sources of error are dominant, $P/U < \beta$; for $0.5 < \beta \leq 0.71$ $P/U < 0.5$; for $0.71 < \beta$ the maximum value of P/U increases from 0.5 at $\beta = 0.71$ to 0.55 at $\beta = 0.75$.

Given the value of P in Table 2.2 and the value of λ associated with the data in the database (and thus with the equation) in Table 2.1, it is possible from Eq. 2.4 to calculate the maximum and minimum values of λ and hence of k/D for use with the equation for each Re_D and β. It was decided to write ISO 5167:2003 in terms of R_a rather than k for specification of pipe roughness, since this is more convenient for users, although k is still mentioned. The maximum and minimum values of R_a/D for pipes upstream of orifice plates are given in Tables 1 and 2 of ISO 5167-2:2003.

It is stated in ISO 5167-2:2003 that the roughness requirements are satisfied in both the following cases:

$$1\ \mu\text{m} \leq R_a \leq 6\ \mu\text{m},\ D \geq 150\ \text{mm},\ \beta \leq 0.6 \text{ and } \text{Re}_D \leq 5 \times 10^7.$$

$$1.5\ \mu\text{m} \leq R_a \leq 6\ \mu\text{m},\ D \geq 150\ \text{mm},\ \beta \leq 0.6 \text{ and } \text{Re}_D \leq 1.5 \times 10^7.$$

These are not additional restrictions, just a method of removing the need to use Tables 1 and 2 of ISO 5167-2:2003 in some cases.

The tables prescribe $k/D \leq 0.005$ even if the calculated value is higher. Although the limits for very large Re_D are much tighter than those in ISO 5167-1:1991, for $Re_D = 3 \times 10^5$ they are very similar; this is unsurprising since the limits in ISO 5167-1:1991 were probably derived from data collected at around that Reynolds number. The fact that the original orifice meters were not hydraulically smooth means that in some situations a hydraulically smooth pipe is too smooth for the Reader-Harris/Gallagher (1998) Equation to be used without additional uncertainty. The minimum values for relative pipe roughness in Table 2 of ISO 5167-2:2003 represent exceedingly smooth pipes.

It might be argued that an equation with a friction factor term included explicitly should have a lower uncertainty. Such an equation was rejected on the grounds that it would be difficult to use.

2.3.3.2 Rough Pipes with a Smooth Portion Immediately Upstream of the Orifice

The tests of Clark and Stephens (1957) included the effect of extremely rough pipes (e.g. ones with peas stuck to them) and also the effect of cleaning: $5D$ of clean pipe upstream of the orifice plate reduced the error due to roughness to about a tenth of what it would have been. Where significant change in roughness over time might be expected a possible strategy might be to use stainless steel pipe for the pipe immediately upstream of the plate, carbon steel further upstream.

To examine the effect of a smooth section of pipe immediately upstream of an orifice plate tests were undertaken at SwRI (Morrow et al. 2002) in a 4″ pipe.

Baseline data were taken with the whole upstream pipe of roughness $R_a = 1.27$ μm (50 μin.). The pipe from 10*D* to 45*D* upstream of the orifice plate was then roughened: the pipe roughness increased from 1.27 to 18.3 μm (50 to 719 μin.). With $\beta = 0.67$ the results were very similar to the baseline (a shift in discharge coefficient of 0.15 % in magnitude at most). This suggests that cleaning the last 10*D* of a pipe will in many cases be sufficient.

When two high-performance flow conditioners were included in turn 10*D* upstream of the orifice plate in the baseline pipe the data were very close to the baseline; when the same flow conditioners were included in turn 10*D* upstream of the orifice plate with roughened pipe from 10*D* to 45*D* upstream of the orifice plate the data were again very close to the baseline. This suggests that rough pipe upstream of a flow conditioner does not spoil its performance.

The final set of tests were carried out with the pipe rough [$R_a = 18.3$ μm (719 μin.)] all the way from 45*D* upstream to the orifice plate: with $\beta = 0.67$ the shift in discharge coefficient was an increase of about 1.7 %. Inserting the two high-performance flow conditioners in turn 10*D* upstream of the orifice plate with rough pipe upstream and downstream of the flow conditioner gave a shift from the smooth baseline at least as large as had been obtained with a rough pipe and no flow conditioner and in some cases almost twice as large. So whereas rough pipe upstream of the flow conditioner has very little effect, rough pipe downstream of the flow conditioner may have a very significant effect.

2.3.3.3 Non-uniform Roughness

In practice roughness is often non-uniform, because in gas pipelines liquid is particularly present on the bottom of the pipe. Before undertaking research in this area video recordings of the inside of 10″ pipes (230 mm ID) were kindly lent to NEL by BP Amoco, and from these it was concluded that it would not be unreasonable to assume that values of roughness height *k* of the order of 0.5 mm would be possible. Accordingly experiments were undertaken with the same level of relative roughness as had been found in these pipes (Reader-Harris et al. 2003).

To achieve the same relative roughness in a 4″ pipe as had been found by BP Amoco gave *k* of the order of 0.2 mm. This corresponds to a P80 Sandpaper (actually silicon carbide paper, commonly called wet and dry paper) with a grain size of 197 μm. It was decided to work also with a smoother sandpaper (P240) with a grain size of 60 μm. Sandpaper of each roughness was obtained and glued to the inside of the pipe from ½*D* upstream of the orifice plate to 10*D* upstream of it. As shown in Sect. 2.3.3.2 the discharge coefficient is largely affected by the pipe roughness over the 10*D* upstream of the orifice plate, and it was necessary to leave 25 mm clean upstream of the flange tappings as the sandpaper was simply glued to the surface. Flange Tappings D were at the bottom of the pipe (see Fig. 2.16) and except where the whole pipe was rough the sandpaper was at the bottom of the pipe and symmetrical about the bottom of the pipe. It was not simple to ensure that the sandpaper remained stuck to the surface, but with the rougher sandpaper data were

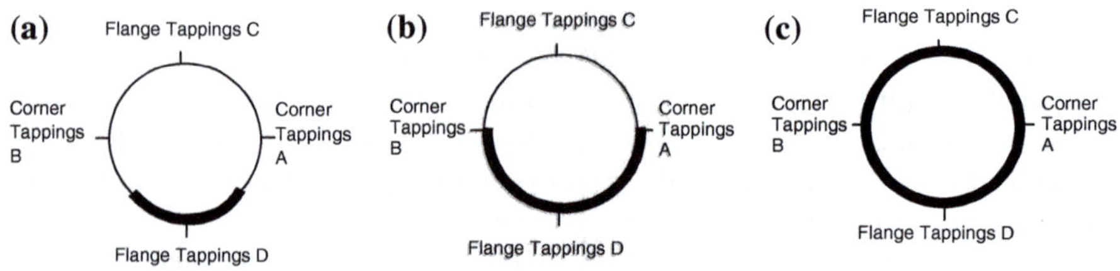

Fig. 2.16 Roughened portions of pipes relative to the pressure tappings. **a** Quarter pipe roughened. **b** Half pipe roughened. **c** Whole pipe roughened

obtained in which the sandpaper remained satisfactorily stuck to the walls. The smoother sandpaper was generally weaker, and so in addition to its tendency to become unstuck it also became wavy in use, and no satisfactory data could be obtained.

The data obtained with the rougher sandpaper are presented in Fig. 2.17. It is interesting that similar data are obtained with each pair of tappings and that the shift is approximately proportional to the fraction of the pipe which is rough. According to Eq. 2.3 for uniformly rough pipe the predicted shift in discharge coefficient at $Re_D = 2.74 \times 10^5$ was 1.26 %, using the Colebrook-White Equation (Eq. 1.21) to determine the friction factor and assuming that the smooth pipe had $R_a = 0.8$ μm and that $k = \pi R_a$. Additionally it would be expected that for entirely rough pipe the thickness of the roughness elements and the paper might increase the shift in discharge coefficient by 0.4 %. So the measured shift in discharge coefficient of 1.98 % compares quite well with the predicted value of 1.66 %.

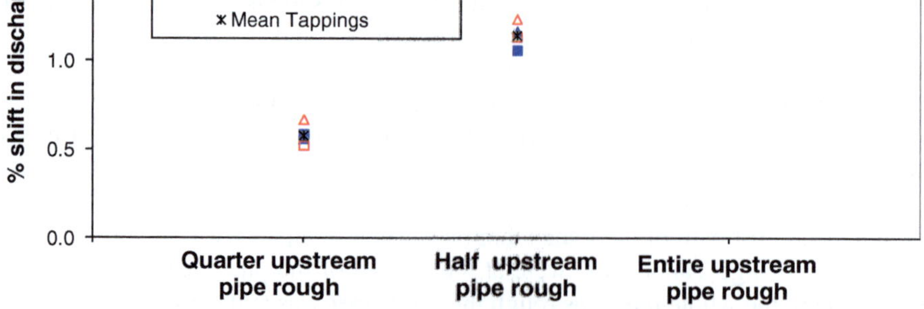

Fig. 2.17 % shift in discharge coefficient due to rough pipe upstream of an orifice plate ($\beta = 0.67$)

2.3.4 Steps and Misalignment

Over the first 2*D* of the pipe upstream of the orifice no measured diameter of the pipe is permitted to differ by more than 0.3 % from *D*, the mean diameter of the pipe (see 6.4.2 of ISO 5167-2:2003 and Sect. 2.4). To be consistent with ISO 5167-2:2003 a protruding weld should not cause a change in measured diameter greater than 0.3 %; if it is greater than this value the weld bead should be removed.

The basis on which this requirement was derived is not clear, but Teyssandier (1985) considered the effect of steps or gaps *D*/4 wide adjacent to an orifice plate or 2*D* upstream of it in a 2″ pipe. As far as these data are concerned a limit of 0.3 % in maximum permissible step as in ISO 5167-2:2003 appears conservative.

Beyond 2*D* from the orifice the standard is less clear. The rules on steps are clear, if rather complicated. They result from NEL tests (Reader-Harris and Brunton 2002): Fig. 2.18 shows the shift in discharge coefficient where pipes of different schedule were placed at different distances upstream of a schedule 40 pipe: the effect of a contraction by about 5 % of *D* (Schedule 10 to Schedule 40) is much less than the effect of an expansion by about 5 % of *D* (Schedule 80 to Schedule 40). The data point for a Schedule 40 pipe shown in the figure was taken using an ordinary Schedule 40 pipe, whereas the baseline pipe was honed. 6.4.3 of ISO 5167-2:2003 permits steps of up to 0.3 % of *D* beyond 2*D* from the orifice plate and up to 2 % of *D* beyond 10*D* from the orifice plate. If the step is a contraction it permits steps of up to 6 % of *D* beyond 10*D* from the orifice plate. It also permits steps (expansions or, of course, contractions) of up to 6 % of *D* beyond the closest permissible location for the insertion of an expander provided that the step is also at

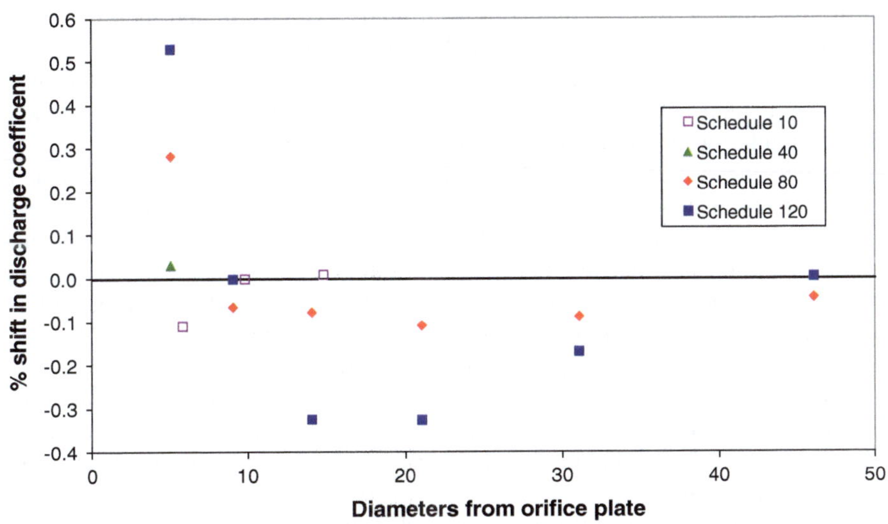

Fig. 2.18 Upstream steps: % shift in discharge coefficient due to pipes at various distances upstream of an orifice plate ($\beta = 0.67$) in a Schedule 40 pipe (Reader-Harris and Brunton 2002)

least $10D$ from the orifice plate. Although larger steps could be used satisfactorily for sufficiently small β it is generally wise to design a system so that β may be easily changed if the flow is higher or lower than the design flow.

The aim of these changes to ISO 5167 was to avoid the requirement to use machined pipes beyond $10D$ from the orifice. Moreover, a change of 6 % of D generally permits a change from one common pipe schedule to the next. The NEL tests provide justification for common practice in calibration laboratories where upstream pipes of the same nominal diameter but larger internal diameter than the customer's meter are used $25D$, say, upstream of the meter.

The steps in Reader-Harris and Brunton (2002) were concentric steps, but at steps there may be offsets as well: 6.4.3 of ISO 5167-2:2003 states that the actual step caused by misalignment and/or change in diameter shall not exceed the permitted diameter step at any point of the circumference. To meet these restrictions will probably require the use of dowels or the equivalent to ensure alignment. If it is intended to machine the pipe immediately upstream of the orifice plate but not to machine the pipe upstream of that then the latter pipe should be selected to be sufficiently cylindrical that the former pipe can be machined to a diameter such that where the flanges meet there is a sufficiently small step. If the pipe immediately upstream of the orifice plate is manufactured first then the pipe upstream of that may require machining, and it may be necessary to buy a thicker walled pipe to machine.

What is not clear are the rules where the internal diameter of the upstream pipe changes not suddenly (as with a step) but gradually. This is not permitted in the first $2D$ upstream of the orifice plate. If the diameter change occurs more than $2D$ away it is not clear what is permitted. The standard is not clear because of a lack of experimental data. What often results in poor practice is to buy a pipe for installation immediately upstream of the orifice, measure it, find it insufficiently cylindrical, machine the first $2D$ until it is sufficiently cylindrical, and then blend the machined and unmachined diameters at $2D$. The problem with this is that it leaves an expansion $2D$ upstream of the orifice. It is much better to work the other way round, i.e. to start with an upstream pipe a little shorter than is required, to weld a piece of thicker-walled pipe to the upstream pipe (the thicker-walled pipe is to be adjacent to the orifice plate) and to machine the thicker-walled pipe so that it approximately matches the rest of the upstream pipe and there is very small deviation from constant diameter and concentricity where the machined and unmachined sections are blended. For 2″ orifice meters it is reasonable to machine the whole upstream length; for 24″ meters it is desirable either to select the pipework so that no machining is required or to have it manufactured to the required tolerance; for 8″ meters machining of a part of the upstream length is probably required.

Goodson et al. (2004) studied the effect of recesses corresponding to those formed where an RTJ-flanged orifice fitting is joined to an upstream meter tube: tests in 4″ ($\beta = 0.2$, 0.4 and 0.67) and 8″ ($\beta = 0.67$) showed that the effect of the recesses was negligible.

If gaskets (rather than ‘O’ rings) are used between pipes or at the orifice it is important that they do not protrude into the pipe; therefore, in practice there will be a small recess. 5.2.6.6 of ISO/TR 9464:2008 recommends that at the plate a gasket be not thicker than 0.03D; such a gasket will have a negligible effect on the discharge coefficient.

Straightness of the upstream pipe is defined by 7.1.3 of ISO 5167-1:2003, and is normally checked visually.

The downstream pipe diameter is much less critical: 6.4.6 of ISO 5167-2:2003 states that it only needs to be within 3 % of that of the upstream pipe: Teyssandier (1985), using a 2″ orifice meter with flange tappings, looked at the effect of a protrusion or a recess $D/4$ wide at the orifice plate on its downstream side and found that for $\beta = 0.3$ or 0.5 its effect is less in magnitude than 0.1 % for protrusions or recesses up to 0.0625D in height (or depth) and less in magnitude than 0.2 % for protrusions or recesses up to 0.125D in height (or depth); for $\beta = 0.7$ its effect is less in magnitude than 0.1 % for a protrusion up to 0.03125D in height or a recess up to 0.0625D in depth. From those data it can be seen that provided that a protrusion is less than 20 % of the dam height its effect is less than 0.1 %; it is not surprising that the effect increases rapidly when the protrusion is a large percentage of the dam height (at 83 % of the dam height for $\beta = 0.7$ the effect is 5.25 %); at 100 % of the dam height a thick plate is obtained. The dam height is the distance from the sharp edge of the orifice plate to the nearest point of the pipe wall, i.e. $\frac{D(1-\beta)}{2}$. Teyssandier (1985) also looked at the effect of a protrusion or a recess $D/4$ wide 2D downstream of the orifice plate and found that even a protrusion 0.125D in height for $\beta = 0.7$ had an effect of only 0.1 %.

In conclusion, the rules on steps are given in 6.4 of ISO 5167-2:2003. They are rather complicated; an explanation for much of their complexity has been given in this section.

2.3.5 Eccentricity

The orifice plate and the pipe need to be close to concentric. The effect of eccentricity has been measured by Norman et al. (1984), Husain and Teyssandier (1986a) and Miller and Kneisel (1968). The eccentricity (i.e. the distance between the centre-line of the pipe and that of the orifice) is generally non-dimensionalized with $D/(0.1 + 2.3\beta^4)$. This non-dimensionalization was used in ISO 5167:1980 (ISO 1980). When an orifice plate with a single tapping upstream and downstream is used the effect of eccentricity is greater in the direction parallel to the tappings than in the direction perpendicular to them. The limit on eccentricity in ISO 5167-2:2003 parallel to the tappings is

$$\frac{0.0025D}{0.1 + 2.3\beta^4} \tag{2.5}$$

and that in the direction perpendicular to the tappings is twice that.

NOTE The limit in ISO 5167:1980 was $0.0005D/(0.1 + 2.3\beta^4)$ in all directions. In ISO 5167-1:1991 (ISO 1991) it was increased to $0.0025D/(0.1 + 2.3\beta^4)$ in all directions. In ISO 5167-2:2003 it remained unchanged parallel to the tappings but was increased perpendicular to the tappings.

In 2.6.2.1 of API MPMS 14.3.2 the same limit as in Eq. 2.5 is applied parallel to the pressure tappings, but in the direction perpendicular to the pressure tappings the limit is four times that in Eq. 2.5. The ISO limit according to Fig. 6 of ISO/TR 12767:2007 gives shifts in the perpendicular direction up to about 0.08 % in magnitude, the API up to about 0.15 %. API MPMS 14.3.2 allows twice as large a limit if tappings 180° apart are connected; however, Fig. 6 of ISO/TR 12767:2007 shows that sometimes there is no cancellation. The ISO limits appear reasonable.

If the eccentricity is close to the limit, determining whether the eccentricity is inside or outside the limit is not trivial: if the plate is held between three dowels the deviation between the centreline of the orifice and that of the outside of the orifice plate, the deviation between the centreline of the pipe and the centre of the circle formed by the inside of the dowels, and the possible variation in the location of the plate within the dowels will all have to be considered.

On one occasion an orifice plate was installed so eccentrically that there was flow past the outside of the plate: this problem was investigated by Barton et al. (2005).

According to 6.5.2 of ISO 5167-2:2003 the plate shall be perpendicular to the pipe centreline within 1°. Whether this is based on engineering judgment or on experiment is not clear. This requirement should be stated when specifying alignment of flanges being welded during fabrication.

2.4 Dimensional Measurements

Where an orifice plate is being used without flow calibration the dimensional measurements are crucial to the calculated flowrate. The value for orifice diameter used in Eq. 1.15 is the mean of at least four measurements of diameter in different orientations (5.1.8.2 of ISO 5167-2:2003). 6.4.2 of ISO 5167-2:2003 requires that at least four measurements of diameter be made in each of three planes in the pipe. ISO 5167-2:2003 gives limits on the variation in diameters permitted, but measured uncertainties lower than the maximum permitted in ISO 5167-2:2003 can be used in the uncertainty of the mass flowrate (see Sect. 4.8). Modern coordinate measuring machines often do not give individual diameters but give a mean diameter and circularity; nevertheless, it can easily be checked whether the measurements meet the intention of ISO 5167-2:2003. For the pipe the mean diameter is obtained from the average of measurements made in at least three planes over a distance of $0.5D$ upstream of the upstream pressure tapping: one plane must be at $0D$, one at $0.5D$, from the upstream tapping and one in the plane of the weld in the case of a weldneck construction.

The measurements of the orifice are considerably more important than those of the upstream pipe (see Eq. 4.13 in Sect. 4.8): no orifice diameter may differ by more than 0.05 % from the mean, whereas in the pipe the tolerance is 0.3 %. In practice this is a minimum requirement. One of the advantages of orifice metering is that the diameter in the discharge-coefficient equation is that of the orifice plate. Where β is small a low quality pipe is acceptable, provided that the orifice plate is of high quality. For larger β the effect of any error in pipe diameter becomes more significant: from Eq. 4.13 the effect can be approximately calculated.

2.5 Orifice Fittings

The orifice fitting, which enables the operator to change or remove an orifice plate easily (i.e. without undoing flange bolts), was introduced by Paul Daniel. Tests were undertaken in 1949–51 in Rockville by the joint AGA-ASME committee to compare measurement uncertainty using orifice fittings with that obtained using plates held between flanges (AGA 1951, 1954a). Tests were also carried out as part of the Refugio tests in 1952/3: an orifice plate held between flanges (an orifice flange union) and an orifice fitting were placed in a 30″ line, the former upstream of the eight 10″ reference meters and the latter downstream, and 'it was unanimously concluded that the equations for computing coefficients of discharge in AGA Gas Measurement Report No 2 may be used, within the tolerances given, for large diameter meter tubes (e.g., 30-in.) using either orifice flange unions or orifice fittings' (AGA 1954b).

A picture of an orifice plate emerging from an orifice fitting is shown as Fig. 2.19.

Some systems are designed so that an orifice plate may be changed without depressurizing the line.

Fig. 2.19 Orifice plate emerging from an orifice fitting

2.6 Pressure Loss

The pressure loss is required for hydraulic system design, but a more accurate value is required if the pressure loss ratio (the ratio of the pressure loss to the differential pressure) is to be used for diagnostic purposes (see Sect. 4.10).

The pressure loss for an orifice plate is calculated as follows. The reader is warned that this section is more mathematically complex than the rest of the book.

The momentum theorem is obtained by integrating the equation of motion over a fixed volume so that

$$\rho \frac{Du_i}{Dt} = \rho F_i + \frac{\partial \sigma_{ij}}{\partial x_j}, \tag{2.6}$$

where ρ is the density, u_i the velocity in the x_i-direction, F_i the body force, σ_{ij} the stress tensor and D/Dt the derivative following the motion of the fluid, becomes, on expanding the derivative following the motion of the fluid and using mass conservation and the divergence theorem,

$$\iiint\limits_V \frac{\partial(u_i\rho)}{\partial t} dV = -\iint\limits_A \rho u_i u_j n_j dA + \iiint\limits_V F_i \rho dV + \iint\limits_A \sigma_{ij} n_j dA \tag{2.7}$$

where the fixed volume V is bounded by surface A.

The stress tensor consists of two terms: the pressure term is sufficient for the approximation here, so $\sigma_{ij} = -p\delta_{ij}$. The flow is steady and the body force (gravity) makes a contribution to the pressure which will make no contribution to the pressure loss.

Equation 2.7 is applied to the volume marked V on Fig. 2.20. Equations are supplied for incompressible flow.

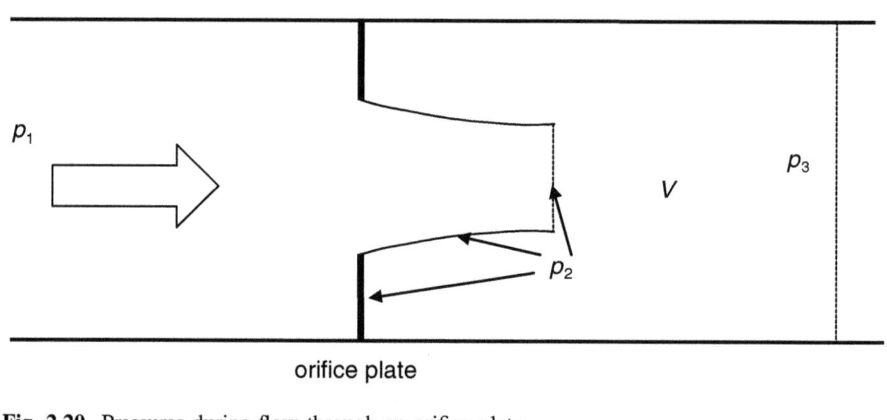

Fig. 2.20 Pressures during flow through an orifice plate

Then, assuming that the pressure has the same value on the back of the orifice plate, on the edge of the orifice jet as far as the *vena contracta* (i.e. the point of maximum convergence) and at the *vena contracta*, and using the divergence theorem

$$0 = \rho A_c \bar{u}_c^2 - \rho A_p \bar{u}_p^2 - p_3 A_p + p_2 A_p \tag{2.8}$$

where A_c and A_p are the area of the *vena contracta* and of the pipe, respectively, $\bar{u}_c$ and $\bar{u}_p$ are the mean velocity in the *vena contracta* and in the pipe, respectively, and p_2 and p_3 are the pressure immediately downstream of the orifice (e.g. at a downstream pressure tapping) and after pressure recovery, respectively.

In addition Bernoulli's Equation (see Sect. 1.3) applies between an appropriate location upstream of the orifice plate and the *vena contracta*:

$$p_1 + \frac{1}{2}\rho \bar{u}_p^2 = p_2 + \frac{1}{2}\rho \bar{u}_c^2 \tag{2.9}$$

where p_1 is the pressure around $1D$ upstream of the orifice.

The basic equation for differential-pressure meters (Eq. 1.15, for incompressible flow) can be expressed as:

$$\bar{u}_o = \frac{C}{\sqrt{1-\beta^4}}\sqrt{\frac{2(p_1 - p_2)}{\rho}} \tag{2.10}$$

where $\bar{u}_o$ is the mean velocity in the orifice. Mass conservation between upstream of the orifice plate, the orifice and the *vena contracta* gives:

$$\rho \bar{u}_p A_p = \rho \bar{u}_o \beta^2 A_p = \rho \bar{u}_c A_c. \tag{2.11}$$

Equations 2.8–2.11 give

$$\frac{\Delta\varpi}{\Delta p} = \frac{p_1 - p_3}{p_1 - p_2} = \frac{\sqrt{1-\beta^4(1-C^2)} - C\beta^2}{\sqrt{1-\beta^4(1-C^2)} + C\beta^2}, \tag{2.12}$$

where $\Delta\varpi$ is the pressure loss (from D upstream to $6D$ downstream).

Of the three standard tapping pairs D and $D/2$ tappings are the most suitable for determining C for use in Eq. 2.12.

Equation 2.12 is given as Eq. (7) in 5.4.1 of ISO 5167-2:2003. The inclusion of this equation in ISO 5167 resulted from the publication of Urner (1997).

A simpler but less accurate approximation is

$$\frac{\Delta\varpi}{\Delta p} = 1 - \beta^{1.9}, \tag{2.13}$$

which is given in 5.4.2 of ISO 5167-2:2003.

Equation 2.12 has been checked by looking at data from Steven et al. (2007). Strictly speaking Eq. 2.12 is applicable to the situation where the upstream tapping is at D upstream. Since Steven's tappings were in the flanges his measured pressure loss ratio $\frac{\Delta\varpi_{up}}{\Delta p_{up}}$ was not equal to the true pressure loss ratio $\frac{\Delta\varpi}{\Delta p}$.

The measured pressure loss ratio $\frac{\Delta\varpi_{up}}{\Delta p_{up}}$, where the upstream tapping is a flange or a corner tapping, is related to the true pressure loss ratio $\frac{\Delta\varpi}{\Delta p}$ by

$$\frac{\Delta\varpi_{up}}{\Delta p_{up}} = \frac{p_{up} - p_3}{p_{up} - p_2} = \frac{\frac{\Delta\varpi}{\Delta p} + \frac{p_{up} - p_1}{p_1 - p_2}}{1 + \frac{p_{up} - p_1}{p_1 - p_2}} \tag{2.14}$$

where p_{up} is the pressure at the upstream pressure tapping. It is not easy, given the assumptions in the use of the momentum theorem, to include a term for the location of the downstream flow-measurement tapping: its effect is small and neglected in Eq. 2.14.

Moreover, from Eq. 1.15,

$$p_{up} \approx p_1 + 2\frac{C_{D-D/2} - C_{up-D/2}}{C_{D-D/2}}(p_1 - p_2) \tag{2.15}$$

where the difference in discharge coefficient between that for a plate with tappings D upstream and $D/2$ downstream, $C_{D\text{-}D/2}$, and that for a plate with tappings at a different upstream location and $D/2$ downstream, $C_{up\text{-}D/2}$, can be obtained from the Reader-Harris/Gallagher (1998) Equation (see Eq. 5.22).

So,

$$\frac{p_{up} - p_1}{p_1 - p_2} \approx 2\frac{C_{D-D/2} - C_{up-D/2}}{C_{D-D/2}} \tag{2.16}$$

Then the measured pressure loss ratio in Steven et al. (2007) can be compared with calculated values of the pressure loss ratio, obtained by substituting Eqs. 2.12 and 2.16 into Eq. 2.14: see Fig. 2.21. The equation in 2.4.5.1 of API MPMS 14.3.2:2000 is also shown.

It appears possible to predict pressure loss remarkably accurately, but the population of data is very small.

The formula for pressure loss applies where the downstream pressure loss tapping is about $6D$ downstream of the orifice plate, where the pressure has recovered (to the maximum extent that it does recover). No term for friction loss has been included: it might be appropriate to include one, but, where the upstream and downstream tappings are $1D$ and $6D$ respectively from the orifice plate, the friction loss will be significantly less than the friction loss in fully developed flow through $7D$ of straight pipe.

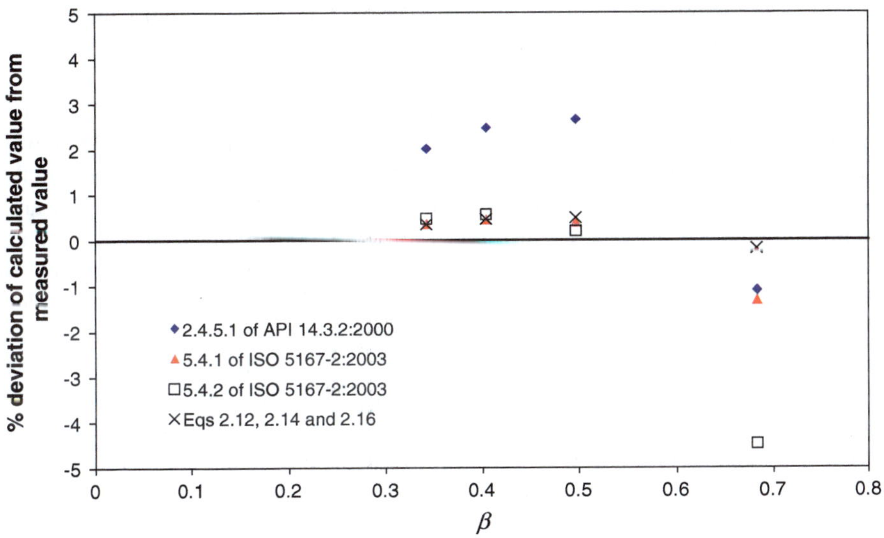

Fig. 2.21 Pressure loss ratio: deviation of calculated values from measured CEESI data (taken in dry gas for their wet-gas JIP): only 'Eqs. 2.12, 2.14 and 2.16' takes specific account of the upstream tapping location

In conclusion, Eq. (7) in ISO 5167-2:2003 for pressure loss ratio (Eq. 2.12) appears remarkably accurate: if the upstream tapping is not located 1*D* upstream of the orifice plate and the pressure loss ratio that will actually be measured is required, the necessary additional term has been derived.

2.7 Reversed Orifice Plates

The effect of reversing an orifice plate is very large. An increase in discharge coefficient of the order of 15 or 20 % is typical. Much of the published literature relating to the discharge coefficients of reversed orifice plates has notable deficiencies when considered for general application. In most papers the orifice plates are not fully described: some papers do not give the bevel angle, and the size of the bevel is rarely stated.

8.6.2 of ISO/TR 12767:2007 (ISO 2007c) provides information on the variation of discharge coefficient with bevel size. However, as no orifice diameter is given, the data cannot be used to determine the magnitude of error in a specific case of interest. 6.1 of ISO/TR 15377:2007 (ISO 2007b) describes the 'conical entrance' orifice, which is considered to have a constant discharge coefficient of 0.734. However, the conical entrance orifice is not sufficiently representative of reversed square-edged orifice plates to give results of general application. In SPE Note 22867 (Ting 1991) the author describes the results of a number of tests with air at low Reynolds numbers. The tests were carried out for various β, but neither bevel angle nor bevel size is given in the paper. When consulted, the author stated that he believed that the bevel angle was 30° to the face of the plate.

Witte (1997) published a paper on the effect of reversing orifice plates in a 10″ pipe in gas at 60 barg. The plates had thickness, E, of 6.35 mm (¼″) with 45° bevel. The orifice (bore) thickness, e, was 4.76 mm (3/16″), although this is not stated in the paper. Witte carried out the work at the Gas Research Institute Metering Research Facility at Southwest Research Institute. Morrow (2000) reanalysed Witte's data. CFD was undertaken by Brown et al. (2000). Morrow also measured the effect of reversing a series of plates in a 4″ line; these plates had a bevel angle of 45°, a plate thickness, E, of 3.18 mm (1/8″) and an orifice (bore) thickness, e, of 1.59 mm (1/16″). George and Morrow presented the data listed above and both 2″ data and additional 4″ data and presented conclusions in George and Morrow (2001).

Even where the bevel angle is 45° no simple formula has achieved universal acceptance. The correlation in George and Morrow (2001), also quoted in Morrow et al. (2002), is

$$\%\,\mathit{flowrate\ error} = -18.93 + 12.91\beta - 34.04\frac{E}{D_{nom}} - 8.900\frac{e}{E} + 13.64\left(\frac{e}{E}\right)^2 \tag{2.17}$$

This does not have the correct performance in the limit as b/E tends to 0, where b is the bevel width ($=E - e$), (i.e. e/E tends to 1).

The error should be a function of b/d (similar to edge rounding), but not of b/d alone: β is significant too. Figure 2.22 shows the experimental data for bevel angle 45° from SwRI including those of Witte plus two points from NEL. The lines described as 'experiment' join up the experimental data for each value of β. If the bevel is treated as a rounded edge and the best-fit radius determined as in Sect. 2.2.4, then the best–fit radius is $1.57b$ (presumably $\pi b/2$). Then the expected flowrate error for a bevelled edge can be calculated as in Sect. 2.2.4. This has been plotted for a maximum ratio of edge radius to diameter of 0.01, since the data of Burgin (1971) show that above that value the data deviate increasingly from the fit in Sect. 2.2.4.

The data from Witte appear different (especially for $\beta = 0.2$) from the other data sets. This could be due to e/E being equal to 0.75. So the other data points (i.e. those for $e/E \le 0.5$) in Fig. 2.22 were fitted with the following equation:

$$\%\,\mathit{flowrate\ error} = -(a - b\beta^c)\left(1 - \exp\left(\frac{-f\frac{b}{d}}{a - b\beta^c}\right)\right)$$

This behaves correctly as b/d tends to 0.

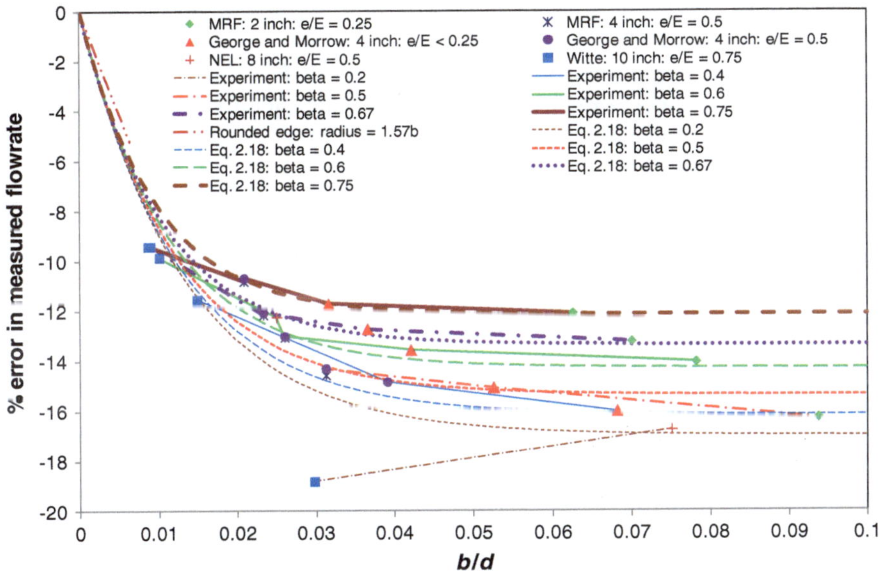

Fig. 2.22 Error in measured flowrate using a reversed orifice plate

With the fit

$$\%flowrate\ error = -\left(17.2 - 10.4\beta^{2.5}\right)\left(1 - \exp\left(\frac{-1270\frac{b}{d}}{17.2 - 10.4\beta^{2.5}}\right)\right) \quad (2.18)$$

the r.m.s. deviation of the equation from the data is 0.35 %. Equation 2.18 is shown on Fig. 2.22.

To avoid having a problem with a reversed plate, where possible a plate should be marked externally in such a way that correct orientation can be verified without looking into the pipe.

2.8 Conclusions

This chapter has described the main design specifications for orifice plates: it has pointed the reader to important parts of ISO 5167 and given reasons for the requirements in the standard. It has covered the orifice plate (the circularity of the bore, the flatness, the parallelism of the two faces, the surface condition of the upstream face, and, above all, the edge sharpness), the pipe (the pressure tappings, the pipe roughness, the effect of upstream steps, the concentricity of the orifice plate in the pipe and the circularity and cylindricality of the pipe), the measurements of both the orifice plate and the pipe, and the pressure loss. A very significant incorrect installation of an orifice plate within the pipe, a reversed orifice plate, has also been covered. Appendix 2.A considers the use of orifice plates of diameter smaller than that permitted in ISO 5167-2:2003.

The basic instruction remains to follow ISO 5167. For some important deviations from ISO 5167 the errors in discharge coefficient can be calculated using what is described in this chapter.

Appendix 2.A: Orifice Plates of Small Orifice Diameter

2.A.1 Introduction and Test Work

Orifice plates of small orifice diameter, d, are often used for measuring low gas flowrates in declining wells in order to increase the available differential pressure. However, some flowrates may result in the orifice diameter being sufficiently small that the orifice diameter is outside ISO 5167-2:2003, i.e. less than 12.5 mm (0.492″).

Twelve orifice plates were obtained and tested by NEL for ConocoPhillips, four of them from a manufacturer that specializes in spark erosion techniques, eight of them from a manufacturer that specializes in orifice plates. The aim was to determine the probable errors that might be obtained in using orifice plates of very small sizes.

Accordingly four orifice plates were manufactured for the project by ATM (Advanced Tool Manufacture), East Kilbride, Scotland and used electrical discharge machining (EDM—or spark erosion) to machine the orifice bore; eight were manufactured by Kelley Orifice Plates, Texarkana, Texas using conventional machining techniques. From ATM, the orifice diameters were 6.35 mm (1/4″), 3.18 mm (1/8″) (2 off) and 1.59 mm (1/16″); from Kelley they were 9.52 mm (3/8″), 6.35 mm (1/4″), 3.18 mm (1/8″) and 1.59 mm (1/16″) (2 off in each case). In the case of ATM $e/d = 0.1$ was specified, where e is the thickness of the orifice, and d the orifice diameter (see Sect. 2.2.5).

All the orifice plates were calibrated in water in a 4″ line ($D = 101.80$ mm) over a range of Reynolds number. The Reynolds number was below the minimum pipe Reynolds number of 5000 permitted by ISO 5167-2:2003 for all the data, except for most of the d = 9.52 mm (3/8″) data. The data are shown in Reader-Harris et al. (2008). The smallest value of pipe Reynolds number was 116; so the deviations here are presented from the complete orifice plate discharge-coefficient equation:

$$
\begin{aligned}
C = {} & 0.5961 + 0.0261\beta^2 - 0.216\beta^8 + 0.000521\,(10^6\beta/Re_D)^{0.7} \\
& + (0.0188 + 0.0063A)\beta^{3.5}\max\{(10^6/Re_D)^{0.3},\ 22.7 - 4700\,(Re_D/10^6)\} \\
& + (0.043 + 0.080\,e^{-10L_1} - 0.123\,e^{-7L_1})(1 - 0.11A)\frac{\beta^4}{1-\beta^4} \\
& - 0.031\,(M_2' - 0.8M_2'^{1.1})\{1 + 8\max(\lg(3700/Re_D), 0.0)\}\beta^{1.3} \\
& + 0.011\,(0.75 - \beta)\max(2.8 - D/25.4, 0.0)
\end{aligned}
\tag{5.21}
$$

where D is the pipe diameter in mm (following ISO 80000 2:2009, $\log_{10}$ is written lg).

The Reader-Harris/Gallagher (1998) Equation in ISO 5167-2:2003 (ISO 2003b) (given here as Eq. 5.22) is the special case of Eq. 5.21 for $Re_D \geq 5000$. The differences between Eq. 5.21 and the extrapolated Reader Harris/Gallagher (1998) Equation are very small, less than 0.06 % in magnitude, over the range of the data here.

For each orifice plate the mean deviation of the data from Eq. 5.21 over its range of Reynolds number is given in Fig. 2.A.1: the mean deviations are plotted against orifice diameter. To use a simple additive correction to Eq. 5.21 for each orifice plate it is necessary that the deviation be almost constant over its range of Reynolds number: for each orifice plate the standard deviation of the deviations from Eq. 5.21 over its range of Reynolds number is given in Fig. 2.A.2.

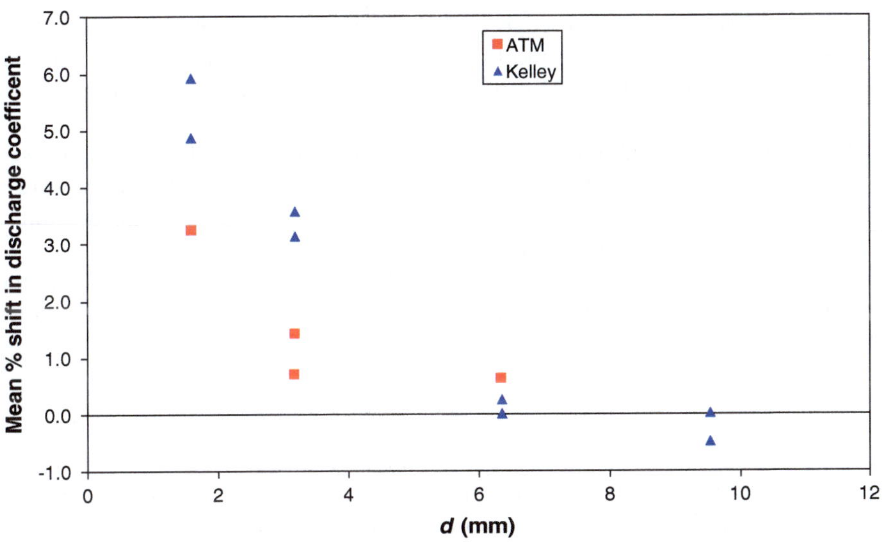

Fig. 2.A.1 Mean percentage shift in discharge coefficient from Eq. 5.21 for each orifice plate over its range of Reynolds number

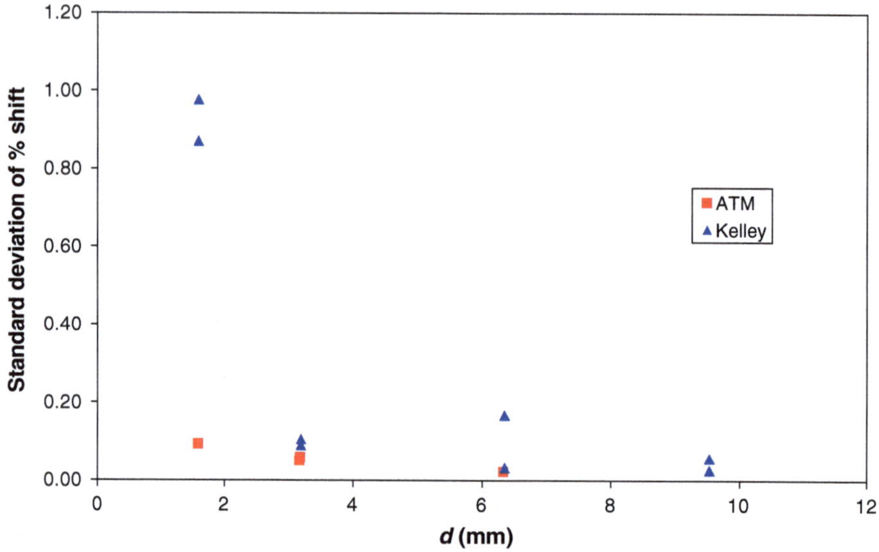

Fig. 2.A.2 Standard deviation of percentage shift in discharge coefficient from Eq. 5.21 for each orifice plate over its range of Reynolds number

From Fig. 2.A.2 for all the ATM data and for most of the Kelley data the deviation from Eq. 5.21 (and hence from the Reader-Harris/Gallagher (1998) Equation) was fairly constant and so Eq. 5.21 could be used with a simple additive correction, presumably due to edge rounding. The fact that there appears to be a simple rounding correction for most of the sets supports the view that, given such an edge-rounding term, Eq. 5.21 can be used with general success for orifice plates that have both small d and small β.

If the orifice plates calibrated here were to be used in service they would be used with a simple additive shift as given in Fig. 2.A.1. Since, however, they were a sample of plates used to determine how other plates from the same manufacturers would perform in service then further analysis was necessary.

If the edge radius of an orifice plate is r, and the increase in edge radius is Δr, the percentage increase in discharge coefficient, S, is given by Hobbs and Humphreys (1990) (see Sect. 2.2.4):

$$S = 550\frac{\Delta r}{d}. \tag{2.A.1}$$

So given a shift in discharge coefficient it is possible to calculate the increase in edge radius that gives rise to it. It is not clear on what relative edge radius Eq. 5.21 is based at the low throat Reynolds numbers obtained with these plates. If it were

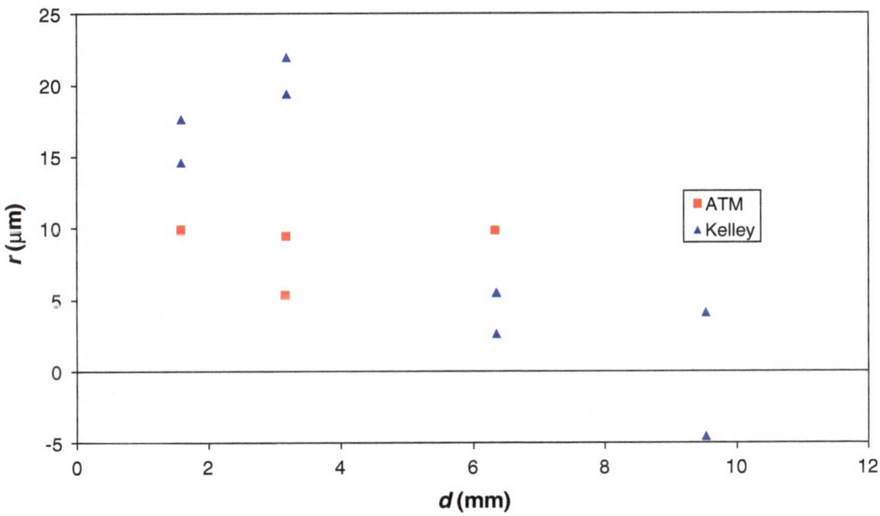

Fig. 2.A.3 Calculated edge radius of orifice plates v *d*

assumed that it is based on $r/d = 0.0004$, that is the maximum permitted edge radius in ISO 5167-2:2003 (at high Reynolds numbers with larger plates Eq. 5.21 (and thus the Reader-Harris/Gallagher (1998) Equation) is based on relatively sharper plates), then the edge radii for the different plates would be as shown in Fig. 2.A.3.

The mean edge radius of the ATM plates is calculated to be 8.7 µm.

NOTE If the PR 14 Equation (Eq. 5.C.2) were used instead of Eq. 5.21 the mean edge radius of the ATM plates would be calculated to be 10.9 µm. If Eq. 5.21 were used but it were assumed that at these low throat Reynolds numbers Eq. 5.21 was based on $r/d = 0.0003$ the mean edge radius of the ATM plates would be calculated to be 0.36 µm smaller than if Eq. 5.21 were based on $r/d = 0.0004$.

When the following term (based on an edge radius of 8.7 µm and Eq. 2.A.1) is used

$$\Delta C = 3.3\left(\frac{0.0087}{d} - 0.0004\right)\ (d:\ \text{mm}) \tag{2.A.2}$$

95 % of the ATM data lie within 0.65 % of Eq. 5.21 with term 2.A.2 added.

In Fig. 2.A.4 the data from the spark-eroded plates are compared with Eq. 5.21 (essentially the Reader-Harris/Gallagher (1998) Equation) with an additional term representing an edge radius of 8.7 µm.

The Stolz Equation in ISO 5167:1980 (ISO 1980) (given here as Eq. 5.24) does not give good performance as β tends to 0.

It is more difficult to analyse the data from the Kelley plates: the analysis is provided in Reader-Harris et al. (2008).

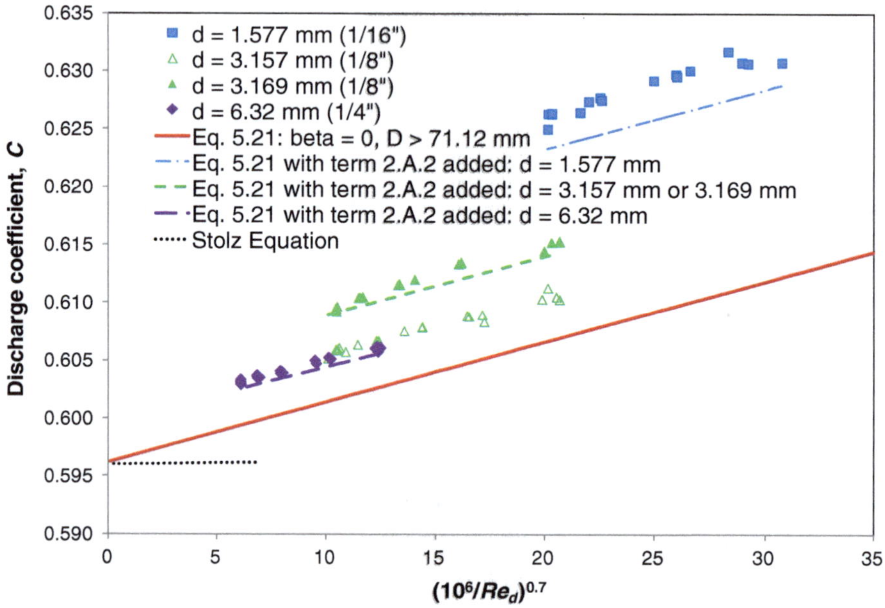

Fig. 2.A.4 The discharge-coefficient data from the spark-eroded orifice plates

2.A.2 Conclusions

Orifice plates can be a surprisingly good way of measuring small gas flows. The Reader-Harris/Gallagher Equation appears to work well for small orifices if an additional term is added to allow for edge rounding (from Sect. 2.2.5 it would be wise also to specify that $e/d \leq 0.1$). The edge radius for the spark-eroded plates appears to be fairly constant as d decreases, and so for uncalibrated spark-eroded orifice plates from one manufacturer good results may be obtained by adding term 2.A.2 arithmetically to the Reader-Harris/Gallagher (1998) Equation. Results presumably depend on the manufacturer. An orifice plate with edge radius equal to the average edge radius of the plates manufactured by ATM will meet the requirements of ISO 5167-2:2003 provided that d is greater than 22 mm. On this basis in a 24″ line an orifice plate with $\beta = 0.05$ would still have a sharp edge and not require an additional term added to the discharge-coefficient equation.

To be able to use orifice plates of very small orifice diameter in an existing installation in a declining gas field may be much more economical than to replace the metering or shut the field. From the beginning of the flow measurement to its conclusion a 4″ orifice meter with different orifice plates (including those described in this appendix) might measure a range of around 3000:1 in terms of mass flowrate (if the static pressure were the same throughout the period of measurement).

References

AGA (1951) Investigation of orifice meter installation requirements. American Gas Association
AGA (1954a) Investigation of orifice meter installation requirements. American Gas Association
AGA (1954b) Large diameter orifice meter tube tests. Final report of Supervising Committee. Research Project NX-4. American Gas Association
API (2000) Natural Gas Fluids Measurement. Part 2—Specification and Installation Requirements —Concentric, Square-edged Orifice Meters. API MPMS (Manual of Petroleum Measurement Standards) Chapter 14.3.2:2000. American Petroleum Institute, Washington, DC
Barton N, Hodgkinson E, Reader-Harris MJ (2005) Estimation of the measurement error of eccentrically installed orifice plates. In: Proc 23rd North Sea Flow Meas Workshop, Norway: Paper 4
Bean HS, Murdock JW (1959) Effects of pipe roughness on orifice meter accuracy, Report of Supervising Committee on two-inch tests. American Gas Association Research Project NW-20, American Gas Association, New York
Beitler SR (1935) The flow of water through orifices. A study in 1-in., 1½-in., 2-in., 3-in., 6-in., 10-in, and 14-in. lines. Engineering Experiment Station Bulletin No 89. Ohio State University
Benedict RP, Wyler JS, Brandt GB (1974) The effect of edge sharpness on the discharge coefficient of an orifice. Trans ASME J. Eng. Power: paper No 74-WA/FM-4
Botros KK, Studzinski W, Barg P (1992) Results of NOVA's gas metering research. Canadian Gas Association, Measurement School, Vancouver, May 26–29
Brain TJS, Reid J (1973) Measurement of orifice plate edge sharpness. Meas Control 6:377–384
Brennan JA, McFaddin SE, Sindt CF, Wilson RR (1989) Effect of pipe roughness on orifice flow measurement. NIST Technical Note 1329. National Institute of Standards and Technology, Boulder, Colorado
Brown GJ, Reader-Harris MJ, Gibson JJ, Stobie GJ (2000) Correction of readings from an orifice plate installed in reverse orientation. In: Proceedings of 18th North Sea Flow Meas Workshop, Gleneagles: Paper 2.2. National Engineering Laboratory, East Kilbride, Glasgow
Burgin EJ (1971) Factors affecting accuracy of orifice measurement (primary element). In: Proceedings of International School Hydrocarbon Measurement
Clark WJ, Stephens RC (1957) Flow measurement by square edged orifice plates: pipe roughness effects. Inst Mech Eng 171(33):895–904
Crocket KA, Upp EL (1973) The measurement and effects of edge sharpness on the flow coefficients of standard orifices. Trans ASME J Fluids Eng (June): 271–275 (also paper No 72-WA/FM-4, 1972)
Dall HE (1958) The effect of roughness of the orifice plate on the discharge coefficient. Instrument Engineer: 2(5) (April)
Gallagher GR (1968) Measuring edge sharpness of orifice plates. The Engineer, 17 May 1968
George DL, Morrow TB (2001) Orifice meter calibration for backwards-facing orifice plates. GRI Report No 01/0074 on SwRI Project No 18-8890. Gas Research Institute, Chicago
Goodson FD, Zanker K, Husain ZD (2004) The effects of upstream recesses on the discharge coefficient of a flange tapped orifice meter. AGA Operations Conference
Gorter J (1978) Deformation of orifice plates: theory and practice. In: Dijstelbergen HH, Spencer EA (eds) Flow measurement of fluids. North-Holland Publishing Company
Herning F (1962) Untersuchungen zum Problem der Kantenunschärfe bei Normblenden und bei Segmentblenden [Experiments on the problem of the edge sharpness of standard and segmental orifice plates]. Brennst-Wärme-Kraft 14(3):119–126
Herning F, Lugt H (1958) Neue Versuche mit Segmentblenden und Normblenden. Brennst-Wärme-Kraft 10(5):219–223
Herning F, Wolowski E (1963) Die Kantenunschärfe von Normblenden und Segmentblenden und das Ähnlichkeitsgesetz [The edge sharpness of standard and segment orifices and the laws of similarity]. Brennst-Wärme-Kraft 15(1):26–30

Hobbs JM, Humphreys JS (1990) The effect of orifice plate geometry upon discharge coefficient. Flow Meas Instrum 1(3):133–140

Husain ZD, Goodson FD (1986) Effects of plate thickness and bevel angle on discharge coefficients of a 50 mm (2 inch) orifice meter. ASME paper 86-WA/FM-4, Winter Annual Meeting, FM-4 Symposium on Flowmeters, Anaheim, California, December 1986

Husain ZD, Teyssandier RG (1986a) Orifice eccentricity effects for flange, pipe and radius (D-D/2) taps. ASME paper 86-WA/FM-1, presented at ASME Winter Annual Meeting, Anaheim, California

Husain ZD, Teyssandier RG (1986b) The effects of plate thickness and bevel angle in a 150 mm line size orifice meter. In: Proceedings of flow measurement in the mid 80s. National Engineering Laboratory, East Kilbride, Glasgow

ISO (1980) Measurement of fluid flow by means of orifice plates, nozzles and Venturi tubes inserted in circular cross-section conduits running full. ISO 5167:1980. International Organization for Standardization, Geneva

ISO (1991) Measurement of fluid flow by means of pressure differential devices—Part 1: orifice plates, nozzles and Venturi tubes inserted in circular cross-section conduits running full. ISO 5167-1:1991. International Organization for Standardization, Geneva

ISO (2003a) Measurement of fluid flow by means of pressure differential devices inserted in circular cross-section conduits running full—Part 1: general principles and requirements. International Organization for Standardization, Geneva. ISO 5167-1:2003

ISO (2003b) Measurement of fluid flow by means of pressure differential devices inserted in circular cross-section conduits running full—Part 2: Orifice plates. Geneva: International Organization for Standardization. ISO 5167-2:2003

ISO (2007a) Fluid flow in closed conduits—Connections for pressure signal transmissions between primary and secondary elements. International Organization for Standardization, Geneva. ISO 2186:2007

ISO (2007b) Measurement of fluid flow by means of pressure differential devices—Guidelines for the specification of orifice plates, nozzles and Venturi tubes beyond the scope of ISO 5167. International Organization for Standardization, Geneva. ISO/TR 15377:2007

ISO (2007c) Measurement of fluid flow by means of pressure differential devices—Guidelines on the effect of departure from the specifications and operating conditions given in ISO 5167. International Organization for Standardization, Geneva. ISO/TR 12767:2007

ISO (2008) Guidelines for the use of ISO 5167:2003. International Organization for Standardization, Geneva. ISO/TR 9464:2008

Jepson P, Chipchase R (1973) The effect of plate buckling on orifice meter accuracy. Report No ERS.R.467, British Gas Engineering Research Station

Jepson P, Chipchase R (1975) Effect of plate buckling on orifice meter accuracy. J Mech Eng Sci 17(6):330–337

Johansen WR (1966) Effects of thin films of liquid coating orifice plate surfaces on orifice flowmeter performance. GRI Report No GRI-96/0375. Gas Research Institute, Chicago

Johansen W, Seidl W, Kegel T (1966) The effects of oil coating on the measurement of gas flow using sharp-edged orifice flowmeters. American Gas Association: AGA Operating section Conference, Montreal

Lansverk NB (1990) Effects of abnormal conditions on accuracy of orifice measurement. ISHM. (sent by Loy Upp)

Mason D, Wilson MP, Birkhead WG (1975) Measurement error due to the bending of orifice plates. ASME Winter Annual Meeting, Houston, Texas: Paper 75-WA/FM-6. American Society of Mechanical Engineers

McVeigh JC (1962) Further investigations into the effect of roughness of the orifice plate on the discharge coefficient. Instrument Engineer, pp 112–113

Miller RW, Kneisel O (1968) Experimental study of the effects of orifice plate eccentricity on flow coefficients. ASME J Basic Eng, pp 12–131, March 1969, previously presented at ASME Winter Annual Meeting, New York, December 1–5, Paper 68-WA/FM-1, 1968

Morrison GL, DeOtte RE, Moen M, Hall KR, Holste JC (1990) Beta ratio, swirl and Reynolds number dependence of wall pressure in orifice flowmeters. Flow Meas Instrum 1:269–277

Morrow TB (1998) Effects of oil on orifice plates in gas flow measurement. In: Proc ASME Fluids Eng. Div. Summer Meeting, paper FEDSM98-M284. American Society of Mechanical Engineers, Washington, DC

Morrow TB (2000) Private communication. GRI MRF, San Antonio, Texas

Morrow TB, Morrison GL (1999) Effect of meter tube roughness on orifice C_d. In: 4th international symposium on fluid flow meas, Denver

Morrow TB, George DL, Nored MG (2002) Operational factors that affect orifice meter accuracy. MRF Topical Report GRI-00/0141 on SwRI Project No 18-8890. Gas Research Institute, Chicago

Norman R, Rawat MS, Jepson P (1983) Buckling and eccentricity effects on orifice metering accuracy. International Gas Research Conference

Norman R, Rawat MS, Jepson P (1984) An experimental investigation into the effects of plate eccentricity and elastic deformation on orifice meter accuracy. In: Spencer EA (ed) Proceedings of International Conference Metering of Natural Gas and Liquefied Hydrocarbon Gases, London, UK, 1–2 Feb 1984, paper 3.3. Oyez, London

Pritchard M, Niazi A, Marshall D (2003) Assessment of the effect of contamination on orifice plates. In: Proceedings of 11th Flomeko conference on flow measurement of gas and liquid, Groningen, 12–14 May 2003: Paper 6.2. Gasunie Research, Groningen (on CD-ROM)

Pritchard M, Marshall D, Wilson J (2004) An assessment of the impact of contamination on orifice plate metering accuracy. In: Proceedings of 22nd North Sea Flow Meas Workshop, St Andrews: Paper 2.2. National Engineering Laboratory, East Kilbride, Glasgow

Reader-Harris MJ (1990) Pipe roughness and Reynolds number limits for the orifice plate discharge coefficient equation. In: Proceedings of 2nd International Symposium Fluid Flow Meas, Calgary, pp 29–43

Reader-Harris MJ (1991) The effect of plate roughness on orifice plate discharge coefficients. Flow Measurement Memo. FL/421. National Engineering Laboratory Executive Agency, East Kilbride, Glasgow

Reader-Harris MJ (1993) The effect of plate roughness on orifice plate discharge coefficients—Further work. Flow Measurement Memo FL/438. National Engineering Laboratory Executive Agency, East Kilbride, Glasgow

Reader-Harris MJ, Brunton WC (2002) The effect of diameter steps in upstream pipework on orifice plate discharge coefficients. In: Proceedings of 5th International Symposium Fluid Flow Meas, Washington, DC

Reader-Harris MJ, Keegans W (1986) Comparison of computation and LDV measurement of flow through orifice and perforated plates, and computation of the effect of rough pipework on orifice plates. In: Proceedings of International Symposium Fluid Flow Meas, pp 653–666, Washington, DC

Reader-Harris MJ, Brunton WC, Nicholson IG, Rushworth R (2003) Ageing effects on orifice metering. In: Proceedings of 21st North Sea Flow Meas Workshop, Norway

Reader-Harris MJ, Stobie G, Kelley T (2008) The discharge coefficient of orifice plates with diameters from $^1/_{16}''$ (1.6 mm) to $^3/_8''$ (9.5 mm). In: Proceedings of Production and Upstream Flow Meas Workshop, Houston, Texas

Reader-Harris MJ, Barton N, Hodges D (2010) The effect of contaminated orifice plates on the discharge coefficient. In: Proceedings of 15th FLOMEKO, Taipei

Schlichting H (1960) Boundary layer theory. McGraw-Hill, New York

Spencer EA (1987) Study of edge sharpness effects measured during the EEC orifice plate coefficient programme. BCR Report EUR 11131, Commission of the European Communities, Brussels

Spencer EA, Calame H, Singer J (1969) Edge sharpness and pipe roughness effects on orifice plate discharge coefficients. NEL Report No 427. National Engineering Laboratory, East Kilbride, Glasgow

Steven R, Britton C, Stewart D (2007) CEESI Wet Gas JIP Data Release. EI/CEESI Wet Gas Metering Seminar Aberdeen. CEESI, Colorado

Stewart IA (1989) The EEC 600 mm orifice meter project. Part II Inspection of the meter run and orifice plates. Report No ERS.R.4255, British Gas Engineering Research Station

Studzinski W, Berg D (1988) Effects of orifice surface conditions on accuracy of flow rate measurements. In: Proceedings of 2nd International Conference on Flow Meas, London. BHRA, Cranfield, Bedfordshire

Studzinski W, Berg D, Bell D, Karwacki L (1990) Effect of meter run roughness on orifice meter accuracy. In: Proceedings of 2nd International Symposium Fluid Flow Meas, Calgary, pp 1–15

Teyssandier RG (1985) The effect of symmetric steps and gaps on orifice measurement. In: Proceedings of 3rd North Sea Flow Meas Workshop, Norway

Thibessard G (1960) Le coefficient de débit des diaphragmes, la rugosité et le nombre de Reynolds. Chaleur et Industrie 415:33–50

Ting VC (1991) Effects of non-standard operating conditions on the accuracy of orifice meters. SPE 22867. Society of Petroleum Engineers

Ting VC, Corpron GP (1995) Effect of liquid entrainment on the accuracy of orifice meters for gas flow measurement. Int Gas Research Conf

Urner G (1997) Pressure loss of orifice plates according to ISO 5167-1. Flow Meas Instrum 8 (1):39–41

Whetstone JR, Cleveland WG, Baumgarten GP, Woo S (1989) Measurements of coefficients of discharge for concentric, flange-tapped, square-edged orifice meters in water over a Reynolds number range of 1000–2,700,000. NIST Technical Note TN-1264. NIST, Washington, DC

Witte R (1953) Neue Beiträge zur internationalen Normung auf dem Gebiete der Durchflußmessung. Brennst-Wärme-Kraft 5(6):185–190

Witte JN (1997) Orifice meter error with reversed beveled plates. In: Proceedings of AGA operating conference, Nashville, Tennessee: paper 97-OP-062

Zedan MF, Teyssandier RG (1990) Effect of errors in pressure tap locations on the discharge coefficient of a flange-tapped orifice plate. Flow Meas Instrum 1(3):141–148

Chapter 3
Venturi Tube Design

Abstract This chapter describes the main design specifications for classical Venturi tubes: it points the reader to important parts of ISO 5167 and gives reasons for the requirements in the standard. It covers the different types: their shape and their discharge coefficient. It includes the tappings, the dimensional measurements, the effect of roughness and the pressure loss. Gas flow at high Reynolds number and the effect of upstream fittings are not covered here: they are in Chaps. 7 and 8 respectively. The basic instruction remains to follow ISO 5167-4, probably with only one tapping, instead of four tappings, in each plane.

3.1 Introduction

A Venturi tube is a differential-pressure meter consisting of an entrance cylinder, a convergent section, a throat and a divergent section. It has a discharge coefficient quite near to 1, and a ratio of pressure loss to differential pressure generally in the range 0.05–0.25. It is more expensive to manufacture than an orifice plate, but has a smaller pressure loss and requires less straight upstream pipe.

The main recommendations for classical Venturi tubes are to be found in ISO 5167-1:2003 and ISO 5167-4:2003 (ISO 2003a, b). This chapter points the reader to important parts of ISO 5167 and gives reasons for the requirements in ISO 5167. The basic equations are covered in Chap. 1 and standards are covered in Chap. 12. Gas flow at high Reynolds number and the effect of upstream fittings are not covered here: they are in Chaps. 7 and 8, respectively.

The profile of the classical Venturi tube is shown in Fig. 3.1. The upstream parallel section is called the entrance cylinder; the narrow parallel section is the throat; between them is the convergent, and downstream of the throat is the divergent. Figure 3.2 is an external view of an 8″ Venturi tube machined out of a solid block: the external shape gives no indication of the internal shape.

M. Reader-Harris, *Orifice Plates and Venturi Tubes*,
Experimental Fluid Mechanics, DOI 10.1007/978-3-319-16880-7_3

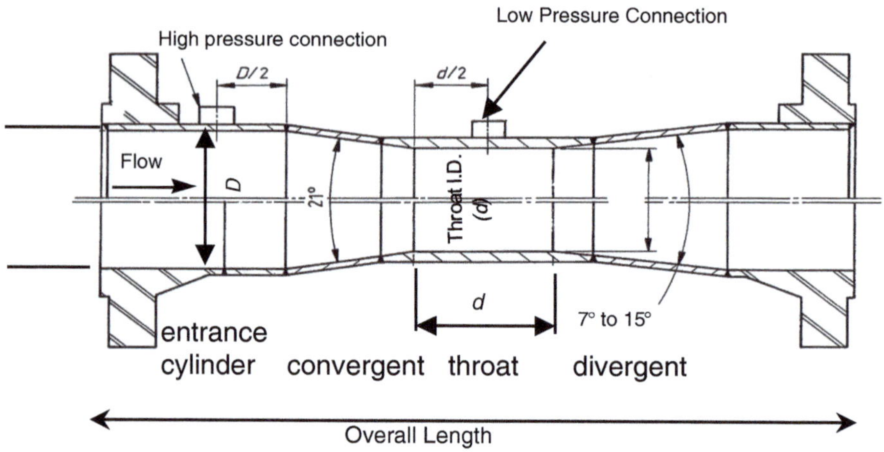

Fig. 3.1 The classical Venturi tube

Fig. 3.2 An 8″ Venturi tube at NEL

To design a Venturi tube, given the range of mass flowrates, the density, the desired range of differential pressures (discussed in Sect. 4.3), the discharge coefficient (from Sect. 3.2), the expansibility factor (from Eq. 1.13) and the desired diameter of the entrance cylinder (D), the throat diameter (d) can be determined from Eq. 1.15:

$$q_m = \frac{C\varepsilon}{\sqrt{1-\beta^4}} \frac{\pi d^2}{4} \sqrt{2\rho_1 \Delta p}. \tag{1.15}$$

If this gives the diameter ratio β (=d/D) greater than the maximum permitted value in Sect. 3.2 (most commonly 0.75) then a larger value of D is required (or a second Venturi tube in parallel could be used); if it gives β less than the minimum permitted value in Sect. 3.2 (most commonly 0.4) then a smaller value of D is required. If there is no preferred value of D then it would not be unreasonable to choose a value of D so that β is around 0.6; however, the choice of β within the permitted range is not critical.

The diameter of the entrance cylinder should be equal to that of the upstream pipe.

3.2 Type

3.2.1 General

There are three types of Venturi tube standardized in ISO 5167-4:2003:

machined convergent in 5.2.9 of ISO 5167-4:2003
rough-welded sheet-iron convergent in 5.2.10 of ISO 5167-4:2003
'as cast' convergent, in 5.2.8 of ISO 5167-4:2003.

For each type the range of Reynolds number over which a specified discharge coefficient is achieved is limited (the limits are given in 5.5.3, 5.5.4 and 5.5.2 of ISO 5167-4:2003 respectively). The differences in discharge coefficient between the types are due to minor differences in geometry and to the fact that data used to derive the predictions have been collected at different times by different researchers. It is not surprising that a rough-welded sheet-iron Venturi tube should have greater losses and thus a lower discharge coefficient than one with a machined convergent; the former's greater variability is reflected in its greater uncertainty. The restrictions on diameter are to some extent descriptions of what can be manufactured. The ranges of β given in ISO 5167-4:2003 reflect what has been tested. Roughness limits are covered in Sects. 3.2.2–3.2.5 and 3.7; discharge coefficients, their uncertainties, and the Reynolds number limits are also covered in Sects. 3.2.2–3.2.5.

For all three types the divergent has very little effect on the discharge coefficient (see also Sect. 3.3).

3.2.2 Machined Convergent (5.2.9, 5.5.3 and 5.7.2 of ISO 5167-4:2003)

The entrance cylinder, convergent and throat are machined. The surface roughness R_a shall be less than 10^{-4} d for all these three parts of the Venturi tube (see Sect. 3.7). The radii of curvature at joins should be as small as possible.

The main design limits are:

$50 \text{ mm} \leq D \leq 250 \text{ mm}$
$0.4 \leq \beta \leq 0.75$
$C = 0.995 \pm 1\ \%$
$2 \times 10^5 \leq Re_D \leq 10^6$

3.2.3 Rough-Welded Sheet-Iron Convergent (5.2.10, 5.5.4 and 5.7.3 of ISO 5167-4:2003)

R_a shall be less than 10^{-4} d for the throat; it may need machining. R_a should be approximately 5×10^{-4} d for the entrance cylinder and the convergent. There shall be no curvature at joins except the effect of welding.

The main design limits are:

$200 \text{ mm} \leq D \leq 1200 \text{ mm}$
$0.4 \leq \beta \leq 0.7$
$C = 0.985 \pm 1.5\ \%$
$2 \times 10^5 \leq Re_D \leq 2 \times 10^6$

3.2.4 'As Cast' Convergent (5.2.8, 5.5.2 and 5.7.1 of ISO 5167-4:2003)

R_a shall be less than 10^{-4} d for the throat. R_a shall be less than 10^{-4} D for the entrance cylinder and the convergent. The radii of curvature at joins are specified in ISO 5167-4:2003 and are quite large.

The main design limits are:

$100 \text{ mm} \leq D \leq 800 \text{ mm}$
$0.3 \leq \beta \leq 0.75$
$C = 0.984 \pm 0.7\ \%$
$2 \times 10^5 \leq Re_D \leq 2 \times 10^6$

Table 3.1 Discharge coefficient and uncertainty for Venturi tube with a machined convergent from (informative) Annex B of ISO 5167-4:2003

Re_d	C	Uncertainty (%)
5×10^4	0.970	3
1×10^5	0.977	2.5
2×10^5	0.992	2.5
3×10^5	0.998	1.5
5×10^5 to 10^6	0.995	1
10^6 to 2×10^6	1.000	2
2×10^6 to 10^8	1.010	3

3.2.5 Wider Range of Reynolds Number

For Reynolds numbers outside those given in Sects. 3.2.2–3.2.4 there are values of discharge coefficient and uncertainty given in (informative) Annex B of ISO 5167-4:2003. In particular, for a Venturi tube with a machined convergent, the discharge coefficients and uncertainties are given in Table 3.1.

Information on discharge coefficients at high Reynolds number is given in Chap. 7. The increase in discharge coefficient at high Reynolds number is partly due to static-hole error, and the increase in uncertainty at high Reynolds number is due to the spread in values of static-hole error at high Reynolds number and to acoustic effects (see Sect. 3.8 and Chap. 7).

NOTE Static-hole error is the effect that pressure tappings of finite size do not measure the pressure which would have been measured using an infinitely small hole: it is discussed in Chap. 7.

In Table 3.1 it is worth noting that the discharge coefficient does not increase continuously with Reynolds number: there is a hump around $Re_d = 3 \times 10^5$. This hump is often observed in calibration and can be seen, for instance, in Spencer and Thibessard (1962) or in Fig. 3.7.

Where a smaller uncertainty can be obtained using Sect. 3.2.2 the value of C (and its associated uncertainty) should be used.

Some text on Venturi tubes of the other types besides machined convergent is provided in Annex B of ISO 5167-4:2003, but the statement that the discharge coefficient does not appear to change for Re_D greater than 2×10^6 for a Venturi tube with a rough-welded sheet-iron convergent and 3.2×10^6 for one with an ‘as cast’ convergent is inconsistent with the work described in Chap. 7.

3.3 Angles, Pressure Loss and Truncation

For all Venturi tubes the convergent angle is 21° ± 1° (5.2.3 of ISO 5167-4:2003). This may be optimum for water flows, but may not be optimum for high-pressure gas (see Chap. 7). The divergent (or diffuser) included angle is in the range 7°–15° (5.2.5 of ISO 5167-4:2003).

There are graphs in Annex C of ISO 5167-4:2003 giving overall pressure loss across the Venturi tube as a fraction of the differential pressure. The effects of relative roughness, Reynolds number and divergent angle are all included. These losses were calculated by measuring the loss between a tapping at least 1 *D* upstream of the upstream flange of the Venturi tube and a tapping at least 6 *D* downstream of the downstream flange. Next, the loss that there would have been if there had only been a pipe between the two tappings is subtracted from the measured loss. Consequently only the additional permanent loss due to the Venturi tube is given in Annex C of ISO 5167-4:2003, and the location of the tappings is no longer critical, since increases in distance between them will lead to equal increases in loss both with and without a Venturi tube. A divergent angle of 7° gives an overall pressure loss of around 10 % of the differential pressure (from upstream of the Venturi tube to its throat); one of 15° gives an overall pressure loss of around 15 % of the differential pressure.

For many applications, for example, the measurement of wetness in wet-gas flow or diagnostic measurements, it is more useful simply to consider the pressure difference (commonly called the pressure loss) between the upstream pressure tapping and an additional tapping about 6 *D* downstream of the downstream end of the divergent. The pressure loss ratio, the ratio of pressure loss to differential pressure, is often used. Values for the pressure loss ratio (ratio of pressure loss to differential pressure) of four Venturi tubes with a divergent angle of 7.5° obtained at NEL are given in Fig. 3.3.

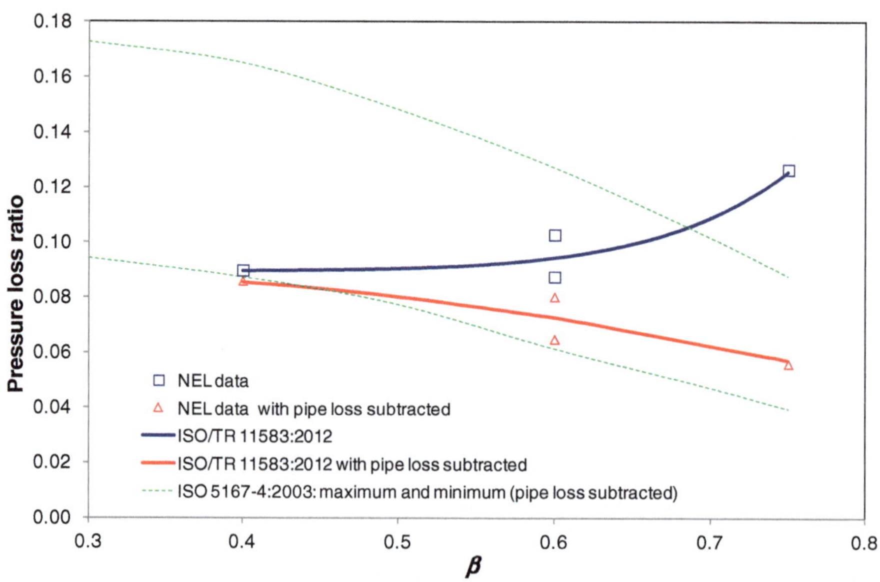

Fig. 3.3 Pressure loss ratio: data across four Venturi tubes at NEL without subtracting the pipe loss; the same data with the pipe loss subtracted; the equation in ISO/TR 11583:2012; the same equation with the pipe loss subtracted; the maximum and minimum values in ISO 5167-4:2003 (which already has the pipe loss subtracted)

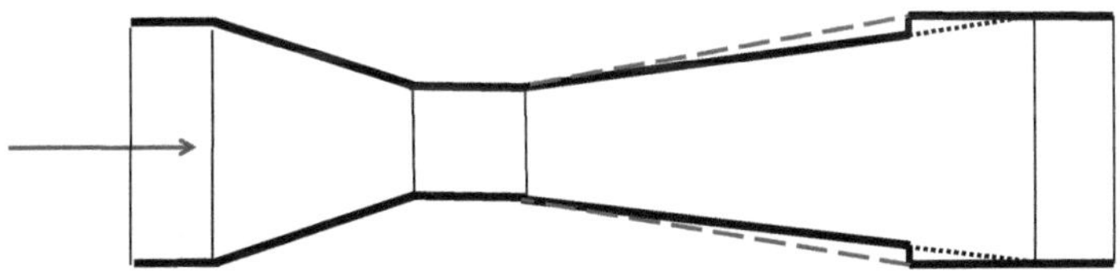

Fig. 3.4 Truncated Venturi tube

It will be noticed that the NEL data in Fig. 3.3 appear to be significantly different from the values in ISO 5167-4:2003, which expects the pressure loss to reduce with increasing β, but this apparent difference is due to the fact that the NEL data are in fact the total loss without subtracting the pipe loss. The NEL data have also been recalculated with the pipe pressure loss subtracted, and the recalculated data are also shown in Fig. 3.3. ISO/TR 11583:2012 (for reasons of ease of use) also does not subtract the pressure loss (the definition is in its Table 3.1): its values are shown in Fig. 3.3. The equation in ISO/TR 11583:2012 has also been recalculated with the pipe pressure loss removed (using a pipe friction factor $\lambda = 0.012$ (see Sect. 1.5 for definition) and taking the loss to the mid-point of the permitted range in 6.4.5 of ISO/TR 11583:2012), and the recalculated values are shown in Fig. 3.3. When the equation in ISO/TR 11583:2012 is used with the pipe loss removed it is near the bottom of the band in Fig. C.1 of ISO 5167-4:2003 (and in Fig. 3.3).

Warren (1951) gives measurements of losses in a Venturi tube which is non-standard in the convergent and throat sections, but standard in terms of the divergent. Most of the loss takes place in the divergent. For a divergent angle of 7.5°, he obtained ratios of pressure loss (including the pipe loss) to differential pressure of 0.147, 0.101 and 0.120 for diameter ratios of 0.25, 0.5 and 0.75 respectively.

The divergent can be truncated by up to 35 % of its length with no effect on the discharge coefficient and very little on the pressure loss (5.2.6 of ISO 5167-4:2003). A truncated Venturi tube is depicted in Fig. 3.4: its pressure loss is almost the same as that of the same Venturi tube without truncation (in which the Venturi wall follows the dotted lines) and less than that of the Venturi tube of the same length but with larger divergent angle (following the dashed lines). So it is more efficient to shorten the length of a Venturi tube by truncating it than by increasing its divergent angle. Details of measurements of pressure loss in both truncated and non-truncated Venturi tubes are given in Warren (1951).

3.4 Dimensional Measurements

Particularly where a Venturi tube is being used without flow calibration the dimensional measurements are crucial: the values of d and D for use in Eq. 1.15 are the values in the planes of the tappings. 5.2.2 and 5.2.4 of ISO 5167-4:2003 require that at least four measurements of diameter be made in each plane and that measurements also be taken in other planes. Best practice is to measure near the beginning and the end of the throat (but not so close to a corner that the diameter is affected by a joining curvature to the convergent or divergent).

ISO 5167-4:2003 gives limits on the permitted variation in measured diameters; if the measured diameters justify it, a measured uncertainty lower than the maximum permitted in ISO 5167-4 can be used in the uncertainty of the mass flowrate (see Sect. 4.8). Modern methods of measurement using coordinate measurement machines give not individual diameters but a mean diameter and circularity, but it can easily be checked whether the measurements meet the intention of ISO 5167-4:2003.

The measurements of the throat are considerably more important than those of the upstream cylinder (see Sect. 4.8). The machining of the throat can be demanding as it can be a considerable distance from the centre of the machine, particularly where β is small.

3.5 Steps and Straightness

It is not only the geometry of the Venturi tube itself that affects the measured flowrate. The upstream pipe and its join to the upstream cylinder are particularly significant. ISO 5167-4:2003 states that the upstream pipe shall be sufficiently cylindrical that all its measured diameters are within 2 % of their mean (see 6.4.1.1 of ISO 5167-4:2003) and at the join between the upstream pipe and the entrance cylinder the pipe diameter shall be within 1 % of measured D (see 6.4.1.2 of ISO 5167-4:2003). The offset between the centrelines of the pipe and the entrance cylinder shall be less than 0.005 D, and there are restrictions on angular alignment and the sum of the offset and half the diameter deviation (see 6.4.3 of ISO 5167-4:2003). To meet these requirements will probably require dowels or some equivalent method to achieve alignment. If it is intended not to machine the upstream pipe it should be selected to meet the cylindricality requirement and the Venturi tube manufactured to match it. If the Venturi tube is manufactured first, the upstream pipe may require machining, and it may be necessary to buy pipe of a higher schedule than the Venturi tube to machine. For the equivalent problem with orifice plates see Sect. 2.3.4.

Straightness of the upstream pipe is defined in 7.1.3 of ISO 5167-1:2003, and is normally only checked visually.

The basis for these requirements is not stated in the standard.

3.6 Pressure Tappings

The tapping locations are 0.5 D upstream of the upstream end of the convergent and 0.5 $d \pm 0.02$ d downstream of the downstream end of the convergent. Full definitions, including tolerances for the former, are given in 5.4.7 of ISO 5167-4:2003.

5.4.2 of ISO 5167-4:2003 requires that for $d \geq 33.3$ mm the tapping diameter shall be in the range from 4 mm to 10 mm, but not greater than 0.1 D for the upstream tappings and not greater than 0.13 d for the throat tappings. The requirement in terms of relative diameter is required for dimensional similarity. Because static-hole error depends on the tapping diameter a requirement in terms of absolute diameter is reasonable: very small tappings would give static-hole error smaller than was obtained with the Venturi tubes on which ISO 5167-4:2003 is based; moreover, tappings should not be so small that they might block. Requirements for Venturi tubes with $d < 33.3$ mm are also given in 5.4.2 of ISO 5167-4:2003. Static-hole error and its consequences are discussed in Chap. 7: in Sect. 7.5 it is recommended that, where possible, the throat-tapping diameter not be greater than 0.07 d.

5.4.3 of ISO 5167-4:2003 requires four tappings in each tapping plane. If this requirement is met the tappings should be arranged in a triple-T as in Fig. 4.8. Using a triple-T reduces the effect of upstream installation if there is an asymmetric flow profile, and has the advantage that a burr on one tapping in four has less effect on the flow measurement than the same burr would have if a single tapping with that burr were used. However, if there is any risk of the presence of liquid in a gas flow tappings should only be in the top half of the pipe. If there is any risk of gas in a liquid flow tappings should only be in the bottom half of the pipe. Therefore, a single tapping in each tapping plane is usual, i.e. it is common to disregard 5.4.3 of ISO 5167-4:2003. It is possible to have two (or conceivably more) tappings in the same half of the pipe. For the circumferential location of the tappings see Sect. 4.2.2.

At least some of the minimum upstream straight lengths were determined using four tappings at right angles connected in a triple-T; so especially if single tappings are used it would be wise to use a longer upstream straight length than the minimum stated in ISO 5167-4:2003.

Owing to the high velocity of the fluid passing the throat tapping, the largest source of error in measurement using Venturi tubes lies in the tappings. Tappings, where they enter the pipe or throat wall must be of the correct diameter, round and have a sharp clean edge. Even very small burrs can create significant errors in C. Errors of up to 5 % are not impossible. The throat tapping is more important than the upstream tapping because the velocity past it is greater. One solution is spark erosion of the tapping where it enters the pipe wall, which has proved to give very consistent results (see Sects. 7.A.6 and 7.A.7). The additional cost in manufacture can be easily saved in not having to modify poorly performing Venturi tubes later.

Protruding tappings or unintended recesses have large effects [see Sect. 7.3.6 and e.g. Allen and Hooper (1932)]. Further details regarding flow past tappings are given in Sect. 7.3. Great care should be taken if tappings are formed by welding a boss to a pipe or throat wall that the welding heat does not create an internal bulge which alters the local velocity at the tapping as well as altering the pipe or throat diameter at the tapping. Welding of bosses should be carried out prior to machining a bore, or an alternative method of preventing this problem employed.

The importance of correct throat tapping location can be seen from Fig. 3.5, which gives an example of a computed pressure profile through a Venturi tube. The requirement for the tapping locations is given in 5.4.7 of ISO 5167-4:2003. The pressure profile, with its spikes on the wall, is more complex than might have been expected. Other computational examples are given by Sattary and Reader-Harris (1997). Experimental measurements were made by Lindley (1966). The pressure dip at the beginning of the throat can lead to cavitation at that point even when the pressure at the throat pressure tapping is high enough to suggest that cavitation will be avoided (see Sect. 4.9).

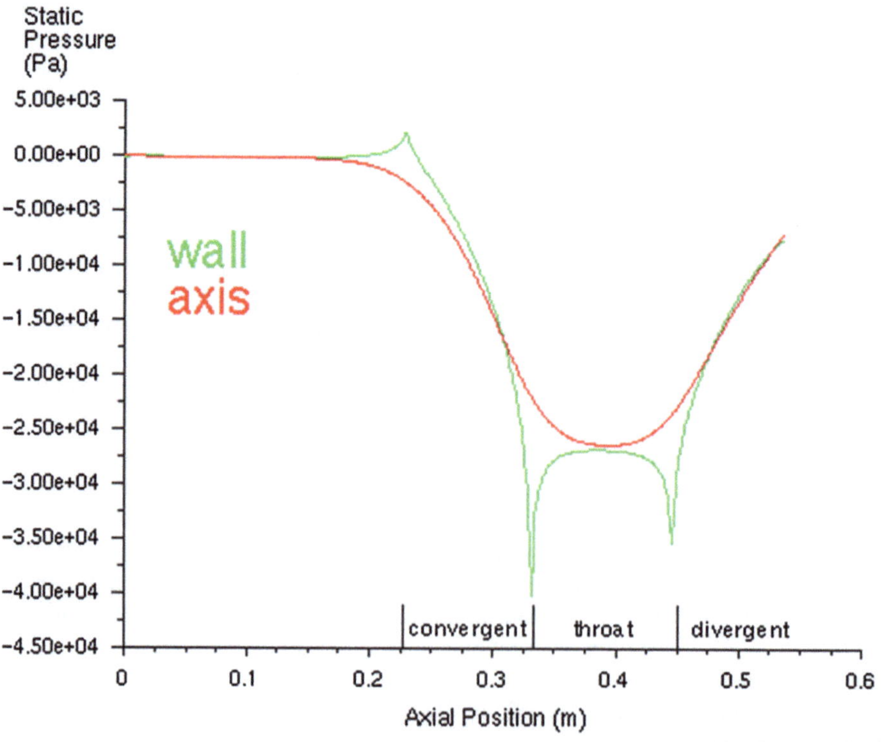

Fig. 3.5 Example of a pressure profile through a Venturi tube ($\beta = 0.75$): CFD by Sattary at NEL

3.7 Effects of Roughness and Reynolds Number

The effect of the roughness of a Venturi tube is addressed in 9.6.3 of ISO/TR 12767:2007 (ISO 2007). This is based on the work of Hutton (1954).

The limits on acceptable roughness are given in ISO 5167-4:2003: for the upstream pipe $R_a/D < 3.2 \times 10^{-4}$ (6.4.2 of ISO 5167-4:2003); for the throat $R_a/d < 10^{-4}$ (see 5.2.7 of ISO 5167-4:2003); for the other parts of the Venturi tube the roughness depends on the type of Venturi tube (5.2.8–5.2.10 of ISO 5167-4:2003). See also Sects. 3.2.2–3.2.4. It is normal to measure R_a, the arithmetical mean deviation of the roughness profile. To obtain R_a from the uniform equivalent roughness k, it is assumed that $k \approx \pi R_a$ (Annex B of ISO 5167-1:2003): this is based on assuming that the surface takes the form of a sine wave.

One practical issue here is that the roughness can change with time. Some information on this is given in Reader-Harris et al. (2000). It is rare to check the roughness in use but it may be necessary. A visual method to check that it has not changed may be a good option.

Computational Fluid Dynamics (CFD) work on the effect of roughness and Reynolds number is described in Appendix 3.A.

NOTE One of the strange features of the classical Venturi tube is that there is no correlation between the discharge coefficients of the population of standard Venturi tubes described in Chap. 7 and the relative roughness of these Venturi tubes, whereas there is, as would be expected, a strong correlation between the discharge coefficients of the population of Venturi tubes with a convergent angle of 10.5° and their relative roughness.

3.8 High or Low Reynolds Number

In practice there are interesting results that may be experienced at high Reynolds number: these are very important and are described in Chap. 7.

A different area in which Venturi tubes might be used is low Reynolds number flow (Pushkarny and Kar 1979; Reddy 1973; Pal 1993; Stobie et al. 2007), which occurs for oils of higher viscosity.

At NEL Marshall (2012) and Mills et al. (2013) calibrated 4″–8″ Venturi tubes with β values of 0.4–0.75 in oils of kinematic viscosity 100 and 200 cSt and also in a higher viscosity oil (Aztec) over the range 550–1450 cSt: see Fig. 3.6. For larger values of β there was a distinct hump in C at transition from laminar to turbulent flow. Marshall also used an additional pressure tapping downstream of the Venturi tube in order to correlate the discharge coefficient and the pressure loss ratio.

It is reasonable to identify several Reynolds number regions:

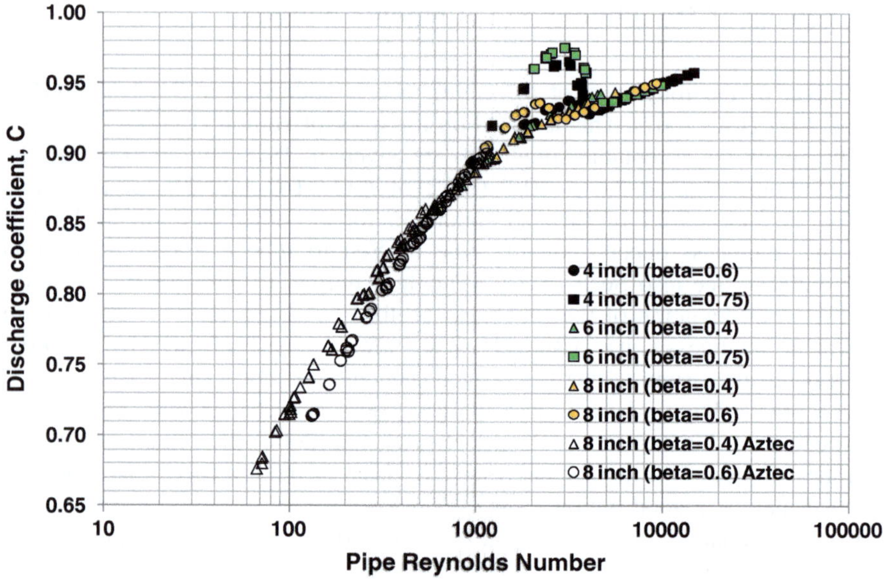

Fig. 3.6 Calibration of 4″–8″ Venturi tubes with β from 0.4 to 0.75 in oil at 100 and 200 cSt (sets of data not marked 'Aztec') and in a higher viscosity oil (Aztec) over the range 550–1450 cSt [see Marshall (2012)]

- At pipe Reynolds numbers below about 2000 the flow upstream of the Venturi tube was laminar: the discharge coefficient depended strongly on Reynolds number.
- As the pipe Reynolds number increased by about 1000 from a value in the range 2200–3200 (different for each pipe) the flow upstream of the Venturi tube became turbulent, and because the upstream velocity profile became flatter the discharge coefficient for large β dropped: the drop increased rapidly with β.
- At pipe Reynolds numbers above about 5000 the flow upstream of the Venturi tube was fully turbulent, and the discharge coefficient rose with Reynolds number.

There is disagreement about the flow within the Venturi tube (Hall 1959; Lindley 1969). According to Greenblatt et al. (1989) there is laminarization (a change from turbulent flow to laminar) when

$$\frac{\nu}{\bar{u}^2}\frac{d\bar{u}}{dx} \text{ is greater than about } 3.6 \times 10^{-6},$$

where $\bar{u}$ is the mean axial velocity in the duct, x is the distance along the axis of the duct and ν is the kinematic viscosity.

In a classical Venturi tube this occurs at the entrance to the convergent when $Re_D \leq$ about 2×10^5 but will not be true throughout the convergent unless $Re_d \leq$ about 2×10^5. However, Lindley (1969) found that in the particular Venturi tubes in which he made measurements (10″, $\beta = 0.71$, machined but with "as cast" profile on the corners) the boundary layer was fully turbulent at all the locations at the Reynolds numbers at which he made measurements ($Re_d \geq 6 \times 10^4$).

- As the pipe Reynolds number increases from 10^5 to 2×10^5 sometimes there is a drop in the discharge coefficient (see Fig. 3.7 for a striking example): this may be taking place as the Reynolds number becomes too high for laminarization in the convergent; Lindley's measurements, however, did not show laminarization (or a hump in discharge coefficient in this Reynolds number range). He did show that in the particular Venturi tubes in which he made measurements there was a small separation bubble just downstream of the sharp corner at the entrance to the throat for $Re_d < 3 \times 10^5$ but that this bubble was not present at higher Reynolds numbers.
- At pipe Reynolds numbers between about 2×10^5 and 10^6 the flow both upstream of and through the Venturi tube is certainly turbulent; the effect of static-hole error is very predictable and the uncertainty is low.
- At pipe Reynolds numbers above about 10^6 the flow both upstream of and through the Venturi tube is turbulent; the effect of static-hole error and acoustic effects in gas become quite large and less predictable and the uncertainty increases: see Chap. 7.

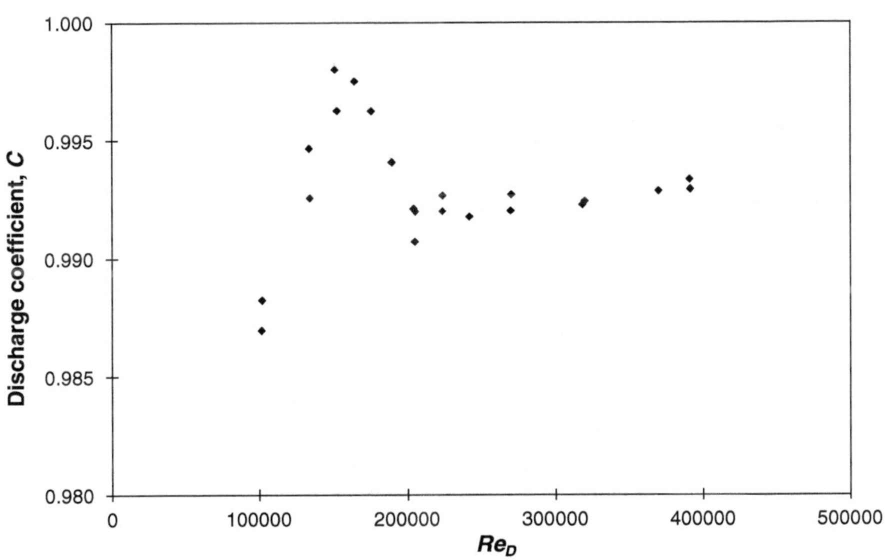

Fig. 3.7 Discharge coefficient for one Venturi tube in water: 4″, $\beta = 0.4$

3.9 Conclusions

This chapter has described the main design specifications for classical Venturi tubes: it has pointed the reader to important parts of ISO 5167 and given reasons for the requirements in the standard. It has covered the different types: their shape and their discharge coefficient. It has included the tappings, the dimensional measurements, the effect of roughness and the pressure loss. Gas flow at high Reynolds number and the effect of upstream fittings are not covered here: they are in Chaps. 7 and 8 respectively.

The basic instruction remains to follow ISO 5167-4:2003, probably with only one tapping, instead of four tappings, in each plane (see Sect. 3.6).

Appendix 3.A: Effect of Roughness: Computational Fluid Dynamics

3.A.1 General

In the work at NEL described in this appendix the effect of the roughness of Venturi tubes and upstream pipes was computed (Reader-Harris et al. 2000). The results are summarized here. The tappings were not included in the computation. In this appendix roughness is described using the arithmetical mean deviation of the roughness profile, R_a, (the average of the absolute values of the deviation from the mean).

3.A.2 Venturi Tube Roughness

3.A.2.1 Effect of Venturi Tube Roughness Height

Except for two points for R_a/d equal to 10^{-2} increasing the roughness of Venturi tubes of diameter ratio 0.4 and 0.75 caused a decrease in the discharge coefficient, as shown in Tables 3.A.1 and 3.A.2 (because the friction loss increases). The pressure spike at the corner between the convergent section and the throat is smaller for rough Venturi tubes, and there is a significant pressure loss in the throat itself. These factors reduce the pressure at the throat tapping, increasing the measured pressure difference across the Venturi tube, thus reducing C. Very rough walls cause difficulty in CFD calculations.

Table 3.A.1 Predicted discharge coefficient for a range of Reynolds numbers and Venturi tube roughness heights: $\beta = 0.4$, inlet pipe $R_a/D = 10^{-5}$

Re_D	R_a/d				
	0	10^{-5}	10^{-4}	10^{-3}	10^{-2}
10^6	0.987	0.987	0.979	0.956	0.961
5×10^6	0.990	0.988	0.978	0.956	0.961
2×10^7	0.992	0.987	0.978	0.956	0.933

Table 3.A.2 Predicted discharge coefficient for a range of Reynolds numbers and Venturi tube roughness heights: $\beta = 0.75$, inlet pipe $R_a/D = 10^{-5}$

Re_D	R_a/d			
	10^{-5}	10^{-4}	10^{-3}	10^{-2}
10^6	0.988	0.979	0.951	0.918
2×10^7	0.988	0.976	0.950	0.921

3.A.2.2 Effect of Reynolds Number

For smooth Venturi tubes, increasing Reynolds number causes an increase in C (because the friction loss as a fraction of the differential pressure reduces): this can be seen in Table 3.A.1 for $\beta = 0.4$. For $\beta = 0.75$ C increases from 0.988 when $Re_D = 10^6$ to 0.992 when $Re_D = 2 \times 10^7$ provided that both the Venturi tube and the upstream pipe are smooth (i.e. for the Venturi tube $R_a/d = 0$ and for the upstream pipe $R_a/D = 0$). However, the discharge coefficient of rough Venturi tubes appears to be independent of Reynolds number over the range tested. The decrease in C shown by one very rough Venturi tube is likely to be due to numerical errors in the CFD.

3.A.2.3 Effect of Venturi Tube Roughness Type

In the Fluent 5 CFD code roughness is specified in terms of an actual roughness height and a shape factor C_{Ks}. The C_{Ks} value is varied to account for the shape and distribution of the roughness. For example, a riveted surface which has rivets of height 2 mm could be represented by setting the actual roughness height to 2 mm and C_{Ks} to an appropriate value. A ribbed surface which has 2 mm high ribs would also be represented by setting the actual roughness height to 2 mm, but a different value of C_{Ks} would be used. Setting C_{Ks} to 0.5 reproduces sand roughness.

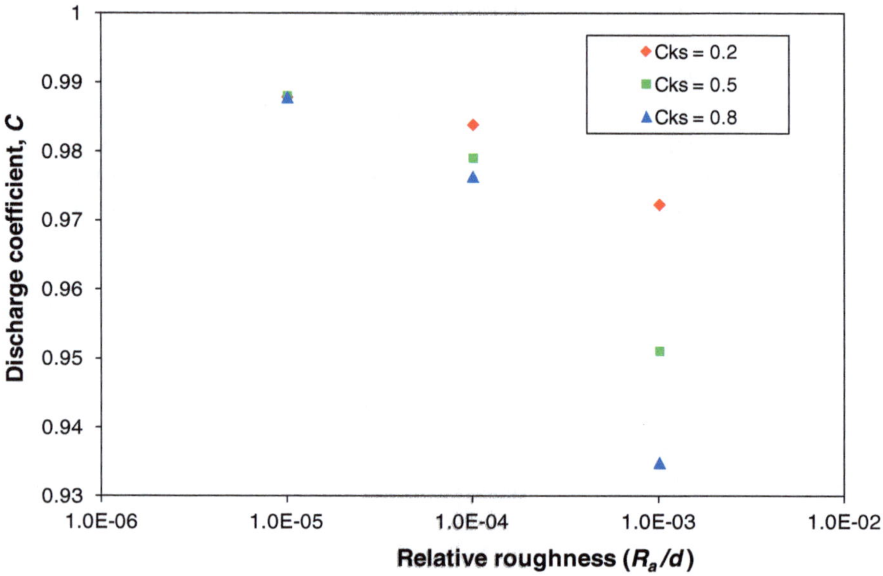

Fig. 3.A.1 Variation of discharge coefficient with roughness shape factor, C_{Ks}: $\beta = 0.75$, $Re_D = 10^6$

Unfortunately, very little information is available which links values of C_{Ks} to particular roughness patterns. Figure 3.A.1 shows that varying C_{Ks} significantly affects the extent to which discharge coefficient varies with roughness height, particularly for very rough Venturi tubes. This implies that the shift in discharge coefficient caused by, say, 100 μm grooves eroded to run parallel to the flow could differ significantly from the shift caused by 100 μm grooves running across the flow.

3.A.3 *Pipe Roughness*

Tables 3.A.3 and 3.A.4 show that increasing the roughness of the pipe upstream of a Venturi tube causes an increase in *C* (a peakier velocity profile results in a smaller differential pressure); this effect is small in magnitude compared with the effect of roughening the Venturi tube itself to an equivalent degree. The effect of pipe roughness increases with β as would be expected.

Table 3.A.3 Predicted discharge coefficient for a range of Reynolds numbers and pipe roughness heights: $\beta = 0.4$, Venturi tube $R_a/d = 10^{-5}$

Re_D	R_a/D				
	0	10^{-5}	10^{-4}	10^{-3}	10^{-2}
10^6		0.987	0.987	0.988	0.993
2×10^7	0.985	0.987	0.985	0.985	0.993

Table 3.A.4 Predicted discharge coefficient for a range of Reynolds numbers and pipe roughness heights: $\beta = 0.75$, Venturi tube $R_a/d = 10^{-5}$

Re_D	R_a/D				
	0	10^{-5}	10^{-4}	10^{-3}	10^{-2}
10^6	0.988	0.988	0.990	0.994	0.996
2×10^7	0.987	0.988	0.989	0.990	1.020

3.A.4 Effect of Rounding the Corner Between the Convergent Section and the Throat

Figure 3.A.2 shows how rounding the corner between the convergent section and the throat reduces the magnitude of the pressure spike upstream of the throat. The figure shows the pressure profile for a $\beta = 0.4$ Venturi tube with a sharp and with a 4.839 d radius corner. In addition to reducing the magnitude of the pressure spike, the rounded corner reduces the straight length of throat upstream of the throat tapping. This causes the spike to recover later in the throat, reducing the pressure at the throat tapping and thus increasing the measured pressure drop and decreasing the discharge coefficient.

Figure 3.A.3 shows that rounding the corner of a smooth Venturi tube has a more significant effect on small β Venturi tubes. ISO 5167-4:2003 defines an "as cast" Venturi tube as having a radius of 3.625 d at the join of the convergent and the throat. The greatest difference seen between the discharge coefficient of a Venturi tube with a sharp corner and that of a rounded Venturi tube of this radius was −0.3 % (for the $\beta = 0.4$ Venturi tube). This difference in discharge coefficient is only slightly affected by Reynolds number.

NOTE Data collected in 1400 mm Venturi tubes are presented by Han et al. (1998): the 13 Venturi tubes had discharge coefficients in water of 0.985 ± 1 %. There was very little difference in discharge coefficient between rough-welded and machined convergent Venturi tubes.

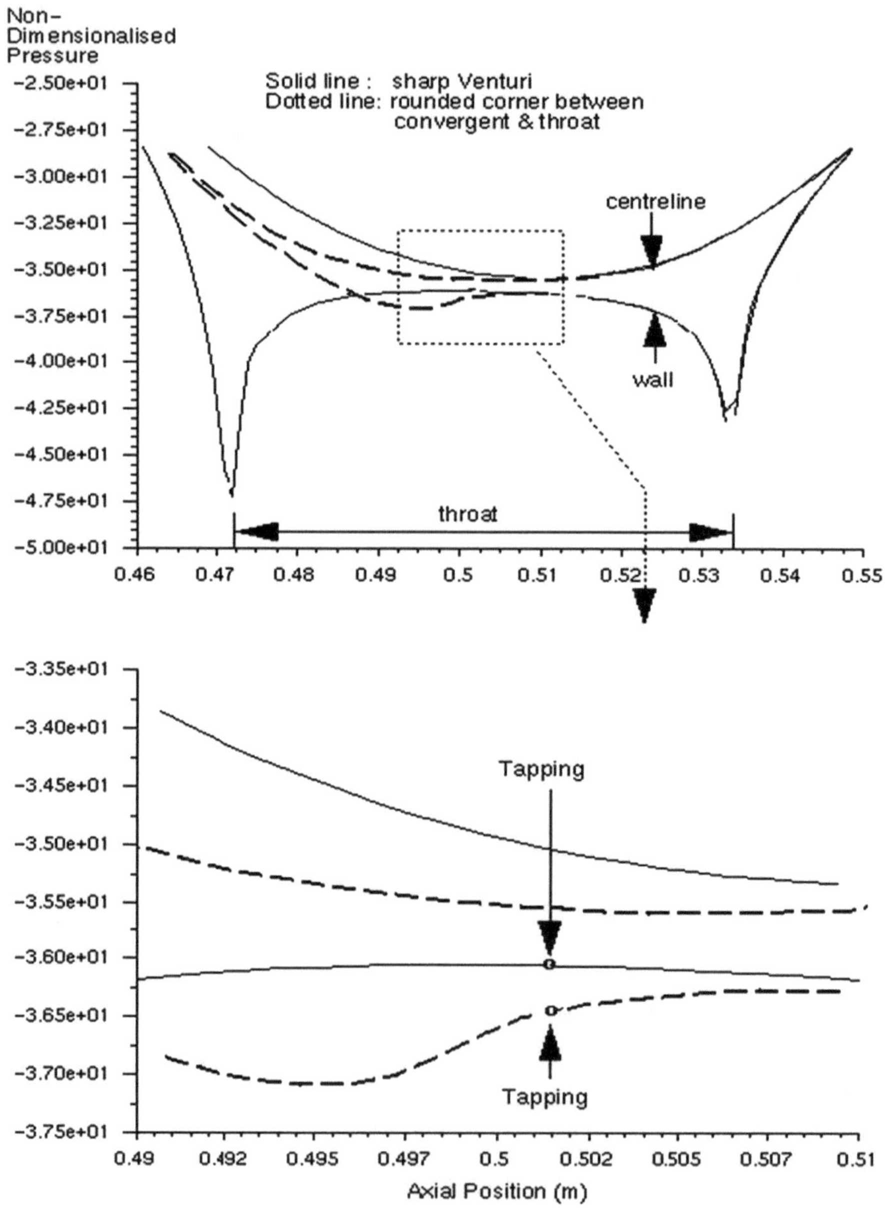

Fig. 3.A.2 The pressure profile through a Venturi tube: $\beta = 0.4$

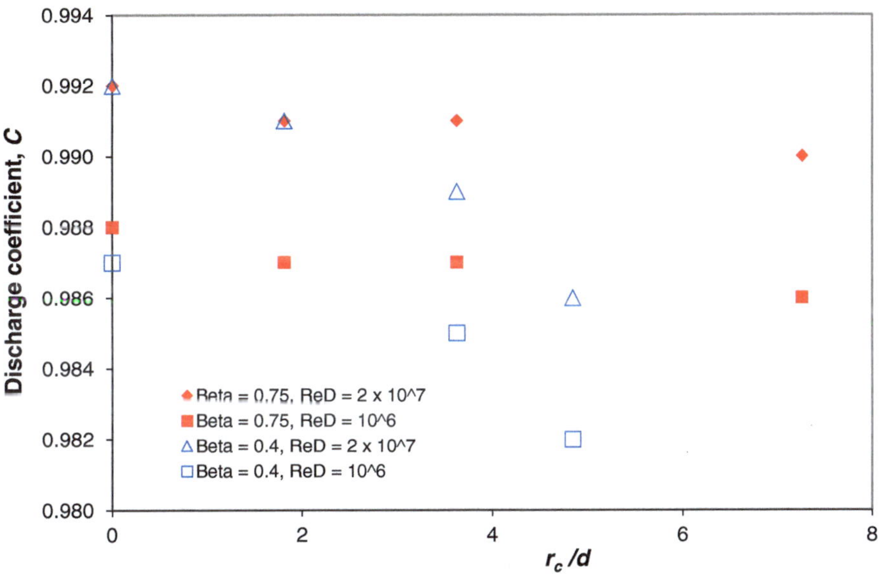

Fig. 3.A.3 The effect of rounding the corner between the convergent section and the throat: r_c is the radius of that corner. The Venturi tube and upstream pipe were both smooth

References

Allen CM, Hooper LJ (1932) Piezometer investigation. ASME Hydraulics Div: HYD-54-1 54:1–16

Greenblatt D, Da Silva DF, Moss EA (1989) A numerical and experimental investigation of temporally accelerated turbulent pipe flow. In: Taylor C, Gresho P, Sani RL, Hauser J (eds) Numerical methods in laminar and turbulent flow (the proceedings of the 6th International Conference on Numerical Methods in Laminar and Turbulent Flow, Swansea, July 1989), part 1, vol 6. Pineridge Press, Swansea, pp 365–375

Hall GW (1959) Application of boundary layer theory to explain some nozzle and Venturi flow peculiarities. Proc Inst Mech Eng 173(36):837–870

Han Y, Han S, Xu L (1998) The practice on DN 1400 mm Venturi tubes. Flow Measur Instrum 9:11–23

Hutton SP (1954) The prediction of Venturi-meter coefficients and their variation with roughness and age. Proc Inst Civ Eng Part 3, 3: 216–241; 922–927

ISO (2003a) Measurement of fluid flow by means of pressure differential devices inserted in circular cross-section conduits running full—part 1: general principles and requirements. International Organization for Standardization, Geneva. ISO 5167-1:2003

ISO (2003b) Measurement of fluid flow by means of pressure differential devices inserted in circular cross-section conduits running full—part 4: Venturi tubes. International Organization for Standardization, Geneva. ISO 5167-4:2003

ISO (2007) Measurement of fluid flow by means of pressure differential devices—guidelines on the effect of departure from the specifications and operating conditions given in ISO 5167. International Organization for Standardization, Geneva. ISO/TR 12767:2007

Lindley D (1966) Venturimeters and boundary layer effects. Ph.D. Thesis, Department of Mechanical Engineering, University College of South Wales and Monmouthshire, Cardiff

Lindley D (1969) An experimental investigation of the flow in a classical Venturimeter. Proc Inst Mech Eng 184(1):133–145 (1969–1970)

Marshall C (2012) Measurement of high viscosity fluids using a smart Venturi meter. Report no 2012/549 on project no FHRE17. NEL, East Kilbride

Mills C, Marshall C, Kay A, MacDonald M (2013) Flow measurement of high viscosity fluids. In: Proceedings of 31st International North Sea Flow Meas Workshop, paper 7, Norway

Pal R (1993) Flow of oil-in-water emulsions through orifice and Venturi meters. Ind Eng Chem Res 32:1212–1217

Pushkarny BH, Kar S (1979) Investigations of discharge coefficient of Herschel type smooth Venturimeters for laminar flow. IE(I) J ME 59: 217–220

Reader-Harris MJ, Barton N, Brunton WC, Gibson JJ, Hodges D, Nicholson IG, Johnson P (2000) The discharge coefficient and through-life performance of Venturi tubes. In: Proceedings of 18th North Sea Flow Meas Workshop, Gleneagles, paper 5.2. National Engineering Laboratory, East Kilbride, Glasgow

Reddy YR (1973) Herschel-type Venturimeter discharge coefficients at low Reynolds number. Int J Mech Sci 15:757–763

Sattary JA, Reader-Harris MJ (1997) Computation of flow through Venturi meters. In: Proceedings of 15th North Sea Flow Meas Workshop, Norway

Spencer EA, Thibessard G (1962) A comparative study of four classical Venturimeters. Flow measurement in closed conduits I (paper C-3). HMSO, Edinburgh, pp 279–316. (Proceedings of symposium at national engineering laboratory, September 1960)

Stobie G, Hart R, Svedeman S, Zanker K (2007) Erosion in a Venturi meter with laminar and turbulent flow and low Reynolds number discharge coefficient measurements. In: Proceedings of 25th International North Sea Flow Meas Workshop, Norway

Warren J (1951) A study of head loss in Venturi-meter diffuser sections. Trans ASME 73: 399–402

Chapter 4
General Design

Abstract This chapter covers the auxiliary equipment required to derive the flowrate from the differential pressure generated by a differential-pressure meter such as an orifice plate or a Venturi tube, described in Chaps. 2 and 3, respectively. The main recommendation has to be to follow ISO 5167 and ISO 2186. This chapter points the reader to important parts of the standards and gives reasons for their requirements. In particular, it covers the appropriate angular location of the pressure tappings, the appropriate slopes of the impulse lines, the need to have the same density in the two impulse lines used to measure the differential pressure, the need to avoid blockage of the impulse lines, suitable ranges for differential pressure, piezometer rings, temperature correction from downstream of the flowmeter to upstream of it, the correct evaluation of uncertainty, the avoidance of cavitation and the use of diagnostics to give increased confidence in the flowrate measurement.

4.1 Introduction

This chapter covers the auxiliary equipment required to derive the flowrate from the differential pressure generated by a differential-pressure meter such as an orifice plate or a Venturi tube, described in Chaps. 2 and 3, respectively. The main recommendation has to be to follow ISO 5167 (ISO 2003a, b, c, d) and ISO 2186 (ISO 2007). This chapter points the reader to important parts of the standards and gives reasons for their requirements.

4.2 Impulse Lines

4.2.1 General

To the tappings through the pipe wall are attached impulse lines. An impulse line is the piping or tubing connecting a pressure tapping to a differential-pressure or a static-pressure transmitter: it may be flexible or rigid. The impulse lines must be

M. Reader-Harris, *Orifice Plates and Venturi Tubes*,
Experimental Fluid Mechanics, DOI 10.1007/978-3-319-16880-7_4

installed in such a way that they transmit the pressure at the tapping accurately to the transmitter. Impulse lines are described in ISO 2186 (ISO 2007) (see also Reader-Harris and McNaught 2005). The use of impulse lines is known to cause a number of problems that can lead to an incorrect measurement. The problems can arise from the following:

1. Condensation of liquid in an impulse line that is intended to be filled with gas: see Sect. 4.2.2.
2. Gas bubbles being trapped in an impulse line that is intended to be filled with liquid: see Sect. 4.2.2.
3. Different temperatures (and therefore different fluid densities) in a pair of impulse lines: see Sect. 4.2.3.
4. The fluid in the impulse lines being of a different composition from the fluid in the pipe in a two-component flow through a flowmeter that is not horizontally installed: see Sect. 4.2.3.
5. Leakage from or between impulse lines, generally at joints: see Sect. 4.2.4.
6. Blockage of an impulse line: see Sect. 4.2.5.
7. Damping of the pressure signal or resonance (this problem is exacerbated if a transient measurement is required or if the impulse lines are of different length): see Sect. 4.2.6.

Figure 4.1 illustrates the potential problems listed above.

It might be thought that many problems could be avoided by installing a pressure transmitter directly on the tapping hole. However, it would then be very difficult to replace the transmitter. Moreover, the discharge-coefficient equations in the standards are based on tappings whose depth is at least equal to 2.5 times the tapping diameter: if there were no tappings the discharge coefficient (particularly for Venturi tubes) would be different (because of a lack of static-hole error: see Sect. 7.3). In practice, a transmitter installed on the wall will form either a very small protrusion or a very small recess, causing an error. Using pressure tappings and impulse lines is the standard method.

Low-pressure impulse lines are often made in transparent tube so that bubbles can be seen. If bubbles are present the impulse lines are bled to remove the bubbles. This is common practice in research and calibration facilities.

Two separate pairs of tappings with separate associated impulse lines and differential-pressure transmitters may be used to provide a check on the measurement. Clearly both measurements use the same orifice plate or Venturi tube; so they are not independent. However, if good agreement between the two measurements is obtained, it is much less likely that there is error in the tappings, the impulse lines or the differential-pressure transmitters. Using one pair of tappings but two differential-pressure transmitters in parallel provides a check on the transmitters but not on the tappings.

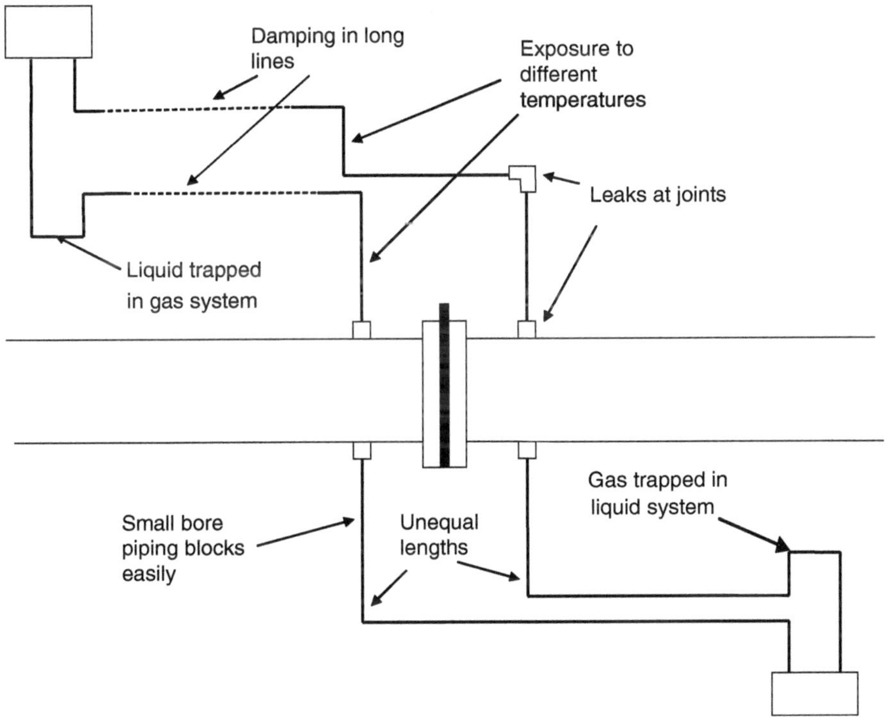

Fig. 4.1 What to avoid in impulse lines

4.2.2 Tapping Locations and Slopes of Impulse Lines

To prevent the problems described in 1 and 2 in Sect. 4.2.1 from occurring the impulse lines should be installed with appropriate slopes to prevent liquid from being trapped in gas lines and gas in liquid lines: i.e. in a gas flow or a wet-gas flow the impulse lines should slope continuously upwards to the pressure transmitter; in a liquid flow they should slope continuously downwards to the pressure transmitter. Pressure transmitters are generally installed with the inlet ports horizontal.

Table 4.1 shows recommended tapping locations for horizontal meters.

Steam-flow applications are treated slightly differently from other gas applications: on a horizontal steam meter the tappings should be horizontal with the impulse lines sloping downwards to the pressure transmitter; in this case the fluid in the impulse lines is condensed water: care may need to be taken at start-up to ensure that the pressure transmitter is not exposed to the steam temperature (see 5.3 of ISO 2186:2007).

Good examples of impulse lines for dry natural gas are given in Figs. 4.2 and 4.3. Figure 4.2 shows impulse lines leading from an orifice plate with horizontal

Table 4.1 Recommended pressure tapping locations for horizontal meters

Fluid	Pressure tapping location
Non-condensing gas	Between the top of the pipe and the horizontal centre line
Wet gas	Usually on the top of the pipe
Liquid	On the horizontal centre line (the bottom half of the pipe can be used if there is no danger of solids accumulating in the tappings)
Condensing vapour (e.g. steam)	On the horizontal centre line

Fig. 4.2 Gas metering installation showing insulated impulse lines [Courtesy of CATS (Central Area Transmission System)]

tappings to an instrument room. The lines are sloped upwards towards the instrument room, and are insulated.

Figure 4.3 is a view of the interior of the instrument room. It shows the impulse lines again correctly sloping upwards towards the differential-pressure transmitters so that any liquids that form in the lines will tend to drain downwards. Since the cabin is at a constant temperature there is no need for insulation. The transmitters are calibrated in situ. There are differential-pressure transmitters in parallel to provide redundancy.

In vertical pipelines there is no restriction on the angular location of the pressure tappings. The recommended slopes of the impulse lines for meters in vertical pipes are generally the same as for horizontal meters. For a vertical steam flow see 6.4 of ISO 2186:2007.

Fig. 4.3 View of interior of instrument room for gas metering installation (Courtesy of CATS)

Other installations besides those described here can be used, but may require more maintenance: for example, drain pots or vent valves may be used (see ISO 2186); they require manual or automatic emptying/opening and must be leak-tight.

4.2.3 Density of the Fluids in Two Impulse Lines to Measure the Differential Pressure

A column of fluid of density ρ (kg/m^3) and height h^* (m) gives rise to a static pressure of ρgh^* (Pa), where g is the acceleration due to gravity (m/s^2). Accordingly if the tappings are at the same height, the transmitter inlet ports are at the same height and the density of the fluid in both the impulse lines is the same, the pressure due to the fluid in the impulse lines will cancel out when the differential pressure is measured. However, if the fluids in the two impulse lines are at different temperatures they will have different densities. Different temperatures can be caused by wind chill or unequal solar gain. Therefore, impulse lines with a significant vertical component should be run close together and insulated as a pair so that they are at the same temperature and thus the same density, and in this way no error to the measured differential pressure is caused. In Fig. 4.4 the measured differential pressure is

$$p_1 + \rho_1 gh^* - (p_2 + \rho_2 gh^*) = \Delta p + (\rho_1 - \rho_2)gh^* \tag{4.1}$$

Fig. 4.4 Differential-pressure transmitter located below a horizontal orifice meter

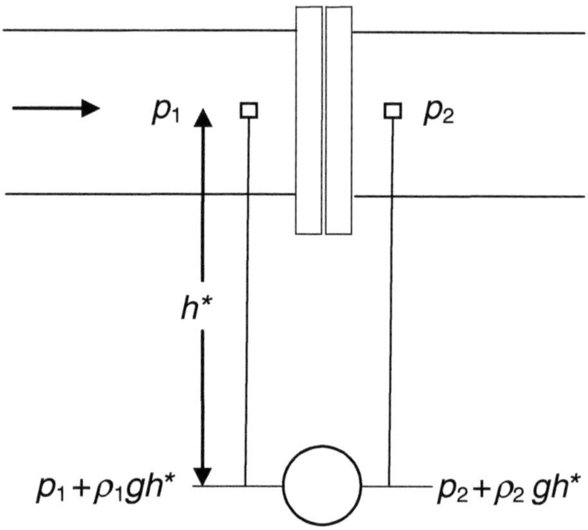

where p_1 and p_2 are the upstream and downstream pressures, respectively, ρ_1 and ρ_2 are the density of the fluid in the upstream and downstream impulse lines, respectively, h^* is the difference in height between the pressure tappings and the differential-pressure transmitter, g is the acceleration due to gravity and Δp is the true differential pressure ($= p_1 - p_2$). If the two impulse lines contain water and are at temperatures of 20 and 30 °C and h^* is 2 m there will be an error in measured differential pressure of 49 Pa (0.49 mbar).

It is necessary to examine vertical flow as well as horizontal flow. Consider a flow vertically upwards. In incompressible flow Bernoulli's Equation (Eq. 1.2) gives

$$\frac{1}{2}\rho u_1^2 + p_1 + \rho g z_1 = \frac{1}{2}\rho u_2^2 + p_2 + \rho g z_2, \tag{4.2}$$

where p, ρ, u, and z are the pressure, fluid density in the pipe, axial velocity and height respectively and the subscripted 1 and 2 represent the upstream and downstream (throat) planes. Equations 4.2 and 1.3 (for horizontal flow) are identical except that p in Eq. 1.3 is replaced by $p + \rho g z$ in Eq. 4.2. From Eqs. 4.2, 1.4 and 1.5 it follows that for incompressible flow

$$q_m = \frac{1}{\sqrt{1-\beta^4}}\frac{\pi d^2}{4}\sqrt{2\rho(p_1 - p_2 + \rho g(z_1 - z_2))}. \tag{4.3}$$

i.e. the differential pressure that is required in calculations of flow, Δp_{req}, is

$$\Delta p_{req} = p_1 - p_2 + \rho g(z_1 - z_2). \tag{4.4}$$

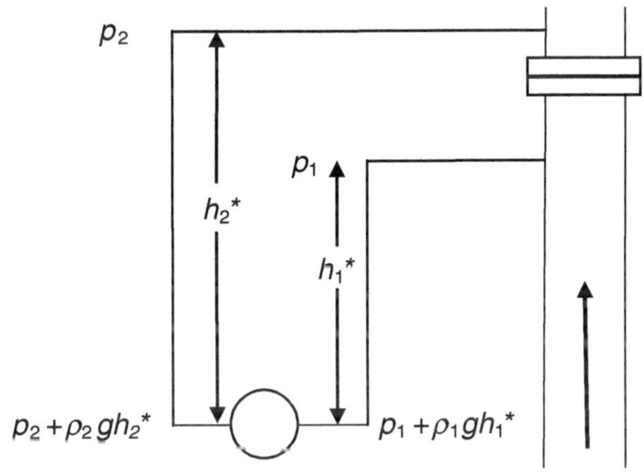

Fig. 4.5 Differential-pressure transmitter located below a vertical orifice meter

In Fig. 4.5 the measured differential pressure, Δp_{meas}, is given by

$$\Delta p_{meas} = p_1 - p_2 + \rho_1 g h_1^* - \rho_2 g h_2^*, \tag{4.5}$$

where ρ_1 and ρ_2 are the density of the fluid in the upstream and downstream impulse lines, respectively, h_1* and h_2* are the differences in height between the upstream and downstream pressure tappings and the differential-pressure transmitter, respectively, and g is the acceleration due to gravity. If $\rho_1 = \rho_2 = \rho$

$$\begin{aligned}\Delta p_{meas} &= p_1 - p_2 + \rho g (h_1^* - h_2^*)\\ &= \Delta p_{req}\end{aligned}$$

So no correction is required to the differential-pressure measurement provided that the differential pressure is measured using a differential-pressure transmitter and the density of the fluid in the impulse lines is equal to that in the flow itself.

NOTE If the density is constant it is clear from the Navier-Stokes Equations (e.g. Equation 3.3.13 of Batchelor 1967) that in a flow of given flowrate through a differential-pressure meter the velocity components are unaffected by gravity and so the discharge coefficient is the same in vertical and in horizontal flow.

Having different densities in the two impulse lines causes a similar problem in vertical flow as it does in horizontal flow. However, in vertical flow if the fluid density in the pipe is different from that in the impulse lines there is also a problem, even if the fluid density in the two impulse lines is the same, i.e. $\rho_2 = \rho_1$. In this case there is an error because

$$\begin{aligned}\Delta p_{meas} &= p_1 - p_2 + \rho_1 g(h_1^* - h_2^*)\\ &= \Delta p_{req} + (\rho_1 - \rho)g(h_1^* - h_2^*)\\ &= \Delta p_{req} + (\rho - \rho_1)g\Delta h^*\end{aligned} \tag{4.6}$$

where $\Delta h^* = h_2{}^* - h_1{}^*$, the difference in height between the tapping points.

Although the flow pattern in wet gas is more symmetrical in vertical flow than in horizontal, the impulse lines will not contain identical fluid to that in the pipe line: therefore, the fluid in the pipe between the two tapping points has a different density from the fluid in the impulse lines. The correction for the difference in height between the tapping points requires a calculation of the mean two-phase density in the pipe line, which is subject to uncertainty. For this reason it was found in Hall and Reader-Harris (1999) that a horizontal orientation of a Venturi tube was preferred in multiphase flow, since the errors in differential-pressure measurement in vertical flow vastly outweighed the errors due to poorer fluid mixing in horizontal flow. ISO/TR 11583:2012 on wet-gas flow through orifice plates and Venturi tubes only covers flow in horizontal pipes.

4.2.4 Length of Impulse Lines

In Annex A of ISO 2186:2007 'it is always recommended that the shortest possible impulse-line lengths be used.' Short impulse lines give a good speed of response, and are better for pulsating flow. They generally reduce the number of joints and leaks. However, longer lines may be needed to reach a convenient cabin where transmitters can be located at a controlled temperature (and possibly calibrated). Small enclosures located close to the meter are less convenient for calibrating the transmitters. Longer lines may also be needed with very hot or cold fluids to ensure that the fluid in the transmitters is at a temperature that will not damage the transmitters.

All joints must be leak-tight. Even a very minor leak in a line will reduce the pressure in that line and cause measurement error.

There is generally a need to include isolating valves between the primary and secondary devices to make possible removal or replacement of the differential-pressure transmitter, to allow connections for calibration and to make it possible to open a path between high- and low-pressure sides to ensure the absence of differential pressure so that the transmitter may be zeroed at line pressure with no flow. A typical arrangement is shown in Fig. 4.6. Additional valves may be required to permit draining or venting.

Isolation and equalization at the transmitter are best achieved using proprietary manifold blocks and needle-type high-integrity valves to ensure a leak-free installation. The manifold block may have valves for drainage, venting and connection to a calibrator.

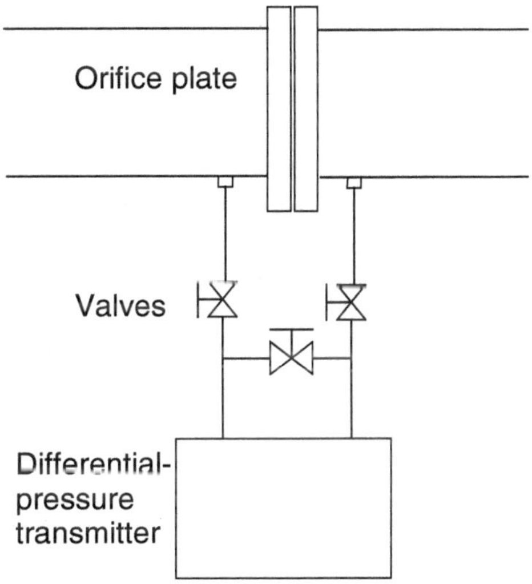

Fig. 4.6 Typical valve arrangement to isolate the differential-pressure transmitter and to provide equalization

Under steady flow conditions without leaks, with the same temperatures in both impulse lines and without other problems such as pulsation both a compact (short-impulse-line) system and a traditional system with long impulse lines will give the same performance.

4.2.5 Blockage

Blockage of an impulse line will cause flowrate information to be lost, with potentially serious consequences in a process-control application. The need to avoid blockages is a primary consideration in the specification of impulse-line diameter. Generally impulse-line diameters in the range 4–25 mm are used. The smallest sizes are used for unsteady flow and for research. ISO 2186:2007 recommends an internal diameter of not less than 6 mm and preferably at least 10 mm if condensation is likely to occur or if gas bubbles may be liberated from a liquid. It recommends a minimum internal diameter of 10 mm for industrial process applications in which reliability is the primary concern. It recommends a maximum internal diameter of 25 mm. If a Venturi tube is used, it is desirable to avoid a change in tapping diameter over as large a distance as possible from the inner wall of the Venturi tube especially for the throat tapping (see Sect. 7.5).

Among the causes of blockage are hydrates, waxing and freezing. Lagging or trace heating may be required to prevent these problems. Blockage of impulse lines due to freezing is addressed by Daiber and Hughes (2003), who estimate that 60 % of trace-heating systems for freeze protection are not working as designed. Useful advice on trace heating using steam is provided by Harrold (1999); however, electrical trace heating is more common. Trace heating must not be so great as to cause liquid to vaporize unintentionally or to prevent steam from condensing.

Solids might be deposited in tappings and impulse lines: if there is a risk of such a deposit occurring tappings below the centre-line of the pipe should not be used.

The use of smart pressure transmitters to detect and predict plugged or frozen impulse lines is described by Menezes (2000). An open impulse line has high frequency noise due to fluid turbulence. If a line becomes plugged its signal becomes flat. A smart transmitter can "learn" the frequency characteristics of open impulse lines and use them to establish an "OK" condition. The different frequency characteristics of a plugged line can then be detected and reported using software. It is possible to detect the plugging of one or both lines. This technique requires the use of a very fast pressure sensor. Tests to assess the ability of smart pressure transmitters to detect an impulse-line blockage in a refinery application are described by Szanyi et al. (2003). Blockage in the impulse lines of pressure transmitters in fluid catalytic cracking (FCC) units can cause very expensive shutdowns. See also Taya and Kuwayama (1995).

4.2.6 Damping of the Pressure Signal or Resonance

The occurrence of flow pulsations introduces the possibility of additional measurement errors. This problem is described in Appendix 4.A.

4.3 Differential Pressure

4.3.1 Differential-Pressure Transmitters

The appropriate range of differential pressure depends on the available transmitters, but some approximate figures that may be useful are given in Table 4.2.

Differential-pressure transmitters (transducers) are affected by outside influences. The magnitude of the effects depends on the design of the device. All are influenced by static pressure and temperature. Some are also influenced by orientation and vibration. Ideally they should be calibrated in situ at the conditions of use. They should be installed with rigid mountings and protected from changes in ambient temperature by using sun or wind shields or better in an enclosure whose temperature is constant.

It is very difficult to calibrate a differential-pressure transmitter at low differential pressure and at high static pressure (e.g. 50 bara), since calibration at high static pressure is carried out using a twin piston pressure balance, which only operates accurately at a minimum differential pressure of about 50 mbar. These pressure balances are high-precision instruments and are not suitable for use in the field or on offshore platforms that are subject to vibration.

Table 4.2 Appropriate range of differential pressure

Fluid	Minimum differential pressure[a]	Maximum differential pressure[c]	
		Venturi tube	Orifice plate
Liquid	c. 15 mbar	Typical values for design might be in the range 500 mbar to 5 bar[d]	Excessive plate bending must be avoided (see Sect. 2.2.2)[f]. Typical values for design will be in the range 500 mbar to 1 bar; a typical plate between flanges can probably be used up to 2 bar
High-pressure gas	c. 50 mbar[b]		
Gas at atmospheric pressure	c. 6 mbar	250 mbar[e] (a smaller value will usually be used)	

[a]One of the reasons why percentage uncertainty in flowrate measurement increases at low differential pressure is that there are greater percentage fluctuations in the data

[b]See the second paragraph below these notes. George and Morrow (2001) recommend a lower limit on differential pressure of around 50 mbar for acceptable accuracy

[c]In liquids it is necessary to avoid cavitation; for gases the uncertainty in the expansibility factor increases with the differential pressure (see Eq. 6.17 or Eq. 1.14, as appropriate); it is generally undesirable to have a large pressure loss in the system

[d]To avoid cavitation in liquid flows through standard Venturi tubes with a machined convergent note Fig. 3.5

[e]p_2/p_1 ($= 1 - \Delta p/p_1$) shall be greater than or equal to 0.75 (5.3.2.2 of ISO 5167-2:2003 or 5.6 of ISO 5167-4:2003)

[f]The maximum differential pressure depends on whether the plate is held in a fitting or between flanges

To maintain such a high minimum differential pressure as 50 mbar will often require multiple flowmeters to be used in parallel and as many flowmeters used as are required. One possible way of proving the performance of differential-pressure transmitters with a low differential pressure at high static pressure is given in Appendix 4.B.

If a differential-pressure transmitter is to be used offshore it will often be calibrated onshore both at high pressure and at atmospheric pressure, the latter calibration being regarded as a "footprint". Then when the transmitter is offshore it can be checked by calibration at atmospheric pressure against the "footprint" and then used at high pressure with its high-pressure calibration. If it is calibrated at low static pressure and used at high, the error may in fact be small, but it is difficult to quantify the uncertainty: Spitzer and Boyes (2003) give a range of approximately 0.001–0.016 % of upper range limit per bar, where the upper range limit is the maximum differential pressure that can be measured by the transmitter: zeroing the differential-pressure transmitter at the actual pressure reduces this effect. The most modern transmitters may have better performance.

The effect of ambient temperature on differential-pressure transmitters can be very significant: Spitzer and Boyes (2003) give a range of approximately 0.001–0.020 % of upper range limit per °C. This is why differential–pressure transmitters are generally used in a location with a controlled temperature so that they can be calibrated and used at the same temperature. The most modern transmitters may have better performance.

It is important to measure the effect of transmitter stability (drift), since this can be significant.

Traditionally differential-pressure transmitters tended to have an uncertainty that was stated as a percentage of full scale. This concealed a statement that the percentage uncertainty increased rapidly as the differential pressure reduced.

Pressure transducers within many transmitters have a significantly higher range and resolution than the electronics interpreting them. For this reason many not only state an upper range limit but can be set to operate across a lower range. In practice the uncertainty quoted might be a percentage of the upper range limit or of the range over which the transmitter had been ranged and calibrated. The best operating range of many older transmitters is around 10:1. Because of the square law relationship between pressure difference and flowrate, the range of a differential-pressure flowmeter with a single differential-pressure transmitter was therefore limited until recently to 3:1 or 4:1. The use of two differential-pressure transmitters of different ranges increased the overall range of the meter. The higher-range transmitter provided a check on the lower-range one. There was normally an overlap to avoid frequent switching between transmitters.

The use of modern 'smart' pressure transmitters significantly increases the range of the meter: 100:1 on differential pressure and thus 10:1 on flowrate is not impossible. Figure 4.7 shows the drift between calibrations of 5 fieldbus

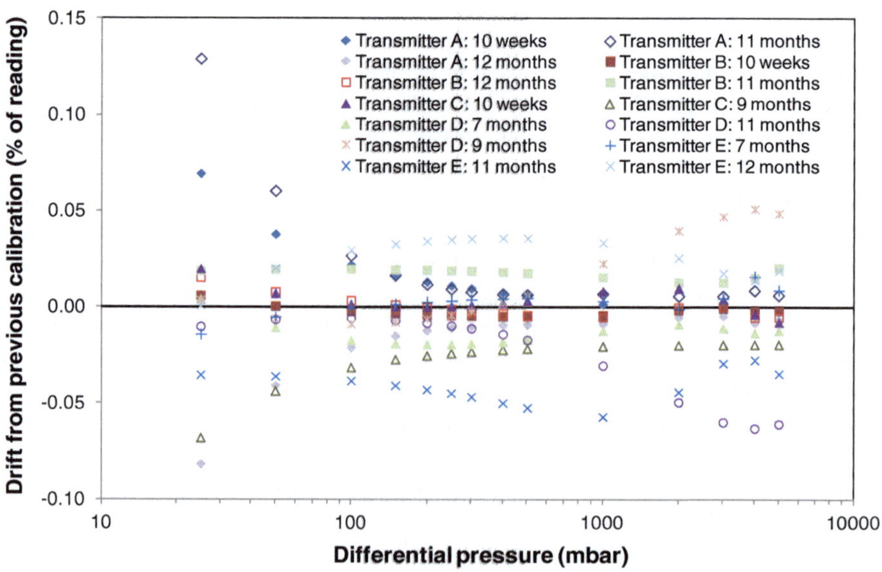

Fig. 4.7 Drift from previous calibration (percentage of reading) of 5 Yokogawa EJX110A differential-pressure transmitters: the time between calibrations is shown

differential-pressure transmitters (all Yokogawa EJX110A), all calibrated at low static pressure at NEL over a range of differential pressure from 25 to 5000 mbar. From 50 to 5000 mbar the largest drift in reading between calibrations was 0.06 %.

The orientation of a differential-pressure transmitter is significant, although its effect can be removed by zeroing the transmitter when it is installed in the field: an arrangement of valves as in Fig. 4.6 will make this possible.

4.3.2 Piezometer Rings

To reduce the effect of a burr on a single pressure tapping it is better in principle to use a piezometer ring connecting multiple tappings together or an annular groove to measure pressure rather than a single tapping. A triple-T (triple-tee) arrangement (Fig. 4.8) has been shown to be the most satisfactory (Blake 1976). Using a triple-T reduces the effect of upstream installation if there is an asymmetric flow profile, and has the advantage that a burr on one tapping in four has less effect on the flow measurement than the same burr would have if a single tapping with that burr were used (this is most useful where burrs have most effect, in the throat of a Venturi tube or a throat-tapped nozzle). However, if there is any risk of the presence of liquid in a gas flow tappings should only be in the top half of the pipe. If there is any risk of gas in a liquid flow tappings should only be in the bottom half of the pipe. In practice a single tapping in each tapping plane is usual (see also Sect. 3.6).

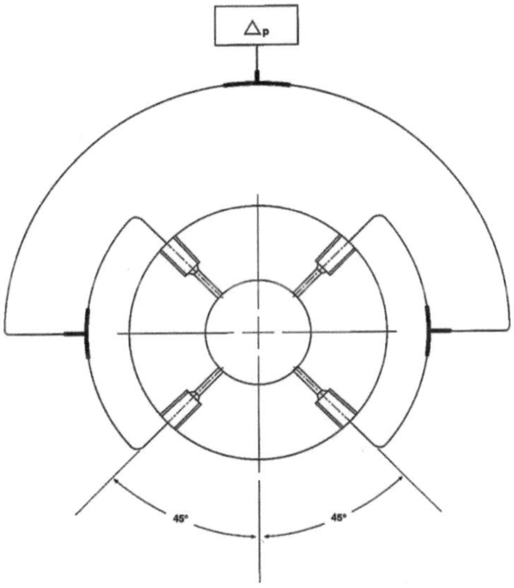

Fig. 4.8 Triple-T arrangement

4.4 Static Pressure

Static pressure is mainly required to calculate the density of the fluid being measured. It is the static pressure on the centreline at the upstream tapping that is required. In practice the vertical height of the static-pressure transmitter will have a substantial effect on the measured static pressure in a liquid but a small effect on the measured static pressure in a non-condensing gas; in both cases this will have a small (but probably not negligible) effect on the measured fluid density. If the fluid is a condensing vapour (e.g. steam) and the impulse lines are filled with liquid, the vertical height of the static-pressure transmitter will have a substantial effect on the measured static pressure and hence on the calculated density of the vapour if no correction is made. In Fig. 4.9 the measured static pressure, p_{meas}, at a transmitter a distance h_1* below the centreline is

$$p_{meas} = p_1 + \rho_1 g h_1{}^*, \tag{4.7}$$

where p_1 is the upstream pressure on the centreline, ρ_1 is the density of the fluid in the impulse line and g is the acceleration due to gravity. p_1 can be simply calculated from p_{meas}.

Since most pressure transmitters read gauge pressure it is normally necessary to measure the ambient (barometric) pressure in order to calculate the absolute pressure in the fluid.

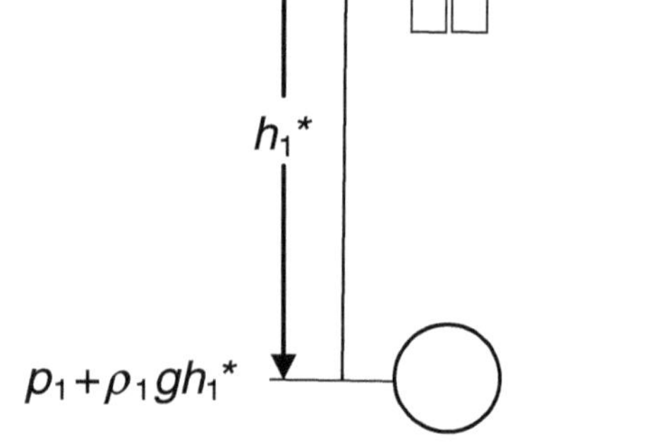

Fig. 4.9 Static pressure transmitter located below a horizontal orifice plate

4.5 Drain and Vent Holes (Through the Pipe Wall)

Gas should not be trapped near a differential-pressure flowmeter in a liquid flow, and liquid should not be trapped near a differential-pressure flowmeter in a gas flow. If drain or vent holes through the pipe wall are required near a differential-pressure flowmeter they should be less than $0.08D$ in diameter, further than $0.5D$ from any pressure tapping, and at least 30° from any pressure tapping (see 7.1.6 of ISO 5167-1:2003).

NOTE Drain holes **through** an orifice plate are covered in Chap. 10.

A gas flowmeter should not be installed where liquids can collect in it.

4.6 Temperature

4.6.1 General

Temperature is mainly required to calculate the density; so it is the temperature at the upstream tapping plane that is required, since it is the density at that location that is required in Eq. 1.15.

The temperature is normally measured downstream of the differential-pressure flowmeter to avoid disturbing the flow into the differential-pressure flowmeter: from 5.4.4 of ISO 5167-1:2003 it should be measured at least $5D$ (and at most $15D$ in gas) downstream of an orifice plate or, for a Venturi tube, at least $5D$ (and at most $15D$ in gas) downstream of the throat pressure tappings and at least $2D$ downstream of the downstream end of the divergent.

A thermometer measures its own temperature; so it is important to ensure that the average temperature of the fluid is very close to the temperature of the thermometer. It may be necessary to mount the thermometer in a thermowell [for more information see 6.3 of ISO/TR 9464:2008 (ISO 2008)]. Using a thermowell protects the thermometer from the process fluid and enables the thermometer to be easily replaced; however, it is important that the thermometer fits tightly inside the thermowell and senses the temperature of the fluid.

The temperature sensor should be in the middle third of the pipe (6.3.4.1 of ISO/TR 9464:2008). According to Ingram and Kimpton (2014) this recommendation is conservative. For accurate flow measurement the temperature of the fluid should not vary significantly across the pipe. If the sun is shining directly on the top of the pipe, there will be a temperature variation across the pipe, especially if the fluid velocity is low; this variation can be avoided by shielding the flowmeter [including its upstream straight length and the downstream straight length to just downstream of the thermowell (and of the densitometer, if present)] from the sun and/or by insulating it. A crosswind can also lead to a temperature variation across the pipe, especially if the fluid velocity is low.

A thermometer in a flowing gas does not record the actual fluid temperature. It will record a temperature $T_{\text{thermometer}}$ higher than the fluid (static) temperature, T, but lower than the stagnation temperature, T_0. The temperature probe recovery factor, R', is given by

$$R' = \frac{T_{\text{thermometer}} - T}{T_0 - T}$$

Therefore, for an ideal gas, using Eq. (1.52) of Curle and Davies (1971),

$$T = \frac{T_{\text{thermometer}}}{1 + \frac{\gamma - 1}{2} R' Ma^2}$$

where Ma is the Mach number of the flowing gas and γ is the ratio of the heat capacities ($=C_p/C_v$, where C_p and C_v are the heat capacities at constant pressure and at constant volume, respectively).

For a typical thermometer R' is around 0.7–0.8.

4.6.2 Temperature Correction from Downstream of the Flowmeter to Upstream of It

It is often assumed that the temperature at the upstream pressure tapping (required for density calculation) is the same as that at a downstream thermowell. However, in a gas as the pressure reduces so the temperature changes. For orifice plates in gas [see Niazi and Thalayasingam (2001) and Gasunie (1988)] it has been shown that there is an isenthalpic expansion between a location upstream of the orifice plate and a location downstream of the orifice plate where the pressure has fully recovered (say 6D downstream of the orifice plate: see Sect. 2.6 for the pressure loss); so

$$\Delta T = \mu_{\text{JT}} \Delta \varpi, \tag{4.8}$$

where

ΔT ($= T_1 - T_3$) is the difference in temperature between that at the upstream tapping, T_1, and that at the pressure recovery location, T_3,

μ_{JT} is the Joule-Thomson coefficient

$\Delta \varpi$ is the pressure loss (i.e. the permanent pressure loss, not the differential pressure—see Fig. 2.11 and also Sect. 1.7)

That the temperature difference between upstream of a Venturi tube and the pressure recovery region should be calculated from an isenthalpic expansion is probably true for Venturi tubes also.

The Joule-Thomson coefficient can be calculated from knowledge of the gas.

For natural gases that are not too rich, a formula for the Joule-Thomson coefficient as a function of temperature and pressure is available in 5.1.5.4.4 of ISO/TR 9464:2008; this gives only a very small increase in uncertainty (see Reader-Harris 2006). Provided that, in the molar composition of the natural gas, methane is greater than 80 %, the temperature is in the range 0–100 °C and the absolute static pressure is in the range 1–200 bar (100 kPa–20 MPa)

$$\begin{aligned}\mu_{\mathrm{JT}} = {} & 0.35 - 0.142\left(\frac{t}{100}\right) \\ & + \left(0.231 - 0.294\left(\frac{t}{100}\right) + 0.136\left(\frac{t}{100}\right)^2\right) \\ & \times \left(0.998 + 0.041\left(\frac{p}{100}\right) - 1.115\left(\frac{p}{100}\right)^2 + 0.3\left(\frac{p}{100}\right)^3\right)\end{aligned} \tag{4.9}$$

where

μ_{JT} is the Joule-Thomson coefficient (K/bar)
t is the temperature of the fluid (°C)
p is the absolute static pressure of the fluid (bar)

The equation was obtained by fitting the Joule-Thomson coefficient of 12 common natural gases, for which in the molar composition methane was in the range 81.2–96.5 %. The differences between Eq. 4.9 and the data can be seen in Figs. 4.10 and 4.11. The uncertainty given in Eq. 4.10 and the Joule-Thomson

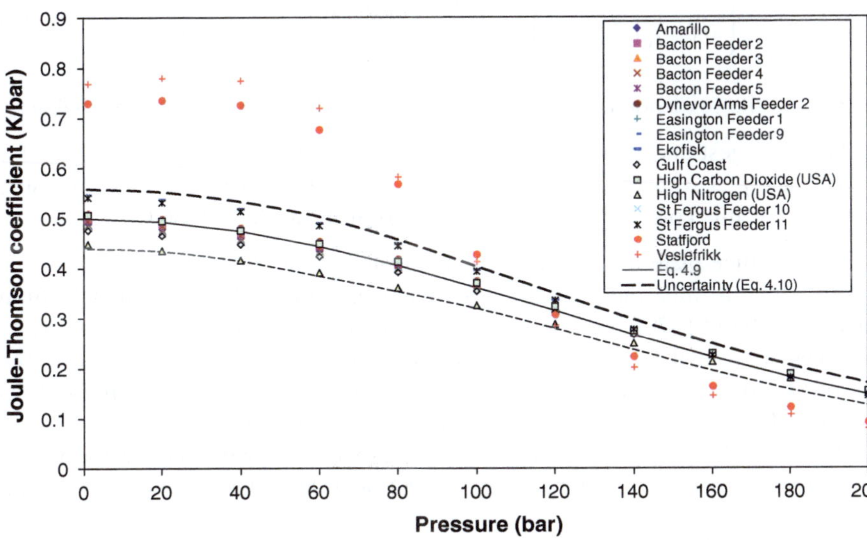

Fig. 4.10 Joule-Thomson coefficient for 14 common natural gases at 20 °C as a function of pressure

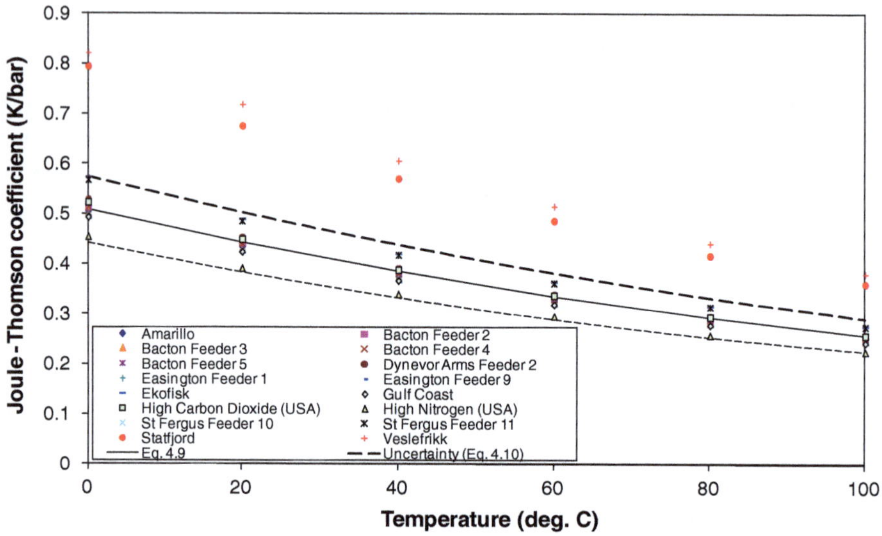

Fig. 4.11 Joule-Thomson coefficient for 14 common natural gases at 60 bara as a function of temperature

coefficient for two natural gases outside the range of use of the equation (Statfjord and Veslefrikk, 73.2 and 66.5 % methane in the molar composition, respectively) are also shown.

$$U = \begin{cases} 0.066\left(1 - \frac{t}{200}\right) & p \leq 70 \text{ bara} \\ 0.066\left(1 - \frac{t}{200}\right)\left\{1 - \frac{(290-t)}{4}\left(\frac{1}{70} - \frac{1}{p}\right)\right\} & p > 70 \text{ bara} \end{cases} \quad (4.10)$$

where U is the (expanded) uncertainty in the Joule-Thomson coefficient (K/bar).

NOTE the largest uncertainty of 0.066 in the Joule-Thomson coefficient occurs for $t = 0$ °C and $p \leq 70$ bara. If an orifice plate with $\beta = 0.6$ has a differential pressure $\Delta p = 0.5$ bar this corresponds to an uncertainty in flowrate of 0.004 %.

For an orifice plate the pressure loss is given in Sect. 2.6.

For a Venturi tube the pressure loss cannot be estimated very accurately from Annex C of ISO 5167-4:2003. It may need to be measured. At present it is probably not necessary to make the temperature correction in Eq. 4.8 for a Venturi tube. The size of the correction can be easily estimated from Annex C of ISO 5167-4:2003 and ISO/TR 9464:2008 to show that in most cases it is very small.

The Joule-Thomson coefficient for carbon dioxide is larger than that for natural gas or for air, and the resulting temperature change may have a more significant effect on the density.

The reason why the temperature must not be measured too close to the orifice plate or Venturi tube is that the temperature must be measured in the pressure recovery region: measurement in this region is required if correction as in Eq. 4.8 is to be used, but is still necessary if no correction is made.

NOTE The required correction will be greater than that in Eq. 4.8 if the temperature probe is installed before pressure recovery is reached: at the downstream flange tapping the temperature drop from the upstream value is greater than (i.e. on the isentropic side of) the value given by an isenthalpic expansion based on the differential pressure (not the pressure loss) (Niazi and Thalayasingam 2001); in some cases the temperature drop from the upstream value to that at the downstream tapping is much greater than the temperature drop from the upstream value to that at the pressure recovery location; at an intermediate point between the downstream flange tapping and the pressure recovery location the temperature drop from upstream of the plate will be hard to predict.

4.6.3 Using a Densitometer

Instead of calculating the density from measurements of pressure and temperature using an equation of state the density may be determined using a densitometer. Temperature measurement (and correction) is still required; so this section is included here.

Where flows of natural gas are measured for fiscal purposes and a densitometer is used, it is mounted in a pocket in the pipe downstream of an orifice plate (see 6.4.2 of ISO/TR 9464:2008): mounting the densitometer in a pocket ensures a known stable temperature of gas within the densitometer. Gas is drawn into the densitometer from a tapping located at least $8D$ downstream of the orifice plate and upstream of the densitometer pocket; presumably the pocket should be not more than $15D$ downstream of the orifice plate so that the gas temperature is still the downstream recovered temperature. Gas is returned, upstream, from the densitometer to a low-pressure return tapping just behind the downstream face of the orifice plate where the downstream static pressure is known (by calculation from the measured upstream static pressure and the measured differential pressure). The return tapping must be the same distance downstream of the orifice plate as the downstream pressure tapping (it is assumed that any variation in pressure round the pipe is insignificant), but the two tappings should be separate to avoid causing any error in the differential pressure measurement.

There are valves on each side of the densitometer to control the flow through it. The valve on the upstream side of the densitometer is opened sufficiently to ensure that there is sufficient (but not excessive) flow through the densitometer; the valve on the downstream side of the densitometer is fully open to ensure that the density of the gas (ρ_m) is measured at p_2 (downstream tapping pressure). A diagram is shown as Fig. 4.12. Because the densitometer is located in the pocket the temperature of the gas is T_3 (downstream recovered temperature); the gas flow from the pipe to the densitometer and the return flow should be included within the thermal lagging. The temperature may be measured upstream or downstream of the

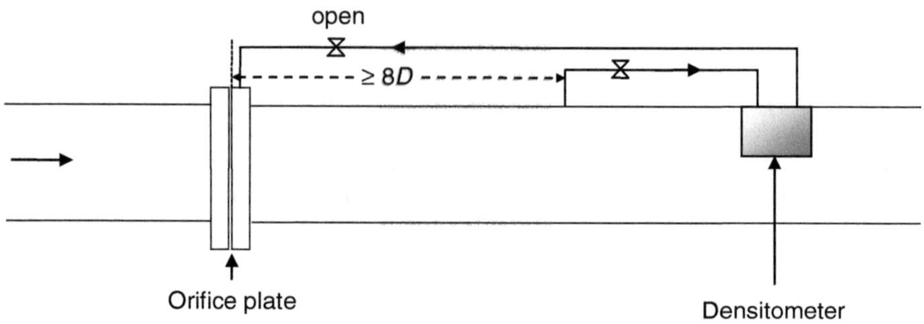

Fig. 4.12 Arrangement of in-line densitometer

densitometer, as long as it is measured in the range 5D to 15D downstream of the orifice plate (as in Sect. 4.6.1). The upstream density is calculated as follows:

$$\rho_1 = \frac{\rho_m p_1 T_3 Z_{(p_2,T_3)}}{p_2 T_1 Z_{(p_1,T_1)}} \tag{4.11}$$

where

ρ_1	upstream density (at p_1, T_1)
ρ_m	measured density (at p_2, T_3) from the densitometer
p_1	upstream pressure
p_2	pressure at the downstream pressure tapping
T_1	upstream temperature
T_3	measured temperature at the downstream recovery position
$Z_{(p_1,T_1)}$	compressibility at p_1, T_1
$Z_{(p_2,T_3)}$	compressibility at p_2, T_3

T_1 is calculated from T_3 as described in Sect. 4.6.2.

4.6.4 Correction of Dimensions for Temperature

Orifice or Venturi throat diameter dimensions should be corrected for temperature following 5.1.6.1 of ISO/TR 9464:2008:

$$d = d_0(1 + \lambda_c(T - T_0)) \tag{4.12}$$

where

d is the orifice/throat diameter at actual temperature, T
d_0 is the orifice/throat diameter at calibration temperature, T_0
λ_c is the linear expansion coefficient for the orifice plate/Venturi tube material

A similar correction is made to the upstream pipe diameter: see 5.2.6.4.2 of ISO/TR 9464:2008. It should be noted that the material for an orifice plate (often stainless steel) may be different from that of the pipe (often carbon steel). Typical values of λ_c for stainless steel and carbon steel are 16.7 $\times 10^{-6}$/°C and 11.2 $\times$ 10^{-6}/°C, respectively.

4.7 Iteration

If C is not a function of Reynolds number then the mass flowrate q_m can be determined directly from Eq. 1.15. If C is a function of Reynolds number then it is necessary to iterate: an assumed value of C is chosen and used to calculate mass flow; from mass flow Re_D is calculated; from this value of Re_D a revised value of C is calculated. This is repeated until there is convergence.

When designing a differential-pressure flowmetering system, it is necessary to iterate, by guessing initial values, calculating others, and recalculating to convergence.

This activity is normally simple and converges in, say, 5 iterations. A more formal method is given in Annex A of ISO 5167-1:2003 with even more detail in Annex A of ISO/TR 9464:2008 (ISO 2008).

4.8 Uncertainty

The basic equation for the uncertainty is given in 8.2 of ISO 5167-1:2003:

$$\frac{\delta q_m}{q_m} = \left[\left(\frac{\delta C}{C}\right)^2 + \left(\frac{\delta \varepsilon}{\varepsilon}\right)^2 + \left(\frac{2\beta^4}{1-\beta^4}\right)^2 \left(\frac{\delta D}{D}\right)^2 + \left(\frac{2}{1-\beta^4}\right)^2 \left(\frac{\delta d}{d}\right)^2 + \frac{1}{4}\left(\frac{\delta \Delta p}{\Delta p}\right)^2 + \frac{1}{4}\left(\frac{\delta \rho_1}{\rho_1}\right)^2\right]^{1/2} \tag{4.13}$$

where δy is the absolute uncertainty in y. Further information on the calculation of uncertainty is available in the GUM (ISO/IEC (2008)).

In most cases the largest source of uncertainty is the uncertainty in the discharge coefficient. The uncertainty of the differential pressure can be very large, especially at small differential pressure. In some situations the uncertainty in the density can be high. Simply using the maximum deviation from the mean in the measurements of d and D to give the uncertainty in d and D as stated in 8.2.2.4 of ISO 5167-1:2003 is incorrect (although this error will usually have a very small effect on the uncertainty of the flowrate): the uncertainty of the reference dimensional measuring device should be included; the maximum deviation is only one part of the diameter uncertainty. The uncertainty in the expansibility factor can be the largest source. It is given as Eq. 1.14 (5.8 of ISO 5167-4:2003) for Venturi tubes, Eq. 6.17 (5.3.3.2 of ISO 5167-2:2003) for orifice plates.

The key is to decide what uncertainty is required and to achieve the required value. This might require calibrating a Venturi tube to reduce the uncertainty in C or using different differential-pressure transmitters to cover different ranges and thereby to reduce the uncertainty in Δp.

4.9 Cavitation

The effect of cavitation on the discharge coefficient of differential-pressure flowmeters has been reviewed by Cousins (1976) (see also Kobayashi 1967). For a Venturi tube the effect of the radius of the corner between the convergent and the throat is significant: with a rounded corner of radius greater than or equal to $3.5d$ cavitation has very little effect on the discharge coefficient; with a corner of smaller radius cavitation appears to cause flow separation at the throat entry, forming a jet at the throat and a sharp decrease in discharge coefficient. Figure 3.A.2 shows that with a sharp corner the minimum pressure is not at a throat tapping but at the intersection between the convergent section and the throat: accordingly care should be taken so that cavitation does not occur at that intersection, even when it appears from the throat tapping pressure that cavitation should not occur. Cavitation prevents much of the pressure recovery in the diffuser. It also causes damage to the flowmeter.

4.10 Diagnostics

One of the advantages of an ultrasonic flow meter is its ability to provide additional diagnostic information on the nature of the flow. This information can be used to check for additional uncertainty in the measured flow caused by, for example, swirl induced by upstream flow disturbances. However, the ability to provide diagnostics is not limited to ultrasonic meters: differential-pressure meters can provide diagnostics too. This has been known for many years: Martin (1986) showed that the use of an additional upstream pressure tapping so that the pressure rise into the upstream corner of an orifice plate could be measured would give the possibility of correcting the measured flowrate to account for the effect of different upstream installations. Recently Steven (2008) and Skelton et al. (2010) have shown the benefits of using an additional pressure tapping around $6D$ downstream of an orifice plate so that the pressure loss can be combined with the differential pressure to give information on the acceptability of the measurement made.

With many but not all distorted upstream velocity profiles change in pressure loss ratio (the ratio of the pressure loss to the differential pressure) is strongly correlated with change in discharge coefficient; so a change in pressure loss ratio gives a warning that there has been a change in discharge coefficient. In some cases, more discrimination may be obtained by measuring the pressure recovery and using the pressure recovery ratio (the ratio of the pressure recovery to the differential pressure) because the pressure recovery is smaller than the pressure loss.

To determine that there has been a change in pressure loss ratio from a baseline value (and a change in discharge coefficient) either the pressure loss ratio has to be measured in good flow conditions for use as a baseline or the value in good flow conditions has to be predicted. For a Venturi tube there is quite a wide spread of values of pressure loss ratio in good flow conditions (see Fig. C.1 of ISO 5167-4:2003; there is one example in Fig. 3.3); so measurement of the baseline is probably required. For an orifice plate there is a good formula for pressure loss ratio in Eq. (7) of ISO 5167-2:2003 (derived as Eq. 2.12 in this book) based on a strong theoretical foundation, but its uncertainty is unspecified. This formula needs the upstream tapping to be about 1*D* upstream of the orifice plate; if the pressure loss is measured from an upstream flange (or corner) tapping a correction is required: use Eqs. 2.12, 2.14 and 2.16. If the pressure loss ratio for an orifice meter is <u>measured</u> in good flow conditions, then a later error in discharge coefficient can be more accurately predicted.

Some problems at or near the orifice plate, e.g. the entry of a wrong pipe diameter, can also be discovered by looking at the value of the pressure loss ratio. Some problems, e.g. the entry of a wrong orifice diameter, can only be discovered if the error in the orifice diameter is large or the pressure loss ratio is compared with a pressure loss ratio obtained in the same orifice meter when the meter was known to have correct inputs.

The pressure loss ratio has been used by DP Diagnostics (Steven 2008, 2009) to identify meter malfunction in many typical operating conditions. This diagnostic method has been developed into a software suite called 'Prognosis', which allows easy monitoring of the health of the measurement. This software and the calculation method behind it are explained in greater detail in Deverapalli et al. (2011).

NOTE If the pressure loss ratio for a meter in good flow conditions is being <u>predicted</u> (e.g. from Sect. 2.6) great care should be taken if tappings are added by welding a boss to a pipe wall that the welding heat does not create an internal bulge which alters the local velocity at the tapping as well as altering the diameter at the tapping. Welding of bosses should be carried out prior to machining a bore, or alternative methods of preventing distortion employed.

4.11 Mixtures

Work on the effect of a small quantity of gas on the performance of a differential-pressure meter in a liquid flow was carried out by Zanker (1996).

Flow of oil-in-water emulsions through orifice plates and Venturi tubes was investigated by Pal (1993). He showed that the single-phase calibration curves apply to surfactant-stabilized emulsions. This applies even for non-Newtonian emulsions provided that a generalized Reynolds number is used.

4.12 Conclusions

Tappings should be appropriately located; impulse lines should have appropriate slopes; differential pressures should be measured over suitable ranges; temperatures may need to be corrected from downstream to upstream; uncertainty should be correctly evaluated; cavitation should be avoided; diagnostics may be used.

Appendix 4.A: Impulse Lines in Pulsating Flows

An extensive research programme on impulse lines in pulsating flows was performed at Southwest Research Institute in the USA, and another major research contributor was the research laboratory of the Nova Chemical Corporation.

In the SwRI research programme, the report by Sparks (1984) describes results of an experimental programme to record and to describe the effect of impulse lines on pressure measurements in pulsating flow and to document pressure-measurement errors. The report is directed towards measurement of the time-average flow.

With a simple impulse line consisting of a straight tube of constant diameter its resonant frequencies will be excited when a pulsation frequency coincides with its quarter-wave resonant frequency or with any odd multiple of that frequency. For this simple geometry the acoustic resonance frequencies $f_{n'}$ are given by

$$f_{n'} = \frac{(2n' - 1)c_s}{4L_p} \text{ (Hz)}$$

where

n' mode number (1, 2, 3,...)

c_s speed of sound in fluid (m/s)

L_p impulse pipe length (m)

The experiments showed that at resonance the pulsation amplitudes in the impulse lines can be much greater than in the main pipe.

Further analysis showed that pulsation data in the impulse lines calculated from the measured differential pressure cannot necessarily be relied upon unless impulse-line lengths are very much shorter than a quarter wavelength for the highest frequency existing in the piping, and then only if there are no constrictions or volumes in the impulse line that would lower its resonant frequency.

Moreover, the existence of pulsations can cause a shift in the mean pressure inside an impulse line. The average pressure transmitted by the impulse line is not necessarily equal to the average pressure in the piping system, even when impulse lines are not at resonance. The basic cause of the shift is stated to be oscillating flow through non-linear resistance elements such as contractions, enlargements, valves and transducer volumes. These elements may have a loss coefficient for flow into the impulse line that is different from the coefficient for outward flow. Oscillating flow is therefore partially rectified, and a net change in the mean pressure in the impulse line results. This flow rectification is greatly amplified if the impulse-line acoustic-resonance frequency coincides with the pulsation frequency.

The main findings of Sparks (1984) are:

- The acoustic response of the impulse lines can cause the pulsations in the impulse line to be many times higher or lower than in the main pipe.
- Impulse lines can distort mean-pressure measurements when there are pulsations in the piping system. These shifts can be either positive or negative, and are most severe when acoustic resonances are excited in the impulse line.
- Pinched valves, volumes and impulse-line filters can be used to provide damping and therefore attenuate the observed pulsations, but they do not totally eliminate the mean pressure shift.
- Differential-pressure measurements are particularly susceptible to impulse-line effects. A positive mean-pressure shift in one line and a negative shift in the other can cause a large shift in the differential pressure.

The main recommendation of the report was that impulse lines should be made as short as possible to give a high acoustic frequency, well above any pulsation frequency in the main piping. The lines should also be of constant diameter, and be free of constrictions and volumes.

In the Nova Corporation Research, Botros et al. (1992) made similar recommendations on impulse-line design to those arising from the SwRI research, i.e.

- Use short impulse lines.
- Avoid abrupt changes in diameter.
- Use a transducer with a very small chamber and with a flat frequency response to the actual frequency range of the pressure oscillation.

ISO/TR 3313:1998 (ISO 1998) (see Sect. 12.3.2.6) provides guidelines on the effects of flow pulsations on flowmeters. It encapsulates and adds to the impulse-line design guidelines in the literature described above, but distinguishes between the use of a slow-response and that of a fast-response transducer. A slow-response transducer is designed to indicate the time-mean differential pressure, whilst a fast-response transducer has to be designed to follow the time-dependent differential pressure so that the square root of the measurement can be averaged to yield a true mean flowrate (i.e. the square-root error is eliminated). Square-root error is discussed in Sect. 8.4.1.

Design guidelines for impulse lines for slow-response transducers to measure time-mean flowrate are:

- A piezometer ring, e.g. a triple-T arrangement as shown in Fig. 4.8, should not be used.
- The distance between pressure tappings should be small compared with the pulsation wavelength.
- The impulse lines should be as short as possible, and of the same bore as the tappings, which should be greater than 3 mm in diameter.
- An impulse-line length close to the pulsation quarter wavelength should not be used.
- For gas-filled lines the sensor cavities or other volumes should be as small as possible.
- Vent points are required for liquid-filled lines to remove any gas bubbles.
- Damping resistances in the connecting tubes and sensing element should be linear. The impulse lines should not be restricted by a partly closed valve.
- The pressure transducer time constant should be about ten times the period of the pulsation cycle.
- If the above rules cannot be observed the secondary measurement system can be effectively isolated from pulsation by the insertion of identical linear-resistance damping plugs into both impulse lines, as close as possible to the pressure tappings.

The above guidelines do not eliminate the square-root error: they just reduce the error in the measurement of the mean differential pressure.

Design guidelines for a fast-response transducer to measure the time-varying flowrate are:

- The mechanical and electronic frequency limits of the secondary measurement system should be at least ten times greater than the pulsation frequency.
- The impulse-line lengths should be as short as possible and less than 10 % of the pulsation quarter wavelength.
- The bore of liquid-filled impulse lines should be greater than or equal to 5 mm.
- Fittings and valves should be of the same bore as the impulse lines.
- The secondary device must be geometrically identical on the upstream and downstream sides.
- Vent points are necessary for liquid-filled lines to remove any gas bubbles.

Pulsation-damping systems are described in 6.1.4 of ISO/TR 3313:1998.

Appendix 4.B: Measuring Low Differential Pressure at High Static Pressure

4.B.1 Introduction

It is desirable to determine how significant are the problems in using a low-static-pressure calibration of a differential-pressure transmitter to measure a small differential pressure at high static pressure. The method suggested here is an idea to be tried; the author is not aware that it has been attempted.

4.B.2 The Problem

As stated in Sect. 4.3.1 calibration of differential-pressure transmitters at high static pressure is carried out using a twin piston pressure balance. At a static pressure of 50 bara this device only operates accurately at a minimum differential pressure of about 50 mbar because of the difficulty of resolving a very small weight difference in a large weight. The availability of calibration of differential-pressure transmitters is as shown in Table 4.B.1.

What is often wanted is a calibration for low differential pressure at high static pressure. The static-pressure correction for $\Delta p \geq 50$ mbar is easily evaluated and quite small; it is desirable to show that that static-pressure correction together with an ambient pressure calibration can be used for $\Delta p < 50$ mbar at high pressure.

4.B.3 A Possible Solution

A possible solution is as follows. In a gas flow laboratory a sonic nozzle is installed upstream of an orifice plate, with a valve installed downstream of the orifice plate. The flow is established and it is ensured that the nozzle is kept choked with the upstream pressure and temperature constant. The valve position is varied. Since the nozzle remains choked, the mass flow remains constant. Then the Reynolds number is (almost) constant and the orifice-plate discharge coefficient constant. Then from Eq. 1.15

Table 4.B.1 Availability of calibration of differential-pressure transmitters

	$\Delta p \geq 50$ mbar	$\Delta p < 50$ mbar
Ambient static pressure	Available	Available
High static pressure	Available	Not available

$$\varepsilon_1\sqrt{\rho_1\Delta p_1} = \varepsilon_2\sqrt{\rho_2\Delta p_2} \tag{4.B.1}$$

where ε, ρ and Δp are the expansibility factor, density (at the upstream pressure tapping) and differential pressure at the orifice plate, and the subscripts represent two static pressures. Then if at conditions 1 and 2 the static pressure were, say, 50 and 5 bara, respectively, and the differential pressure were chosen to be 10 mbar at 50 bara, the differential pressure would therefore be about 100 mbar at 5 bara and the uncertainty in measuring 10 mbar at a static pressure of 50 bara would be around 0.2 %. This is significantly better than the uncertainty claimed by UKAS labs for this measurement, generally 1 % (0.5 % at best).

References

Batchelor GK (1967) An introduction to fluid dynamics. Cambridge University Press, Cambridge

Blake KA (1976) The design of piezometer rings. J Fluid Mech 78(2):415–428

Botros KK, Jungowski WM, Petela G (1992) Gauge line effects and DP transmitter response to flow pulsation through orifice plate. Flow Meas Instrum 3(3):130–144

Cousins T (1976) The effect of cavitation and air/water mixtures on the discharge coefficient of differential pressure flowmeters. IMEKO VII BFL/2411-24110

Curle N, Davies HJ (1971) Modern fluid dynamics. In: Compressible flow, vol 2. van Nostrand Reinhold, London

Daiber J, Hughes A (2003) Freeze protecting your power generation assets O'Brien corporation. http://www.babcockpower.com/pdf/VPI-TP-0002-v1-0000.pdf

Deverapalli V, Glover B, Ayre J, Steven R (2011) A diagnostic system for Venturi meters in single phase and wet gas flow applications. In: Proceedings of 29th International North Sea Flow Meas Workshop, Norway

Gasunie (1988) 600 mm orifice plate project. Part X: temperature profile measurements. Report TP/T 88.R.2050. N V Nederlandse Gasunie, Groningen

George DL, Morrow TB (2001) Metering research facility program, orifice meter operational effects: orifice meter flow measurement at low differential pressures. GRI report No 00/0216. Gas Research Institute, Chicago

Hall ARW, Reader-Harris MJ (1999) Use of Venturi meters in multiphase flow measurement. In: Proceedings of 17th North Sea Flow Meas Workshop, Norway

Harrold D (1999) Protect process instruments to ensure performance. Control Eng

Ingram R, Kimpton S (2014) Experimental research into the measurement of temperature in natural gas transmission metering systems. In: Proceedings of 32nd International North Sea Flow Meas Workshop, St Andrews. NEL, East Kilbride, Glasgow

ISO (1998) Measurement of fluid flow in closed conduits—guidelines on the effects of flow pulsations on flow measurement instruments. International Organization for Standardization, Geneva. ISO/TR 3313:1998

ISO (2003a) Measurement of fluid flow by means of pressure differential devices inserted in circular cross-section conduits running full—part 1: general principles and requirements. International Organization for Standardization, Geneva. ISO 5167-1:2003

ISO (2003b) Measurement of fluid flow by means of pressure differential devices inserted in circular cross-section conduits running full—part 2: orifice plates. International Organization for Standardization, Geneva. ISO 5167-2:2003

ISO (2003c) Measurement of fluid flow by means of pressure differential devices inserted in circular cross-section conduits running full—part 3: nozzles and Venturi nozzles. International Organization for Standardization, Geneva. ISO 5167-3:2003

ISO (2003d) Measurement of fluid flow by means of pressure differential devices inserted in circular cross-section conduits running full—part 4: Venturi tubes. International Organization for Standardization, Geneva. ISO 5167-4:2003

ISO (2007) Fluid flow in closed conduits—connections for pressure signal transmissions between primary and secondary elements. International Organization for Standardization, Geneva. ISO 2186:2007

ISO (2008) Guidelines for the use of ISO 5167:2003. International Organization for Standardization, Geneva. ISO/TR 9464:2008

ISO/IEC (2008) Uncertainty of measurement—part 3: guide to the expression of uncertainty in measurement (GUM: 1995). ISO/IEC Guide 98-3:2008

Kobayashi R (1967) Effect of cavitation on the discharge coefficient of standard flow nozzles. Trans ASME J Basic Eng 677–685

Martin CNB (1986) Effects of upstream bends and valves on orifice plate pressure distributions and discharge coefficients. NEL report no 702. National Engineering Laboratory, East Kilbride, Glasgow

Menezes M (2000) Improving plant safety & availability through advanced measurement diagnostics. Originally presented to ISA/2000 (August), New Orleans. Available from http://www.rosemount.com/support/papers/diagnost0810.doc

Niazi A, Thalayasingam S (2001) Temperature changes across orifice meters. In: Proceedings of 19th North Sea Flow Meas Workshop, Norway

Pal R (1993) Flow of oil-in-water emulsions through orifice and Venturi meters. Ind Eng Chem Res 32:1212–1217

Reader-Harris MJ (2006) ISO standards on differential-pressure flow measurement: developments. In: Proceedings of 5th International South East Asia Hydrocarbon Flow Meas Workshop, Kuala Lumpur

Reader-Harris MJ, McNaught J (2005) Impulse lines for differential-pressure flowmeters best practice guide. NEL, East Kilbride, Glasgow

Skelton M, Barrons S, Ayre J, Steven R (2010) Developments in the self-diagnostic capabilities of orifice plate meters. In: Proceedings of 28th International North Sea Flow Meas Workshop, St Andrews. NEL, East Kilbride, Glasgow

Sparks C R (1984) The effects of gage lines on pressure measurement accuracy. SwRI report no 84-6A. Southwest Research Institute, San Antonio, Texas

Spitzer DW, Boyes W (2003) The consumer guide to differential pressure flow transmitters. Copperhill and Pointer, New York

Steven R (2008) Diagnostic methodologies for generic differential pressure flow meters. In: Proceedings of 26th International North Sea Flow Meas Workshop, St Andrews. NEL, East Kilbride, Glasgow

Steven R (2009) Significantly improved capabilities of DP meter diagnostics methodologies. In: Proceedings of 27th International North Sea Flow Meas Workshop, Norway, paper 1

Szanyi R, Raterman M, Eryurek E (2003) Diagnostic capabilities of FOUNDATION fieldbus pressure transmitters. Hydrocarbon Process 53–59

Taya E, Kuwayama H (1995) Detecting blockage in process connections of differential pressure transmitters. In: SICE '95, Sapporo (paper 315 C-2), pp 1605–1608

Zanker KJ (1996) The influence of air on the performance of differential-pressure water flowmeters. BHRA report RR870. BHRA, Cranfield, Bedfordshire

Chapter 5
Orifice Discharge Coefficient

Abstract The discharge coefficient is required to measure flow using a differential-pressure meter. This chapter considers the discharge-coefficient equation for orifice plates: its history, some older equations, the database collected mostly in the 1980s and the analysis. The orifice discharge-coefficient equation is made up of the following terms: the discharge coefficient using corner tappings for infinite Reynolds number, the slope term which gives the increase in the discharge coefficient (using corner tappings) for lower Reynolds number, the tapping terms, and possibly an additional term for small orifices. The derivation of the Reader-Harris/Gallagher (1998) Equation, which is included in ISO 5167-2:2003, is given. The quality of its performance, both absolutely and relative to the Reader-Harris/Gallagher (RG) Equation in API 14.3:1990 and the Stolz Equation in ISO 5167:1980, is demonstrated and a calculation of its uncertainty provided. Orifice plates with small pipe diameter or with no upstream or with no downstream pipeline or with no upstream and no downstream pipeline are considered in an appendix.

5.1 Introduction

The discharge coefficient, C, is required to measure flow using a differential-pressure flowmeter (called a differential-pressure meter):

$$q_m = \frac{C\varepsilon}{\sqrt{1-\beta^4}}\frac{\pi d^2}{4}\sqrt{2\rho_1 \Delta p}. \tag{1.15}$$

The discharge coefficient can be obtained for any differential-pressure meter and any installation by calibrating it in a flowing fluid: for a particular orifice meter the discharge coefficient is a function of the Reynolds number. However, there are very large savings to be made if the discharge coefficient can be predicted. Over many years of experiment it has been found that the discharge coefficient can be predicted within a defined uncertainty provided that the orifice meter (i.e. the orifice plate and

M. Reader-Harris, *Orifice Plates and Venturi Tubes*,
Experimental Fluid Mechanics, DOI 10.1007/978-3-319-16880-7_5

pipework) are constructed within the standards. If the discharge coefficient is to be used for an orifice meter without calibrating it in a flowing fluid, the discharge coefficient is usually taken from a published discharge-coefficient equation. Therefore, the discharge-coefficient equation is very important for orifice plates: an error of 0.1 % in discharge coefficient gives an error of 0.1 % in many flow measurements of natural gas.

Various discharge-coefficient equations have been used over the years, but currently only four are in common use, the first two of which, listed below, have been largely superseded:

1. The Buckingham Equation in ANSI/API 2530-1985 (API 1985): see Eqs. 5.1a–c.
2. The Stolz Equation in ISO 5167:1980 (ISO 1980) and in ISO 5167-1:1991 (ISO 1991): see Eq. 5.24.
3. The Reader-Harris/Gallagher (RG) Equation in API MPMS 14.3.1:1990 (API 1990) and in API MPMS 14.3.1:2012 (API 2012): see Eqs. 5.23a, b.
4. The Reader-Harris/Gallagher (1998) Equation in ISO 5167-1:1991/Amd.1:1998 (ISO 1998) and in ISO 5167-2:2003 (ISO 2003): see Eqs. 5.22a, b.

How the equations in 3 and 4 above were derived will be described in this chapter.

NOTE 'Orifice plate' can refer just to the plate or to the whole meter; where it is important to be clear that the plate and pipework are meant, 'orifice meter' can be used.

5.2 History

The Buckingham Equation (Eqs. 5.1a–c) was based on the data collected by Beitler in the early 1930s at the Ohio State University (OSU) Engineering Experimental Station (Beitler 1935). Of the tapping arrangements standardized in ISO 5167 it is only applicable for flange tappings:

$$C = K_e\sqrt{1-\beta^4}\,\frac{1+\frac{E_b}{Re_d}}{1+\frac{15E_b}{10^6 d_{in}}}, \qquad (5.1a)$$

where

d_{in} is the orifice diameter in inches
D_{in} is the pipe internal diameter in inches

$$K_e = 0.5993 + \frac{0.007}{D_{in}} + \left(0.364 + \frac{0.076}{D_{in}^{0.5}}\right)\beta^4 + 0.4\left(1.6 - \frac{1}{D_{in}}\right)^5\left(0.07 + \frac{0.5}{D_{in}} - \beta\right)^{5/2}$$
$$- \left(0.009 + \frac{0.034}{D_{in}}\right)(0.5 - \beta)^{3/2} + \left(\frac{65}{D_{in}^2} + 3\right)(\beta - 0.7)^{5/2} \quad (5.1b)$$

and

$$E_b = d_{in}\left(830 - 5000\,\beta + 9000\,\beta^2 - 4200\,\beta^3 + \frac{530}{D_{in}^{0.5}}\right) \quad (5.1c)$$

NOTE In Eq. 5.1b where a term with fractional exponent is negative, its value is to be taken as zero.

To use Eq. 5.1a–c D shall be not less than 40.64 mm (1.6″) and β shall be between 0.1 and 0.75 (preferably between 0.15 and 0.7).

The OSU data are of high quality, but were all collected in water; so any equations based on them require significant extrapolation in Reynolds number when used in high-pressure gas. They were taken in 1″, 1½″, 2″, 3″, 6″, 10″ and 14″ pipes. Tests in 1952/3 (the Refugio tests) measured the discharge coefficient in a 30″ orifice meter in natural gas by passing the flow through eight 10″ orifice meters downstream of a header and showed that the Buckingham Equation gave results in a 30″ pipe that were consistent with those in 10″ pipes (AGA 1954). The OSU data remain valuable for later developments, particularly data from the smallest two sizes, since no data in pipes smaller than 2″ were taken during the European and API tests described in Sect. 5.3.

Stolz (1978) used 303 points from the OSU data together with a table of flow coefficients included in ISO/R 541:1967 (ISO 1967) to establish an equation that is dimensionally correct. The ISO/R 541 table was established by the German VDI and is based on original data of Witte. Unfortunately it is believed that these data are no longer available to researchers. The Stolz Equation was included in ISO 5167:1980 and ISO 5167-1:1991. Inconsistency between ISO standards and AGA (API) standards, which used the Buckingham Equation, was identified as an important problem.

Therefore, to resolve discrepancies between the ISO and AGA (API) equations, data on orifice plate discharge coefficients were collected in Europe [supported by the EEC (now the EU)] from 1980 to 1991 and in the United States around the same time in order to provide a new database from which an improved discharge-coefficient equation could be obtained that would receive international acceptance. Data were collected not only in water, but also in air, natural gas and oil.

In November 1988 a joint meeting of API (American Petroleum Institute) and European flow measurement experts in New Orleans accepted an equation derived by NEL (Reader-Harris and Sattary 1990). At that time the database contained 11 346 points, collected in pipes whose diameters ranged from 2″ to 10″; 24″ data were being collected but had not yet been included in the database from which the discharge-coefficient equation was derived.

The data which were least well fitted by the equation presented at New Orleans were the 2″ data. An additional term was proposed by Gallagher for pipes whose diameter lies below 71.12 mm (2.8″). The resulting equation, the Reader-Harris/Gallagher (RG) Equation, based on the NEL equation accepted at New Orleans with the additional (Gallagher) term, was incorporated in the revised API standard, designated as API MPMS 14.3:1990 (API 1990). Since there was no obvious physical explanation for the additional term for small pipe diameter, additional data were collected in 2″ pipe in water and oil at NEL and included in the database (in addition to the 24″ data now included). Measurements by NEL of the edge radius of the plates used in European tests showed that orifices whose diameter is less than 50 mm tend to have edge radii outside those permitted by ISO 5167. An equation was therefore derived by NEL which included an additional term for small orifice diameter rather than one for small pipe diameter, and put forward in a report to EC BCR numbered PR14 (Reader-Harris et al. 1992a) and thus described as the PR14 Equation. It was also described in papers at the North Sea Flow Measurement Workshop in 1992 (Reader-Harris et al. 1992b) and in Flow Measurement and Instrumentation (Reader-Harris et al. 1995): see Eq. 5.C.2.

However, neither this equation nor the attempt at compromise put forward at a meeting of ISO/TC 30 in Paris in 1993 was acceptable to the ISO/TC 28 delegation, which primarily represented the views of the hydrocarbon industry in the US (API and AGA). It was considered by the ISO/TC 28 delegation that the PR14 Equation gave an insufficiently good fit to the data collected in 3″ pipes. Having already standardized the RG Equation they were unwilling to change the form of the equation: they wanted an equation of the same form as the RG Equation in API MPMS 14.3:1990 (API 1990) for $Re_D \geq 4000$, proposing that only the constants and exponents might be changed to give the best fit to the final database. Moreover, measurements by SwRI of the orifice edge radius for the orifice plates used in 2″ pipe were smaller than those made by NEL.

Since ISO/TC 30 was unwilling to use an equation (the RG Equation) that was not fitted to the data taken in 24″ pipe since they were of particular financial importance but one of the principal objectives of the previous fourteen years of work was to obtain a common equation for use worldwide, an equation of the form of the RG Equation was derived (Reader-Harris and Sattary 1996) and was accepted by ISO/TC 30/SC 2 for inclusion in ISO 5167. It was included in a 1998 amendment to ISO 5167-1:1991 and is thus referred to in ISO 5167-2:2003 as the Reader-Harris/Gallagher (1998) Equation. Given the definition that s is the standard deviation of the data in the database about an equation, the value of s for the Reader-Harris/Gallagher (1998) Equation is very similar to the value of s for the PR14 Equation.

Although chosen by ISO/TC 30 in consultation with the ISO/TC 28 delegation the Reader-Harris/Gallagher (1998) Equation has not been adopted by API [which continues to use the RG Equation (API 2012)], and there are therefore two equations in common use for hydrocarbon products.

This chapter describes the work undertaken to determine the discharge-coefficient equation and will serve as a vital reference for those who wish to improve on the equation. It will also provide important information on orifice performance that can be used for deriving the effects of deviations from good manufacturing or installation.

Further work has been done by others; however, some simplifications fail to cover all the required physics: for example, Keyser's equation is almost constant as a function of Reynolds number for small β (Keyser and Friedman 2005): this is inconsistent with the work described in Appendix 2.A.

5.3 The EEC/API Database

An immense international project was required to determine a database of sufficient quality to determine an international equation. There were two parallel projects, one by API and one by the EEC (now the EU). Data were collected in different laboratories and in different fluids. The tests are listed in Tables 5.1 and 5.2. The European programme included collecting data on the effect of upstream installations.

All European data sets included data with corner, flange and D and $D/2$ tappings. Those sets in Table 5.1 marked * also included data with an additional corner tapping whose design was due to Gasunie and which is simpler to make than the corner tapping depicted in ISO 5167-1:1991: this 'Corner (GU)' tapping was perpendicular to the pipe wall at the corner for a short distance; it then turned through 90° into a short section of tapping parallel to the pipe axis and then through 90° into a longer section of tapping perpendicular to the pipe wall. This type of tapping is described in Sattary and Spearman (1992). The European plates had plate thickness E (see Fig. 2.1) of 2.5, 3.2, 6.1 and 20.0 mm for the plates in 2″, 4″, 10″ and 24″ pipes, respectively; they had orifice thickness e of 0.8, 1.6, 3.2 and 8.1 mm for the plates in 2″, 4″, 10″ and 24″ pipes, respectively.

Where straighteners were tested in good flow conditions the following were tested: two tube bundles of different pattern and length, Sprenkle Flow Conditioners of length $2D$ (see C.3.2.4 of ISO 5167-1:2003) and Zanker Flow Conditioners of length $1D$ (see C.3.2.5 of ISO 5167-1:2003) (for more information on flow conditioners see Sect. 8.2.5.3).

All American sets of data were taken with flange tappings only. The American plates had plate thickness E of 3.2 mm (1/8″) in 2″, 3″, 4″ and 6″ pipes and 6.4 mm (1/4″) in 10″ pipe; the orifice thickness e was not specified.

The American data were taken with an oversized Sprenkle Flow Conditioner followed by a contraction to the pipe size of the orifice meter followed by approximately $45D$ of pipe [see 1.4.7.2 of API MPMS 14.3.1:1990 (API 1990)]. The oversized Sprenkle Flow Conditioner was a Sprenkle Flow Conditioner of diameter D_1 consisting of three perforated plates one pipe diameter D_1 apart: the plates were perforated with a hexagonal pattern of holes (Whetstone et al. 1989). It appears that the ratio of D_1 to the diameter of the orifice meter D was approximately 1.5–2. Oversized Sprenkle Flow Conditioners were not used in the baseline for the

Table 5.1 European tests

Laboratory	Fluid	Pipe size	β	Additional installation effects tests
NEL	Water	4″	0.2, 0.37, 0.5, 0.57, 0.66, 0.75	Straighteners in good flow conditions
NEL	Air	4″	0.2, 0.57, 0.66, 0.75	2 90° bends in perpendicular planes
Gasunie	Natural gas	4″	0.2, 0.57, 0.66, 0.75	Straighteners in good flow conditions
Gaz de France	Natural gas	4″	0.2, 0.57, 0.66, 0.75	2 90° bends in S-configuration (with/without a straightener)
Delft Hydraulics	Water	10″*	0.2, 0.37, 0.5, 0.57, 0.66, 0.75	Straighteners in good flow conditions
Gasunie	Natural gas	10″*	0.57, 0.66, 0.75	Straighteners in good flow conditions
Ruhrgas	Natural gas	10″	0.2, 0.57, 0.66, 0.75	Header (without/with a long tube bundle)
NEL	Water	10″*	0.2, 0.37, 0.5, 0.57, 0.66, 0.75	Straighteners in good flow conditions
British Gas	Natural gas	10″	0.2, 0.57, 0.66, 0.75	2 90° bends in S-configuration (with/without a straightener)
CEAT	Air	10″	0.2, 0.57, 0.66, 0.75	3 90° bends in perpendicular planes (with/without a short tube bundle)
Gasunie	Natural gas	24″*	0.2, 0.37, 0.5, 0.57, 0.66, 0.75	
NEL	Water	24″	0.2, 0.37, 0.5, 0.57, 0.66, 0.75	
Delft Hydraulics	Water	24″	0.2, 0.37, 0.5, 0.57, 0.66, 0.75	
British Gas	Natural gas	24″	0.2, 0.37, 0.5, 0.57, 0.66, 0.75	
NEL	Water	2″	0.24, 0.37, 0.5, 0.57, 0.66, 0.73 or 0.75	
NEL	Gas-oil ($\nu \approx 5$ cSt)	2″	0.24, 0.37, 0.5, 0.57, 0.66, 0.73 or 0.75	

European experiments. Morrow and Park (1992) showed that increasing the distance between the orifice plate and the Sprenkle Flow Conditioner from 45D to 100D made no difference to the measured discharge coefficient for β from 0.2 to 0.67: over this range of β the difference in C was less than 0.1 % for any β and 0.01 % on average; for $\beta = 0.75$ the discharge coefficient with a 100D spacing was 0.17 % lower than that with a 45D spacing, but it is not clear that this difference is significant (with the 100D spacing the measured discharge coefficient was closer to the values calculated by both the Reader-Harris/Gallagher (1998) and the Reader-Harris/Gallagher (RG) Equations than it was with the 45D spacing).

Table 5.2 American tests

Laboratory	Fluid	Pipe size	β
CEESI	Oil ($\nu \approx 7$ cSt)	2″	0.12, 0.24, 0.36, 0.48, 0.54, 0.66, 0.73
CEESI	Oil ($\nu \approx 7$ cSt)	4″	0.06, 0.09, 0.22, 0.37, 0.49, 0.56, 0.65, 0.74
CEESI	Oil ($\nu \approx 7$ cSt)	6″	0.10, 0.20, 0.37, 0.49, 0.57, 0.66, 0.74
NIST Gaithersburg	Water	2″	0.12, 0.24, 0.36, 0.48, 0.54, 0.66, 0.73
NIST Gaithersburg	Water	3″	0.08, 0.12, 0.20, 0.37, 0.49, 0.57, 0.65, 0.73
NIST Gaithersburg	Water	4″	0.06, 0.09, 0.22, 0.37, 0.49, 0.56, 0.65, 0.74
NIST Gaithersburg	Water	6″	0.10, 0.20, 0.37, 0.49, 0.57, 0.66, 0.74
NIST Gaithersburg	Water	10″	0.10, 0.20, 0.37, 0.50, 0.57, 0.66, 0.75
SwRI	Nitrogen	4″	0.20, 0.40, 0.50, 0.60, 0.67, 0.75

The database used to derive equations consisted of 16 376 points: the orifice Reynolds numbers ranged from 1700 to 7×10^7. The data points for which the orifice diameter was less than 12.5 mm (0.49″) and those for which the differential pressure across the orifice plate was less than 6 mbar are very scattered and were excluded from the derivation of the equation.

The American data are given in Whetstone et al. (1989), Britton et al. (1988), API (1988). The complete European data are tabulated in Hobbs (1985, 1987), Hobbs and Sattary (1986), Hobbs et al. (1987a, b), Sattary and Spearman (1992), Sattary et al. (1992); the data sets which have been accepted for analysis are indicated in Spearman et al. (1992). Further details of the 4″ and 10″ European data are available in the reports of the individual laboratories which are listed in Spencer (1993). Both American and European data were used for the purposes of curve fitting and of determining the deviations of the data from the equation. A very small number of points (0.5 % of all the European points) were removed from the European data as outliers; each of those removed was identified as an outlier within its own set of data by the Grubbs' extreme deviation outlier test; details will be found in Reader-Harris et al. (1992a).

American gas data (collected at Joliet) were not included in the final database: the scatter in these data was larger than expected (Whetstone et al. 1988). There are only 146 points from SwRI, Texas (Morrow and Park 1992); their inclusion makes essentially no difference to the final fitted equation.

In the database the discharge coefficients were calculated using the expansibility-factor equation in ISO 5167-1:1991 (ISO 1991); when what is now accepted as a more accurate expansibility-factor equation (Eq. 6.13) replaced the older equation the discharge coefficients should have been recalculated, and there would have been an effect on the calculated discharge-coefficient equation. This did not happen; however, it is shown in Appendix 5.D that to have recalculated the discharge coefficients would have had an insignificant effect on the discharge-coefficient equation.

5.4 The Equation

5.4.1 Introduction

The orifice discharge-coefficient equation is made up of the following terms: C_∞, C_s (the slope term), the tapping terms, and possibly an additional term for small orifices:

- C_∞ is the discharge coefficient using corner tappings for infinite Reynolds number,
- The slope term, C_s, gives the increase in the discharge coefficient (using corner tappings) for lower Reynolds number, so that the discharge coefficient using corner tappings is given by $C_\infty + C_s$;
- The tapping terms are equal to the difference between the discharge coefficient using flange or D and $D/2$ tappings and those using corner tappings (in determining the complete equation the tapping terms were determined first).

5.4.2 The Tapping Terms

5.4.2.1 Introduction

Tapping terms are needed because the pressure changes along the wall upstream and downstream of the orifice. Figure 5.1 gives an outline of the pressure profile albeit with the variation in pressure along the wall expanded relative to the differential pressure across the orifice plate so that the longitudinal variation can be clearly seen. Upstream of the orifice plate the pressure rises into the corner (as the velocity decreases); downstream of the orifice plate it decreases to a minimum; soon after the minimum it rises rapidly as the pressure recovers. The downstream pressure tapping must not be located where the pressure begins to recover rapidly since the discharge-coefficient equations do not apply in the pressure-recovery region.

The tapping terms are expressed as the sum of an upstream and a downstream tapping term. The upstream term is equal to the change in discharge coefficient when the downstream tapping is fixed in the downstream corner and the upstream tapping is moved from the upstream corner to another position. The downstream term is equal to the change in discharge coefficient when the upstream tapping is fixed in the upstream corner and the downstream tapping is moved from the downstream corner to another position.

From the database, as described in Sect. 5.3, only the total tapping term, the sum of the upstream and the downstream terms, is available. To divide the tapping term into two parts so that each can be accurately fitted, measurements of the individual tapping terms collected outside the European and API projects were used to indicate

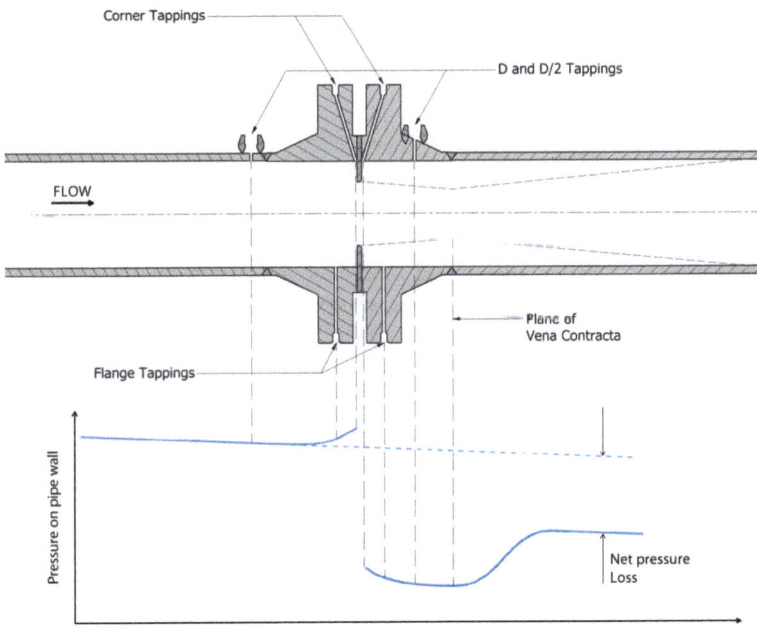

Fig. 5.1 Flow through an orifice plate, showing the pressure tapping positions and approximate pressure profile

the form and approximate value of the upstream and downstream terms; however, the constants in the formulae were obtained to fit the EEC/API database. The European data were collected with several tapping systems; so the total tapping term could be simply obtained (Reader-Harris 1990; Reader-Harris and Sattary 1990).

Although the American data were collected with flange tappings alone, and direct measurements of tapping terms were therefore not available, small adjustments were made to the final tapping terms in order to obtain the optimum fit to the database as a whole.

On examining the measured tapping terms it was shown (Reader-Harris 1990; Reader-Harris and Sattary 1990) that for high Reynolds number (orifice Reynolds number, Re_d, greater than about 80,000) the tapping terms may be considered not to vary with Re_d, but that for low Reynolds number (Re_d smaller than about 80,000) the terms depend on Re_d. To show that for $Re_d > 80{,}000$ the tapping terms do not depend on Re_d the 4″ and 10″ data that had been collected in water at NEL or in air at CEAT (Centre d'Etudes Aerodynamiques et Thermiques, Poitiers) were used, because these consisted of simultaneous measurements of discharge coefficients using different sets of tappings. The tapping terms were calculated for each value of β and plotted against lg Re_D, where Re_D is the pipe Reynolds number (following ISO 80000-2:2009, $\log_{10}$ is written lg). The best-fit straight line and 95 % confidence limits were obtained, and for each value of β the horizontal line representing the mean difference lay within the confidence limits. For some values of β the slope of the best-fit line was small and positive and for other values small and negative. It was established that the tapping terms are independent of Reynolds number for these data.

5.4.2.2 High Reynolds Number Tapping Terms

Total Tapping Terms

For each diameter ratio mean total tapping terms for the European data both for D and $D/2$ tappings and for flange tappings in 2″, 4″, 10″ and 24″ pipe were calculated and used to determine the best constants and exponents. Low Reynolds number data ($Re_d < 80{,}000$) were excluded. The total tapping terms were obtained by considering sets of data collected in the same laboratory with the same orifice in the same pipe on approximately the same day with different pairs of tappings: lines were fitted to these sets of data on the basis that the slope of the data as a function of Reynolds number is the same for each pair of tappings but that the constant depends on the pair of tappings used; the difference between the constant for the data using flange tappings and that for the data using corner tappings is the total tapping term using flange tappings and similarly for D and $D/2$ tappings. From the total tapping terms for individual sets of data mean total tapping terms for the complete European database can then be obtained. Details of the method by which the values of the total tapping terms were calculated are given by Reader-Harris (1990) and Reader-Harris and Sattary (1990).

Upstream Term

As stated above, the form of the upstream tapping term had to be obtained from data sets outside the EEC/API database. The change in discharge coefficient, ΔC, when the downstream tapping is fixed in the corner and the upstream tapping is moved from the corner to a distance one pipe diameter (D) upstream is shown in Fig. 5.2 as a function of $\beta^4/(1-\beta^4)$, the form that would be expected as the velocity reduces to zero into the corner. Data from Bean et al. (1931), BHRA (Martin 1986), Darrow (1954), Engel (1953), Gasunie (1988), Herning (1955), Herning and Bellenberg (1960), NEL (1988), Sens (1977), Stuttgart (Stolz 1988), Teyssandier and Husain (1987), Witte (1931), and Witte and Schröder (Engel 1961) are plotted. The linear dependence on $\beta^4/(1-\beta^4)$ is confirmed. The data from Bean et al. should perhaps not be included since their upstream tapping closest to the corner was about $0.04D$ away from it.

The line on the figure is $\Delta C = 0.043\beta^4/(1-\beta^4)$: the constant was chosen to make it possible to obtain a good fit to the mean total tapping terms from the European data. In theory the total tapping term is the sum not only of the upstream and the downstream term but also of a product term because the discharge coefficient depends on the reciprocal of the square root of the differential pressure (Reader-Harris 1990). In practice the product term is not included in the formula and, to

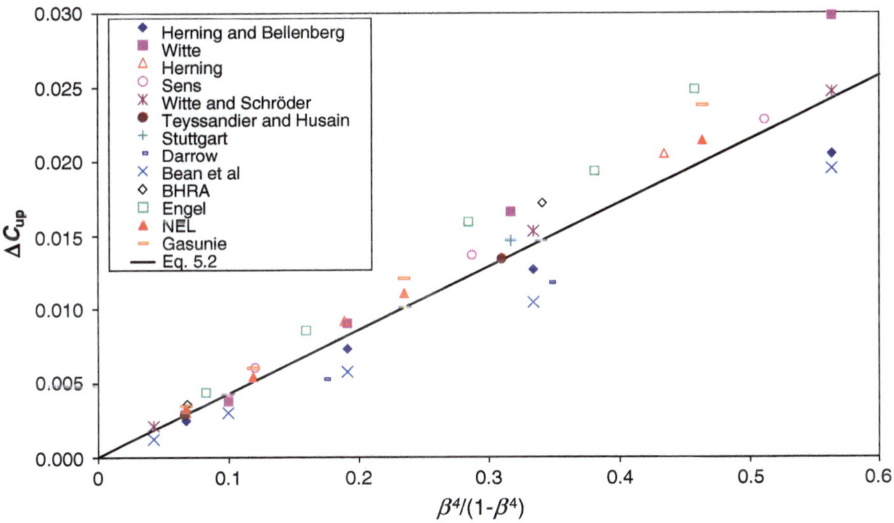

Fig. 5.2 Upstream tapping term one diameter upstream

compensate, the upstream tapping term in the formula is very slightly smaller than the true upstream term.

The dependence of the upstream tapping term on L_1 is also required, where L_1 is the quotient of the distance of the upstream tapping from the upstream face of the plate and the pipe diameter. To determine this dependence, points were taken from most of the experimenters listed above and $\Delta C(1 - \beta^4)/\beta^4$ was plotted against L_1. Since it is much more important to obtain a good fit for large β than for small β, the symbols used relate to the value of β rather than the experimenter. Several forms of equation for the upstream tapping term were tried, and their exponents and constants determined, and the one ultimately used in the Reader-Harris/Gallagher (1998) Equation was the following:

$$\Delta C_{\mathrm{up}} = \left(0.043 + 0.080\mathrm{e}^{-10L_1} - 0.123\mathrm{e}^{-7L_1}\right)(1 - 0.11A)\frac{\beta^4}{1 - \beta^4} \tag{5.2}$$

This equation is physically realistic: it has the required dependence on $\beta^4/(1 - \beta^4)$, is equal to 0 for $L_1 = 0$, does not become negative, tends rapidly to a constant once L_1 exceeds 0.5, and has a continuous derivative. Together with the downstream term it gives a very good fit to the total tapping term data. It is plotted in Fig. 5.3 against many sets of experimental measurements of the upstream tapping term and the quality of the fit is good. A is defined in Eq. 5.9c and is only significant where Re_d is small: in Figs. 5.2 and 5.3 it is taken as 0.

NOTE In Sect. 2.3.2.2 Eq. 5.2 is used to calculate the effect of misplacing the upstream tapping.

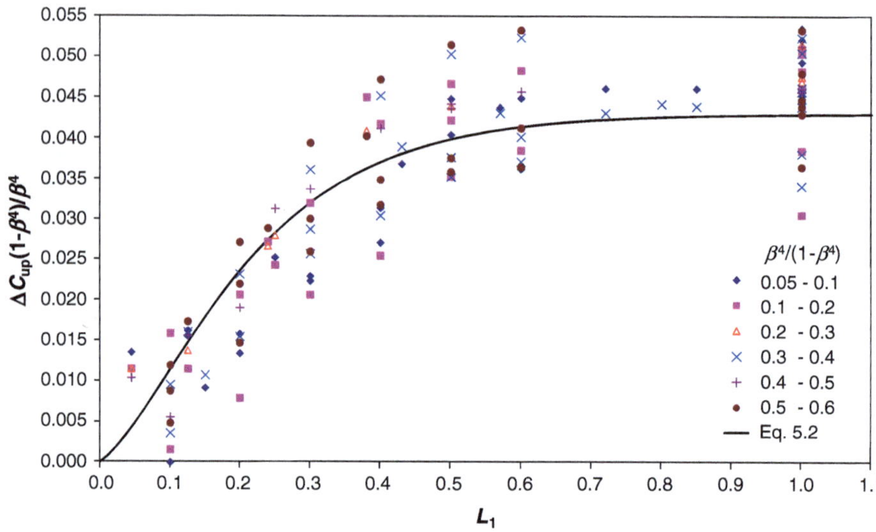

Fig. 5.3 Upstream tapping term as a function of L_1

Downstream Term

Many experimenters have measured the pressure profile downstream of the orifice plate, and, although the data are more scattered than those upstream of the plate, the pattern is clear: the pressure decreases downstream of the plate till it reaches a minimum and then quite a short distance downstream of the minimum it begins to increase rapidly as the pressure recovers.

An important step in the determination of the downstream formula was the work of Teyssandier and Husain (see Teyssandier and Husain 1987; Reader-Harris 1990), who non-dimensionalized downstream distances with the dam height rather than the pipe diameter, where the dam height is the distance from the upstream (sharp) corner of the orifice plate to the nearest point of the pipe wall, i.e. $\frac{D(1-\beta)}{2}$. Instead of working in terms of L_2, the quotient of the distance of the downstream tapping from the upstream face of the plate and the pipe diameter, it is better to use M_2, the quotient of the distance of the downstream tapping from the upstream face of the plate and the dam height, which is given by

$$M_2 = \frac{2L_2}{1-\beta} \tag{5.3}$$

L'_2 and M'_2 are defined identically except that in each case the distance from the downstream face of the orifice plate is used.

As stated above and as seen in Fig. 5.1 the pressure downstream of the orifice plate decreases from its corner tapping value to a minimum and thereafter it

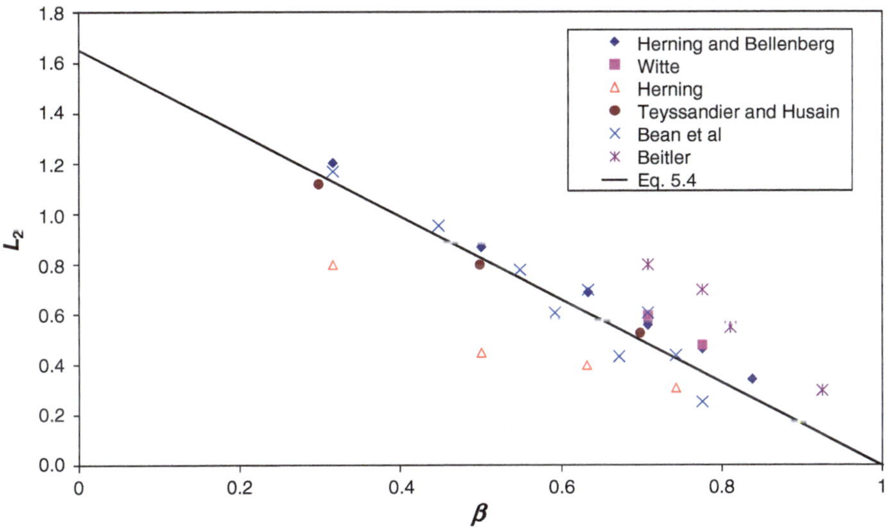

Fig. 5.4 Location of pressure minimum downstream of an orifice plate

increases rapidly; so in Fig. 5.4 the value of L_2 at the pressure minimum is plotted as a function of β. The same symbols as in Fig. 5.2 are used with the addition of data from Beitler (quoted by Teyssandier and Miller 1979). This figure is similar to one in Teyssandier and Miller, except that this one includes a simple fit: at the pressure minimum

$$L_2 = 1.65(1 - \beta) \quad \text{i.e. } M_2 = 3.3 \tag{5.4}$$

Although it is theoretically better to work in terms of M_2 it is more convenient to work in terms of M'_2 since it avoids the need to include the plate thickness in the discharge-coefficient equation. Provided that appropriate restrictions are placed on plate thickness, an equation for M'_2 can be used without introducing significant errors (Reader-Harris 1990; see also Pfeffer 1968; Husain and Teyssandier 1986; see also Sect. 2.2.5.2).

To determine the downstream tapping term, ΔC_{down}, it was desirable next to consider its value at the downstream pressure minimum, $\Delta C_{\text{down,min}}$, and to determine the dependence of $\Delta C_{\text{down,min}}$ on β. Figure 5.5 gives values from measured downstream tapping term data from various experimenters. It is not necessary to determine the best fit, but the following, from Eq. 5.6, gives a good fit to the data:

$$\Delta C_{\text{down,min}} = -0.0101\beta^{1.3}. \tag{5.5}$$

Several forms of equation for the complete downstream tapping term were tried, and their exponents and constants determined, and the one ultimately used in the Reader-Harris/Gallagher (1998) Equation was the following:

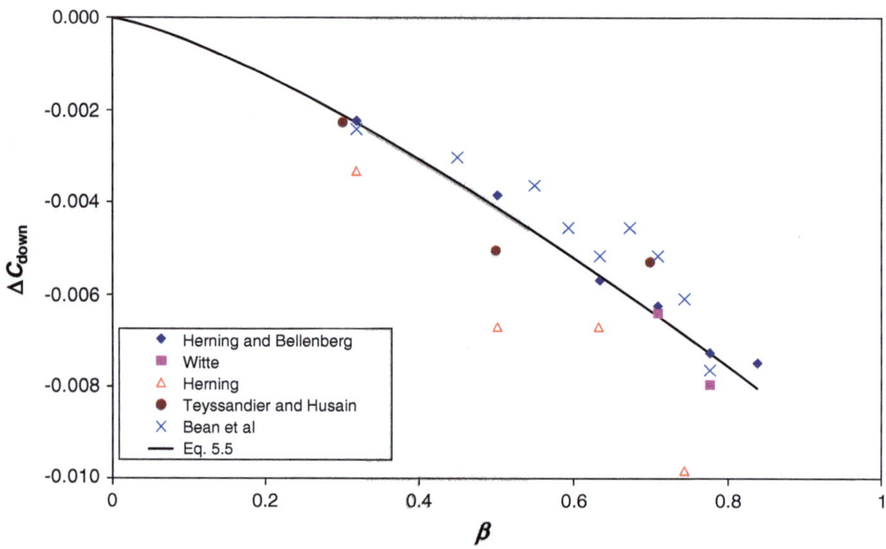

Fig. 5.5 Downstream tapping term at the pressure minimum

$$\Delta C_{\text{down}} = -0.031(M_2' - 0.8M_2'^{1.1})\beta^{1.3} \tag{5.6}$$

This equation is physically realistic: it has an appropriate dependence on β, is equal to 0 for $M'_2 = 0$, has a minimum, and has a continuous and finite derivative. Together with the upstream term it gives a very good fit to the total tapping term data. It is plotted in Fig. 5.6 against many sets of experimental measurements of the downstream tapping term: only data from plates whose thickness, E, does not exceed $0.051D$ are plotted. From the figure it can be seen that an orifice meter should certainly not be used with $M'_2 > 4$ since then its downstream tapping may lie in the region of rapid pressure recovery (the points for large M'_2 are not errors: they show that in some cases pressure recovery is beginning). It is safer to avoid using $M'_2 > 3.75$.

NOTE In Sect. 2.3.2.2 Eq. 5.6 is used to calculate the effect of misplacing the downstream tapping.

5.4.2.3 Low Reynolds Number Tapping Terms

In analysing the European 4″, 10″ and 24″ data, only a small effect of Reynolds number on the tapping terms was found. However, this is not true of the American data since a significant part of the American data were collected in oil (at Reynolds

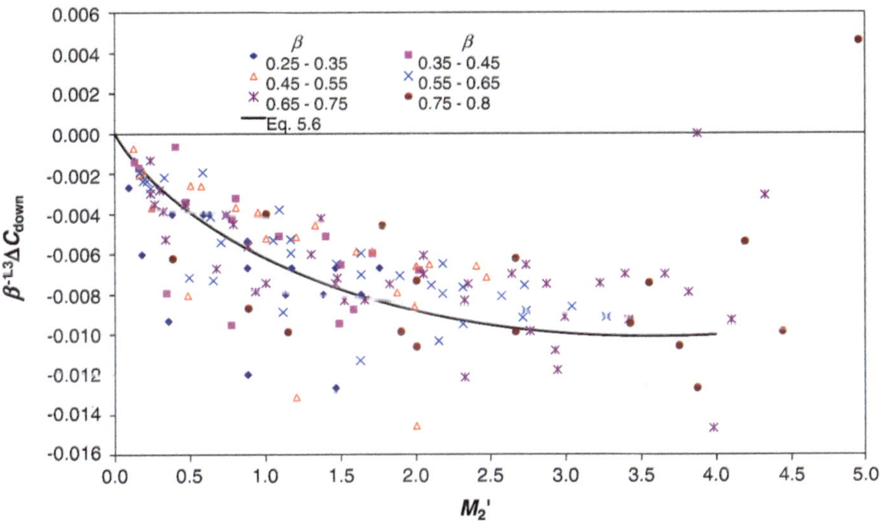

Fig. 5.6 Downstream tapping term as a function of M_2' ($E/D \leq 0.051$)

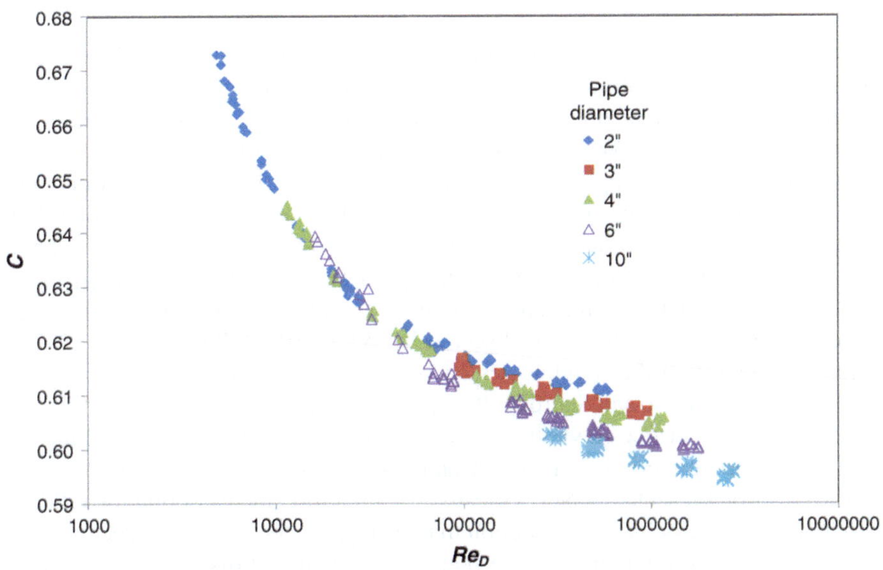

Fig. 5.7 American (CEESI and NIST) data (all with flange tappings) for $\beta \approx 0.735$

numbers below those in water or gas). Figure 5.7 is a graph of the American data for $\beta \approx 0.735$. The high-Reynolds-number tapping terms, which are not functions of Reynolds number and bring the data for high Reynolds numbers on to a common curve, are clearly inappropriate for lower Reynolds number.

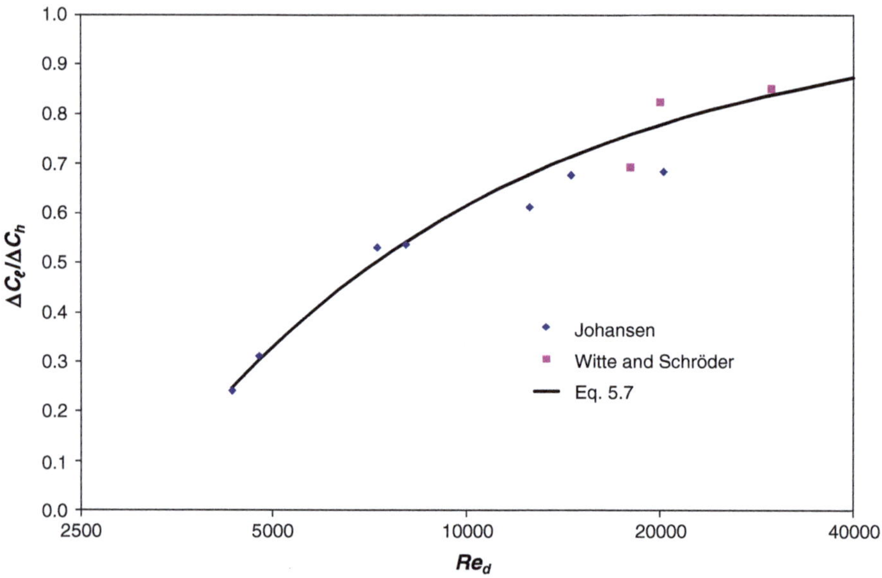

Fig. 5.8 Effect of low Reynolds number on upstream tapping term

Work by Johansen (1930) showed that the upstream tapping term determined at the upstream pressure minimum decreases as Re_d decreases. Similarly the downstream tapping term at the downstream pressure minimum decreases in magnitude as Re_d decreases. Data from Witte and Schröder (quoted in Engel 1961, 1962), who only measured the upstream tapping term, agree with Johansen. The data (excluding those for $\beta = 0.209$ for which the tapping term is small and thus the value of the ratio is both less certain and less important) are presented in Figs. 5.8 and 5.9. ΔC_l is the value of the tapping term at low Re_d and ΔC_h is the appropriate calculated high Re_d tapping term. A reduction in the upstream pressure rise for extremely rough pipes ($\lambda = 0.06$) at moderate Reynolds numbers was observed by Witte (1931), but this effect will be extremely small with the roughnesses encountered in accurate flow measurement. From the work of Witte and Schröder the upstream tapping term becomes effectively constant above an orifice Reynolds number of about 80,000. It is assumed that the downstream tapping term becomes constant at the same orifice Reynolds number.

The equation presented at New Orleans and, therefore, the RG Equation reflected these data on the reasonable assumption that the tapping terms, though decreasing in size as Re_d decreases, retain the same shape as a function of distance from the plate and diameter ratio, since data were not available except at the pressure minima. On Figs. 5.8 and 5.9 the terms in the RG Equation are shown: upstream

$$\frac{\Delta C_\ell}{\Delta C_h} = 1 - 0.23\left(\frac{19,000}{Re_d}\right)^{0.8} \tag{5.7}$$

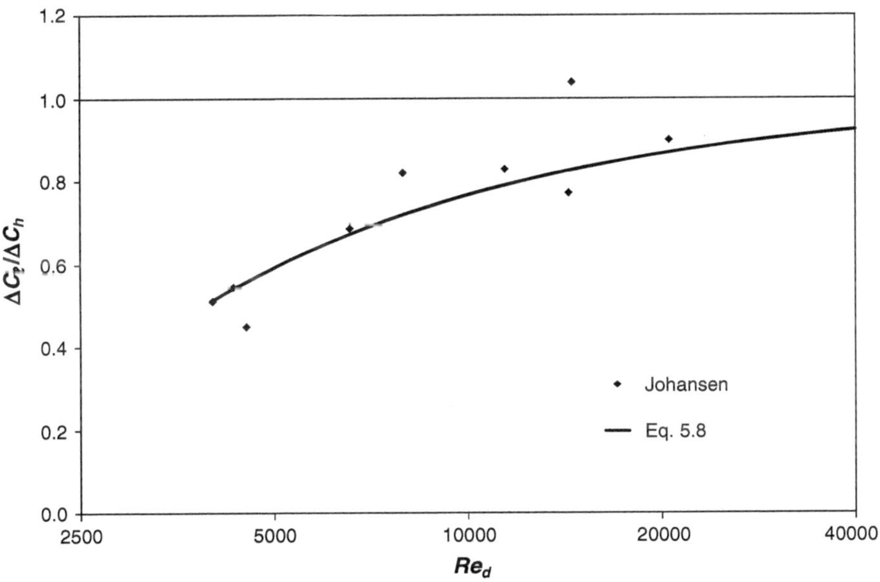

Fig. 5.9 Effect of low Reynolds number on downstream tapping term

and downstream

$$\frac{\Delta C_\ell}{\Delta C_h} = 1 - 0.14\left(\frac{19,000}{Re_d}\right)^{0.8}. \tag{5.8}$$

Further work was carried out after the meeting in New Orleans, and the best option for the low-Reynolds-number tapping terms is described in Appendix 5.A, but since equations using these terms have not been used for commercial purposes, the text that follows here describes the tapping terms in the Reader-Harris/Gallagher (1998) Equation.

To derive the Reader-Harris/Gallagher (1998) Equation the upstream and downstream tapping terms, ΔC_{up} and ΔC_{down}, were taken as:

$$\Delta C_{\mathrm{up}} = (0.043 + 0.080\mathrm{e}^{-10L_1} - 0.123e^{-7L_1})(1 - aA)\frac{\beta^4}{1-\beta^4} \tag{5.9a}$$

and

$$\Delta C_{\mathrm{down}} = -0.031(M'_2 - 0.8M'^{1.1}_2)(1 - bA)\beta^{1.3}. \tag{5.9b}$$

where

$$A = \left(\frac{19,000\beta}{Re_D}\right)^{0.8} \text{ and } M_2' = \frac{2L_2'}{1-\beta}. \tag{5.9c}$$

a and b were not determined to fit the data in Figs. 5.8 and 5.9 but to fit the data in the discharge-coefficient database. When determining the values of a and b to give the best fit of the form of the complete Reader-Harris/Gallagher (1998) Equation to the database, it was found that to permit a non-zero value of b gives a negligible reduction in the standard deviation of the data about the equation. With $b = 0$ the best-fit value of a was 0.11. So that a good fit for the complete database (including data for $Re_D < 4000$) can be obtained it is necessary to include an additional term in ΔC_{down} for $Re_D \leq 3700$ identical to that in the PR14 Equation (see Appendix 5.A). Therefore, the final complete tapping terms for inclusion in the Reader-Harris/Gallagher (1998) Equation are given by

$$\Delta C_{\text{up}} = (0.043 + 0.080\text{e}^{-10L_1} - 0.123\text{e}^{-7L_1})(1 - 0.11A)\frac{\beta^4}{1-\beta^4} \tag{5.10a}$$

and

$$\Delta C_{\text{down}} = -0.031\,(M_2' - 0.8M_2'^{1.1})\{1 + 8\max(\lg(3700/Re_D), 0.0)\}\beta^{1.3}. \tag{5.10b}$$

5.4.3 The C_∞ and Slope Terms

Given the tapping terms it is possible to determine the C_∞ and slope terms. As stated above, C_∞ is the discharge coefficient using corner tappings for infinite Reynolds number, and the slope term, C_s, gives the increase in that discharge coefficient for lower Reynolds number, so that the discharge coefficient using corner tappings is given by $C_\infty + C_s$. There may be an additional term for small orifices. C_∞ increases with β to a maximum near $\beta = 0.55$ and then decreases increasingly rapidly with β. Thus an appropriate form for C_∞ is

$$C_\infty = a_1 + a_2\beta^{m_1} + a_3\beta^{m_2} \tag{5.11}$$

Taking $m_1 = 2$ and $m_2 = 8$ enables a good fit to be obtained.

Reader-Harris and Keegans (1986) found that, for a fixed pipe Reynolds number, as the pipe roughness changed so the change in discharge coefficient was approximately proportional to $\beta^4\Delta\lambda$, where $\Delta\lambda$ is the change in the pipe friction factor. This topic is considered in Sect. 2.3.3.1. This led the author to suppose that the form of the equation for the discharge coefficient (omitting tapping terms) might be

$$C = C_\infty + c'\beta^\ell\lambda \tag{5.12}$$

where ℓ is approximately 4. Rapier (1979) made experimental measurements in very smooth pipes and as a result of his work also proposed that C was best expressed as a function of friction factor. However, equations of the form of Eq. 5.12 do not give good agreement with experiment for small β, since they then give C almost constant. It seems reasonable to suppose that for small β the discharge coefficient should depend only on orifice Reynolds number

$$C = C_\infty + c\left(\frac{10^6}{Re_d}\right)^{n_1}. \tag{5.13}$$

This form is supported by data for β both in the range 0.1–0.375 (Reader-Harris and Sattary 1990) and in the range 0.015–0.06 (Reader-Harris et al. 2008b) (see also Appendix 2.A).

Combining Eqs. 5.12 and 5.13 gives

$$C = C_\infty + c\left(\frac{10^6}{Re_d}\right)^{n_1} + c'\beta^l\lambda \tag{5.14}$$

This form is adequate for high Reynolds number but for practical use it requires three further enhancements. There are no data on the effect of rough pipework on the discharge coefficient for low Reynolds number, and a better fit to the database is obtained by including an additional term proportional to A, on the basis that, as the tapping terms begin to change, so may the dependence on friction factor:

$$C = C_\infty + c\left(\frac{10^6}{Re_d}\right)^{n_1} + (c' + c''A)\beta^l\lambda \tag{5.15}$$

λ is an inconvenient variable with which to work, especially in commercial measurements, but for the pipes used in collecting the data in the EEC/API database a typical pipe roughness as a function of Re_D can be determined; so typical values of λ as a function of Re_D can be calculated, and λ can then be approximated by a constant (which leads to a term to be absorbed into C_∞ and a small term which is neglected) and a reciprocal power of Re_D:

$$C = C_\infty + c\left(\frac{10^6}{Re_d}\right)^{n_1} + (c' + c''A)\beta^l\left(\frac{10^6}{Re_D}\right)^{n_2} \tag{5.16}$$

It may also be necessary to make provision for transition from turbulent to laminar flow since, except for very small β, the gradient of the discharge coefficient as a function of a reciprocal power of Re_D is very different below a transition point in the range Re_D = 3000–5000 from that above it. This change of gradient occurs because the velocity profile and the friction factor change very rapidly as the flow changes from turbulent to laminar. When the velocity profile term is extended so that the equation can be used below the fully turbulent range the discharge coefficient (still omitting tapping terms) becomes:

$$C = C_\infty + c\left(\frac{10^6}{Re_d}\right)^{n_1} + (c' + c''A)\beta^l \max\left(\left(\frac{10^6}{Re_D}\right)^{n_2}, c_1 - c_2\left(\frac{Re_D}{10^6}\right)\right) \quad (5.17)$$

It remains to determine the constants and exponents in Eqs. 5.11 and 5.17. $n_1 = 0.7$ and $n_2 = 0.3$ give the optimum fit to the complete database. Given that the mean Re_D at which the flow becomes fully turbulent was 3700, c_1 is given in terms of c_2. c_2 is obtained by trying appropriate values in turn and obtaining the best overall fit: $c_2 = 4700$ gives an excellent overall fit. Given the tapping terms in Eqs. 5.10a, b and the small pipe diameter term in Eq. 5.20 a least-squares fit of the complete database was performed: on rounding the constants, the C_∞ and slope terms become

$$\begin{aligned} C_\infty + C_s = {} & 0.5961 + 0.0261\beta^2 - 0.216\beta^8 + 0.000521\,(10^6\beta/Re_D)^{0.7} \\ & + (0.0188 + 0.0063A)\beta^{3.5}\max\{(10^6/Re_D)^{0.3},\ 22.7 - 4700\,(Re_D/10^6)\}. \end{aligned} \quad (5.18)$$

5.4.4 A Term for Small Orifice Meters

There is a problem in that Eq. 5.18 when combined with the tapping terms in Eqs. 5.10a, b fits the whole database except for the data from 2″ pipes. Because it failed to fit the original API 2″ data more data were collected at NEL using both the API plates and newly manufactured plates. These data also differ from Eqs. 5.10a, b and 5.18: C was larger in 2″ pipes than in larger pipes, even with corner and D and $D/2$ tappings. The upstream edges of the 2″ plates were then measured by NEL and shown to lie outside the permitted radius (see Appendix 5.B). The simplest solution, proposed by Jim Gallagher, is a small pipe diameter term, ΔC_D, of the following form:

$$\Delta C_D = h_1(h_2 - \beta)\max\left(h_3 - \frac{D}{25.4},\ 0.0\right) \quad (5.19)$$

where D is the pipe diameter in mm.

Although it provides a correction that fits the data, the form of this term has no known physical basis, although this author believes that the need for the term is due to the edge radius of the orifice plates used in the 2″ pipes rather than the pipe diameter itself.

Since there are essentially only discrete values of D in the database h_3 cannot be determined: the value of h_3 was taken to be 71.12 mm (2.8″), 2.54 mm (0.1″) smaller than the internal diameter for 3″ schedule 80 pipe. The quality of fit to the complete database improves as h_2 reduces to 0.75, the smallest value which ensures that this term does not become negative over the permissible range of use. If $(h_2 - \beta)$ were replaced by $\max(h_2 - \beta, 0)$ then the best fit is obtained with h_2 equal to 0.7, but the improvement in the fit is very small and the equation significantly different from and more complicated than the RG Equation. With $h_2 = 0.75$ h_1 was determined by fitting the complete database and is given by 0.011. Therefore, the small pipe diameter term is given by

$$\Delta C_D = 0.011\,(0.75 - \beta)\max\left(2.8 - \frac{D}{25.4},\ 0.0\right) \tag{5.20}$$

where D is as in Eq. 5.19.

NOTE It would be better to work in terms of orifice diameter. This is discussed in Appendix 5.B.

5.4.5 The Complete Equation

Therefore, the complete orifice plate discharge-coefficient equation is as follows:

$$\begin{aligned} C = {} & 0.5961 + 0.0261\beta^2 - 0.216\beta^8 + 0.000521\,(10^6\beta/Re_D)^{0.7} \\ & + (0.0188 + 0.0063A)\beta^{3.5}\max\{(10^6/Re_D)^{0.3},\ 22.7 - 4700\,(Re_D/10^6)\} \\ & + (0.043 + 0.080\,\mathrm{e}^{-10L_1} - 0.123\,\mathrm{e}^{-7L_1})(1 - 0.11A)\frac{\beta^4}{1-\beta^4} \\ & - 0.031\,(M_2' - 0.8M_2'^{1.1})\{1 + 8\max(\lg(3700/Re_D),\ 0.0)\}\beta^{1.3} \\ & + 0.011\,(0.75 - \beta)\max\left(2.8 - \frac{D}{25.4},\ 0.0\right) \end{aligned} \tag{5.21}$$

where D is the pipe diameter in mm.

NOTE Because the velocity profile term (the second line of Eq. 5.21) changes rapidly with Re_D when $Re_D < 3700$ unless β is small, it is wise not to use an orifice plate to measure flow through transition from turbulent flow to laminar flow in the upstream pipe unless β is small.

For $Re_D \geq 5000$ this equation can be written as follows:
For $D \geq 71.12$ mm (2.8″)

$$\begin{aligned} C = {} & 0.5961 + 0.0261\beta^2 - 0.216\beta^8 \\ & + 0.000521\,(10^6\beta/Re_D)^{0.7} + (0.0188 + 0.0063A)\beta^{3.5}(10^6/Re_D)^{0.3} \\ & + (0.043 + 0.080\mathrm{e}^{-10L_1} - 0.123\mathrm{e}^{-7L_1})(1 - 0.11A)\frac{\beta^4}{1-\beta^4} \\ & - 0.031\,(M_2' - 0.8M_2'^{1.1})\beta^{1.3} \end{aligned} \tag{5.22a}$$

Where $D < 71.12$ mm (2.8″) the following term should be added to Eq. 5.22a:

$$+0.011\,(0.75 - \beta)\left(2.8 - \frac{D}{25.4}\right) \tag{5.22b}$$

where D is the pipe diameter in mm.

In this equation

$$A = \left(\frac{19{,}000\beta}{Re_D}\right)^{0.8} \text{ and } M_2' = \frac{2L_2'}{1-\beta}.$$

Equation 5.22, comprising Eq. 5.22a with the additional term 5.22b, is the Reader-Harris/Gallagher (1998) Equation in ISO 5167-2:2003.

When writing computer code for the equation it can be checked using examples from Tables A.1 to A.11 in ISO 5167-2:2003.

Work on the lower Reynolds number limit for the Reader-Harris/Gallagher (1998) Equation as in ISO 5167-2:2003 is included as Appendix 5.F.

5.5 Quality of Fit

The quality of the fit of Eq. 5.21 to the database is very good: the overall standard deviation of the data for $Re_D \geq 4000$ about the equation is 0.259 %; the mean deviations of the data about the equation as a function of β, pipe size, Re_D and pair of tappings used, and of pairs of these independent variables, are both small and well-balanced.

The quality of fit is quantified in Tables 5.3, 5.4 5.5 and 5.6. Table 5.3 gives a description of the meaning of the different lines in Tables 5.4, 5.5 and 5.6. These tables give the deviations of the data in the database about the equation as a

Table 5.3 General information about the analysis of deviations in Tables 5.4, 5.5 and 5.6

For each cell	Line 1—mean percent error
	Line 2—percent standard deviation
	Line 4—number of observations
	Line 5—percent r.m.s. deviation about equation

Table 5.4 Deviations from Eq. 5.21 as a function of β and Re_D

Re_D	10–4000	4000–10^4	10^4–10^5	10^5–10^6	10^6–10^7	10^7–10^8	Summary by β
β							
0.100 (0.0991–0.1028)	0.095	0.108	0.037				0.078
	0.189	0.282	0.305				0.265
	–	–	–				–
	52	49	59				160
	0.211	0.299	0.305				0.276
0.200 (0.1982–0.2418)	−0.027	0.047	−0.050	−0.051	0.102		−0.019
	0.556	0.354	0.249	0.240	0.170		0.300
	–	–	–	–	–		–
	237	238	1190	454	316		2435
	0.556	0.357	0.254	0.245	0.198		0.300
0.375 (0.3620–0.3748)	0.376	0.125	0.008	−0.041	0.049	0.082	0.031
	0.509	0.322	0.175	0.157	0.084	0.083	0.225
	–	–	–	–	–	–	–
	125	133	748	711	325	140	2182
	0.631	0.344	0.175	0.162	0.097	0.116	0.227
0.500 (0.4825–0.5003)	−0.057	−0.093	0.034	0.071	−0.090	−0.057	0.016
	0.760	0.281	0.195	0.163	0.134	0.085	0.205
	–	–	–	–	–	–	–
	33	83	436	788	205	249	1794
	0.750	0.294	0.198	0.177	0.162	0.102	0.205

(continued)

Table 5.4 (continued)

Re_D β	10–4000	4000–10^4	10^4–10^5	10^5–10^6	10^6–10^7	10^7–10^8	Summary by β
0.570 (0.5427–0.5770)	−0.546	−0.334	−0.040	0.002	0.083	0.028	0.009
	0.968	0.459	0.241	0.195	0.273	0.224	0.255
	–	–	–	–	–	–	–
	18	59	502	1430	782	463	3254
	1.087	0.564	0.244	0.195	0.285	0.226	0.255
0.660 (0.6481–0.6646)	−0.087	−0.325	−0.110	0.012	0.025	−0.100	−0.038
	0.116	0.522	0.307	0.203	0.193	0.177	0.233
	–	–	–	–	–	–	–
	5	35	466	1121	475	660	2762
	0.135	0.609	0.326	0.204	0.194	0.203	0.236
0.750 (0.7239–0.7509)		0.242	−0.073	0.038	−0.029	−0.053	−0.004
		0.514	0.404	0.303	0.315	0.355	0.340
		–	–	–	–	–	–
		78	615	1704	1062	476	3935
		0.565	0.410	0.305	0.316	0.359	0.340
Summary by Re_D	0.071	0.019	−0.038	0.013	0.024	−0.041	−0.002
	0.589	0.411	0.273	0.230	0.255	0.239	0.274
	–	–	–	–	–	–	–
	470	675	4016	6208	3165	1988	16,522
	0.593	0.411	0.276	0.231	0.257	0.242	0.274

Table 5.5 Deviations from Eq. 5.21 as a function of β and pipe size ($Re_D \geq 4000$)

Pipe size β	2″ (50 mm)	3″ (75 mm)	4″ (100 mm)	6″ (150 mm)	10″ (250 mm)	24″ (600 mm)	Summary by β
0.100 (0.0991–0.1028)				0.267	−0.004		0.069
				0.277	0.269		0.296
				–	–		–
				29	79		108
				0.381	0.268		0.302
0.200 (0.1982–0.2418)	0.052	−0.047	0.079	0.095	−0.170	0.028	−0.018
	0.341	0.124	0.232	0.079	0.187	0.239	0.258
	–	–	–	–	–	–	–
	324	57	626	83	714	394	2198
	0.345	0.131	0.244	0.123	0.253	0.240	0.258
0.375 (0.3620–0.3748)	0.022	−0.003	0.061	0.081	−0.106	0.037	0.010
	0.190	0.097	0.200	0.175	0.159	0.113	0.174
	–	–	–	–	–	–	–
	344	106	455	122	439	591	2057
	0.191	0.097	0.209	0.192	0.191	0.119	0.174
0.500 (0.4825–0.5003)	−0.028	0.138	0.135	0.098	0.033	−0.063	0.017
	0.211	0.054	0.190	0.103	0.164	0.118	0.179
	–	–	–	–	–	–	–
	365	69	300	109	392	526	1761
	0.213	0.148	0.233	0.142	0.168	0.134	0.180

(continued)

Table 5.5 (continued)

Pipe size β	2″ (50 mm)	3″ (75 mm)	4″ (100 mm)	6″ (150 mm)	10″ (250 mm)	24″ (600 mm)	Summary by β
0.570 (0.5427–0.5770)	−0.086	0.095	0.028	0.009	0.074	−0.090	0.012
	0.322	0.076	0.233	0.143	0.260	0.117	0.242
	–	–	–	–	–	–	–
	330	72	1008	136	1123	567	3236
	0.333	0.121	0.235	0.143	0.270	0.148	0.243
0.660 (0.6481–0.6646)	−0.088	0.101	−0.018	−0.097	0.050	−0.136	−0.038
	0.307	0.100	0.245	0.204	0.197	0.143	0.233
	–	–	–	–	–	–	–
	493	64	642	92	823	643	2757
	0.319	0.142	0.246	0.225	0.204	0.197	0.236
0.750 (0.7239–0.7509)	−0.051	0.092	0.120	0.107	−0.024	−0.251	−0.004
	0.342	0.106	0.322	0.333	0.321	0.338	0.340
	–	–	–	–	–	–	–
	866	101	1024	130	1478	336	3935
	0.345	0.140	0.344	0.349	0.322	0.421	0.340
Summary by *D*	−0.037	0.062	0.064	0.061	−0.013	−0.073	−0.004
	0.305	0.113	0.259	0.217	0.262	0.199	0.259
	–	–	–	–	–	–	–
	2722	469	4055	701	5048	3057	16,052
	0.308	0.129	0.267	0.225	0.262	0.212	0.259

Table 5.6 Deviations from Eq. 5.21 as a function of β and pair of tappings ($Re_D \geq 4000$)

Tappings β	Corner (ISO)	Flange	*D* and *D*/2	Corner (GU)	Summary by β
0.100 (0.0991–0.1028)		0.069			0.069
		0.296			0.296
		–			–
		108			108
		0.302			0.302
0.200 (0.1982–0.2418)	0.018	−0.034	−0.002	−0.098	−0.018
	0.274	0.233	0.272	0.259	0.258
	–	–	–	–	–
	537	925	582	154	2198
	0.275	0.235	0.272	0.276	0.258
0.375 (0.3620–0.3748)	−0.006	0.026	0.006	−0.072	0.010
	0.175	0.178	0.155	0.170	0.174
	–	–	–	–	–
	382	1101	466	108	2057
	0.175	0.180	0.155	0.184	0.174
0.500 (0.4825–0.5003)	0.012	0.020	0.029	−0.037	0.017
	0.242	0.154	0.160	0.178	0.179
	–	–	–	–	–
	369	884	403	105	1761
	0.241	0.155	0.163	0.181	0.180

(continued)

Table 5.6 (continued)

Tappings β	Corner (ISO)	Flange	*D* and *D*/2	Corner (GU)	Summary by β
0.570 (0.5427–0.5770)	−0.002	−0.035	0.084	0.080	0.012
	0.231	0.197	0.274	0.314	0.242
	–	–	–	–	–
	794	1366	845	231	3236
	0.231	0.200	0.286	0.323	0.243
0.660 (0.6481–0.6646)	−0.068	−0.045	0.006	−0.040	−0.038
	0.231	0.215	0.256	0.244	−0.233
	–	–	–	–	–
	671	1200	690	196	2757
	0.241	0.220	0.256	0.246	0.236
0.750 (0.7239–0.7509)	−0.026	0.060	−0.147	0.106	−0.004
	0.324	0.290	0.381	0.381	0.340
	–	–	–	–	–
	963	1765	893	314	3935
	0.325	0.296	0.408	0.395	0.340
Summary by Tappings	−0.017	0.004	−0.011	0.016	−0.004
	0.262	0.229	0.291	0.306	0.259
	–	–	–	–	–
	3716	7349	3879	1108	16,052
	0.263	0.229	0.291	0.306	0.259

function of β, pipe size, Re_D and pair of tappings used, and certain combinations of these. The tappings described as Corner (GU) are tappings in the corners which were designed by Gasunie and are described briefly in Table 5.1. The database used is the complete EEC/API database as described, except that data with $Re_D < 4000$ were excluded in Tables 5.5 and 5.6.

For the ith point in a cell percent error, $P_i = \frac{(C_{im} - C_{ie})}{C_{im}} \times 100$,
where C_{im} is the measured discharge coefficient of the ith point
and C_{ie} is the corresponding discharge coefficient from the equation.

Mean percent error, $\mu = \frac{\sum_{i=1}^{N} P_i}{N}$,
where N is the number of points in the cell.

Percent standard deviation $= \sqrt{\frac{\sum_{i=1}^{N} (P_i - \mu)^2}{N-1}}$.

Percent r.m.s. deviation about equation $= \sqrt{\frac{\sum_{i=1}^{N} P_i^2}{N}}$.
Statistics for the entire population appear in the bottom right hand cell.

5.6 Equations and Comparison Between Them on the Basis of Deviations

5.6.1 The Reader-Harris/Gallagher (RG) Equation as in API 14.3.1:1990

The Reader-Harris/Gallagher (RG) Equation as in API 14.3.1:1990 (API 1990) and in API 14.3.1:2012 (API 2012) is as follows:

$$\begin{aligned} C = {} & 0.5961 + 0.0291\beta^2 - 0.229\beta^8 \\ & + 0.000511\,(10^6\beta/Re_D)^{0.7} + (0.0210 + 0.0049A)\beta^4(10^6/Re_D)^{0.35} \\ & + (0.0433 + 0.0712\,e^{-8.5L_1} - 0.1145\,e^{-6L_1})(1 - 0.23A)\frac{\beta^4}{1-\beta^4} \\ & - 0.0116\,(M_2' - 0.52M_2'^{1.3})(1 - 0.14A)\beta^{1.1} \end{aligned} \tag{5.23a}$$

Where $D_r < 71.12$ mm (2.8″) the following term is added to Eq. 5.23a:

$$+0.003\,(1-\beta)\left(2.8-\frac{D_r}{25.4}\right). \quad (D_r \quad \text{in} \quad \text{mm}) \tag{5.23b}$$

where D_r is the pipe internal diameter at reference temperature (assumed to be 20 °C (68 °F)).

5.6.2 The Stolz Equation in ISO 5167:1980

The Stolz Equation in ISO 5167:1980 (ISO 1980) and in ISO 5167-1:1991 (ISO 1991) is as follows:

$$C = 0.5959 + 0.0312\beta^{2.1} - 0.184\beta^8 + 0.0029\beta^{2.5}(10^6/Re_D)^{0.75} + \min(0.090L_1,\ 0.039)\frac{\beta^4}{1-\beta^4} - 0.0337L_2'\beta^3 \tag{5.24}$$

5.6.3 Comparisons

A direct comparison between the Reader-Harris/Gallagher (1998) Equation (given above as Eqs. 5.22a, b), the Reader-Harris/Gallagher (RG) Equation given in API 14.3:1990 (given above as Eqs. 5.23a, b), and the PR14 Equation in Appendix 5.C for $Re_D \geq 4000$ is given in Table 5.7. To do this the data in Table 5.5 (the table of deviations on the basis of β and pipe size) were analysed: the number of boxes (range of β and pipe size) over which an equation gave a mean deviation greater than 0.1 % was counted and is given in Table 5.7; the number of boxes for which the mean deviation was greater than 0.2 % was also counted and is given in Table 5.7. A similar count was undertaken for Table 5.4 (omitting $Re_D < 4000$) and Table 5.6 and another similar table (not shown), and the results are also given in Table 5.7. These numbers provide a measure of possible bias in an equation. The standard deviation of the data about each equation is also given as a measure of the quality of fit. The quality of fit for the Reader-Harris/Gallagher (1998) Equation is similar to that of the PR14 Equation. They are both better than the Reader-Harris/Gallagher (RG) Equation as in API 14.3:1990.

Since the Stolz Equation in ISO 5167:1980 is applicable over a more limited range of values of Re_D and β than the three equations previously considered, the standard deviation and the number of boxes with mean deviations greater than 0.1 or 0.2 % are given in Table 5.8 for all four equations over the range of applicability of the Stolz ISO 5167:1980 Equation. That there was a need for an improved equation, even over the limited range of Re_D and β, is obvious.

Another way of seeing the difference between equations is to look at the quality of the fit to the 24″ gas data taken with flange tappings (by British Gas and Gasunie),

Table 5.7 Analysis of deviations for the database ($Re_D \geq 4000$)

	Reader-Harris/ Gallagher (1998) Equation (ISO 5167-2:2003)		PR14 Equation		Reader-Harris/ Gallagher (RG) Equation (API 14.3.1:1990)	
Equation	5.22a, b		5.C.2		5.23a, b	
	N_1	N_2	N_1	N_2	N_1	N_2
β versus pipe size	10	2	14	2	18	6
β versus Re_D	8	3	6	5	14	6
β versus tappings	2	0	2	0	6	1
pipe size versus Re_D	10	2	7	2	12	2
TOTAL	30	7	29	9	50	15
s (per cent)	0.259		0.254		0.292	

N_1 is the number of boxes with the mean deviation greater than 0.1 %
N_2 is the number of boxes with the mean deviation greater than 0.2 %
s is the standard deviation of the data in the database about the equation

Table 5.8 Analysis of deviations for the database over the range of the Stolz ISO 5167:1980 Equation ($Re_D \geq 4000$)

	Reader-Harris/ Gallagher (1998) Equation (ISO 5167-2:2003)		PR14 Equation		Reader-Harris/ Gallagher (RG) Equation (API 14.3.1:1990)		Stolz Equation (ISO 5167:1980)	
Equation	5.22a, b		5.C.2		5.23a, b		5.24	
	N_1	N_2	N_1	N_2	N_1	N_2	N_1	N_2
β versus pipe size	8	1	15	2	17	6	26	16
β versus Re_D	4	0	3	1	10	4	16	12
β versus tappings	2	0	2	0	5	0	13	7
pipe size versus Re_D	7	2	10	3	10	3	16	10
TOTAL	21	3	30	6	42	13	71	45
s (per cent)	0.245		0.247		0.277		0.390	

N_1 is the number of boxes with the mean deviation greater than 0.1 %
N_2 is the number of boxes with the mean deviation greater than 0.2 %
s is the standard deviation of the data in the database about the equation

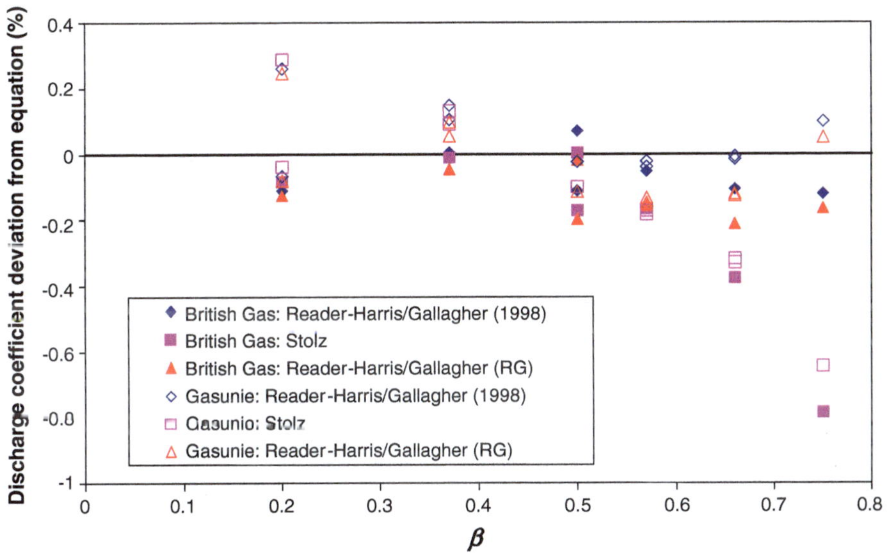

Fig. 5.10 Discharge coefficient deviation from different equations for the 24″ gas data using flange tappings

perhaps the most important sets of data in terms of financial impact. The deviations are shown in Fig. 5.10. The Reader-Harris/Gallagher (1998), the Reader-Harris/Gallagher (RG) (as in API) and the Stolz Equations have mean deviations of −0.01, −0.08 and −0.17 % respectively. The Stolz Equation has a large deviation for large β.

5.7 Uncertainty

It is not simple to calculate the uncertainty for the orifice equation. Evaluating the uncertainty for the orifice equation is different from most evaluations of uncertainty: a normal evaluation of uncertainty is for an instrument which is calibrated and itself subsequently used. However, the orifice discharge-coefficient equation is used for plates other than those used to collect the data. So it is not ideal to evaluate the uncertainty of the orifice discharge-coefficient equation from twice the standard deviation of the data in the database about the Reader-Harris/Gallagher Equation (which is the method used by API 14.3.1).

Twice the standard deviation of the data in the database about the Reader-Harris/Gallagher Equation includes the random uncertainty of the original sets of data collected with equipment whose random scatter may be higher than that available today. To avoid the uncertainty of the equation being increased by the random uncertainty of each data set it is possible to fit each set of data as a linear function of $Re_D^{-0.5}$ and then represent each line by a single point at the middle of its range. Such

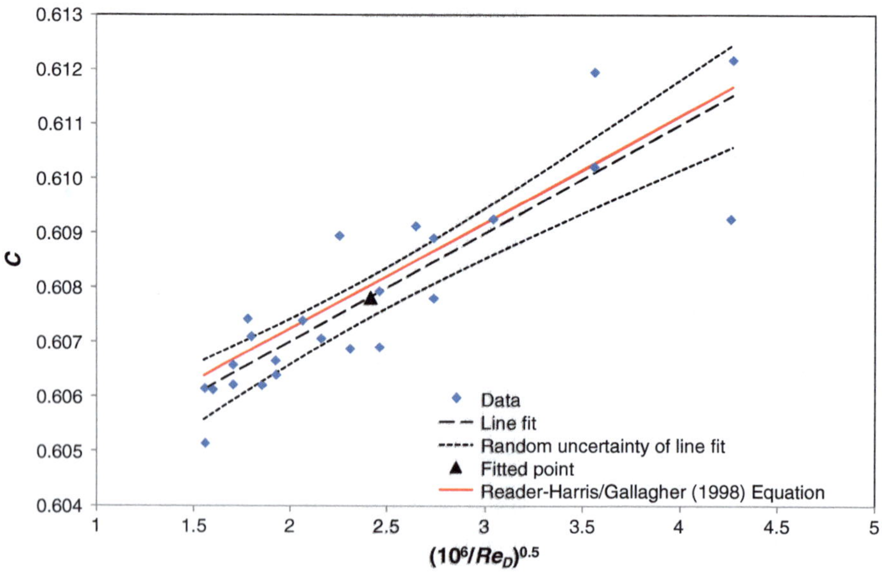

Fig. 5.11 One set of NEL 4″ β = 0.57 water data from the orifice-plate discharge-coefficient database

fitted points have much reduced random uncertainty. Since uncertainties increase for large β, small Re_d and small d, the data for $0.19 < \beta < 0.67$, $Re_d > 30{,}000$ and $d > 50$ mm were analysed as a core population. For this core population of 8515 points the standard deviation, s, about the equation was determined: $2s = 0.42$ %; there are 343 data sets making up the core population: for the 343 mid-range calculated points $2s = 0.36$ %. To illustrate the difference in deviations Fig. 5.11 shows one set of 24 water points from the database. Although twice the r.m.s. deviation of the (24) points about the Reader-Harris/Gallagher (1998) Equation is 0.28 %, the (expanded) random uncertainty of the fitted point is 0.06 %. The deviation of the fitted point from the Reader-Harris/Gallagher (1998) Equation and the mean deviation of the 24 points from the Reader-Harris/Gallagher (1998) Equation are both equal to −0.03 %.

It is necessary to consider the uncertainty contributed by the original laboratories: on the one hand the uncertainty of the mean of the data from the independent laboratories should be added to the uncertainty of the Equation; on the other hand the scatter contributed by calibration in a range of independent laboratories should be removed from the uncertainty of the Equation. It is assumed here that these two contributions cancel each other out.

NOTE For an example of the scatter contributed by calibration in different laboratories consider Fig. 5.10. If twelve points that give the mean deviation for each β for each laboratory from the Reader-Harris/Gallagher (1998) Equation were considered, twice the r.m.s. deviation of these points from the Reader-Harris/Gallagher

(1998) Equation would be 0.16 %; if six points that give the mean deviation for each β were considered, they would have reduced uncertainty: twice their r.m.s. deviation from the Reader-Harris/Gallagher (1998) Equation would be 0.08 %.

The database contains points with a range of values of relative pipe roughness, R_a/D, (creating a variation in C of, say, ±0.2 %) and relative edge sharpness, r/d, (creating a variation in C of, say, ±0.08 %): the fitted equation has effectively been derived for a particular relative pipe roughness (as a function of Reynolds number) and relative edge sharpness (based on the relative pipe roughness and relative edge sharpness of the original data); if the actual orifice meter to be manufactured for a user of ISO 5167-2:2003 had the same relative pipe roughness and the same relative edge sharpness as those on which the fitted equation is based the uncertainty could be considerably reduced, but in fact that relative pipe roughness and that relative edge sharpness are not known exactly. Using the Reader-Harris/Gallagher (1998) Equation with the calculated relative pipe roughness shown in Table 2.1 and a relative edge sharpness based on an approximate relative edge sharpness formula (which is unstated here but is a function of Reynolds number, reducing as Reynolds number increases: see Fig. 5.B.2 for the appropriate EEC data on edge radius for conversion to relative edge radius) $2s$ was approximately obtained by subtracting quadratically 0.16 and 0.06 % to give 0.32 %.

What is required as the uncertainty of the discharge-coefficient equation is the uncertainty of an orifice meter with the variability permitted by the standard (figures here exclude $\beta > 0.6$):

	$2s$ (%)
Base uncertainty for calculated pipe roughness and edge sharpness (see above)	0.32
Add pipe roughness (at most; see Sect. 2.3.3.1)	0.25
Add installation effects (see Chap. 8)[a]	0.22
Add edge sharpness (approximate value; see Sect. 2.2.4 and Fig. 5.B.2)	0.13
Add plate bending at manufacture (see Sect. 2.2.2)	0.10
Add plate bending due to differential pressure (see Sect. 2.2.2)	0.10
Add eccentricity (see Sect. 2.3.5)	0.08
Add steps (see Sect. 2.3.4 and assume 0.1 % for $\beta = 0.67$)	0.07
Add others	0.05
Total (added quadratically)	0.51

[a]A little higher than this for $\beta < 0.5$, but the effect of roughness reduces for $\beta < 0.5$ in such a way that the worst case for the sum of pipe roughness and installation effects is as given above. Alternatively, if a flow conditioner is used, in the resulting non-swirling flow the shift for $\beta \leq 0.6$ should be less than the shift at $\beta = 0.67$, where in the flow-conditioner test (Sect. 8.2.5.3) the shift must not exceed 0.23 %

NOTE It is assumed (reasonably) that the two types of bending, i.e. at manufacture and due to differential pressure, are independent of each other.

It is not generally desirable to round the uncertainty down, but on this occasion where some terms would in practice be omitted or reduced (e.g. there may be no steps, or the plate bending due to differential pressure is calculated to be much less than 0.1 %) rounding down to 0.5 % is reasonable.

The effects of roughness and installation increase the uncertainty for the highest β; the uncertainty increases for lower Reynolds number; moreover for $\beta > 0.5$ and $Re_D < 10{,}000$ the discharge coefficient is changing very rapidly with Reynolds number and the uncertainty increases yet more. In each case see Table 5.4. These effects are included in the statement of uncertainty in ISO 5167-2:2003; where $D < 71.12$ mm (2.8″) it is assumed that an additional uncertainty of half the additional term should be added.

Another way to determine the uncertainty of the equation is to examine how it performs when the discharge coefficients of orifice plates not used to determine the equation are compared with it. Just over 100 runs of calibration data using more than 50 orifice plates have been compared with it: for each set of data a mean deviation has been calculated. The following reports have been used to provide data: Morrow and Park (1992), George and Morrow (2001), Studzinski et al. (2001a, b), Morrow (2004, 2005), Reader-Harris (2002, 2003, 2008a, 2010). These sets of data include both water data and high-pressure gas data; so there is a useful range of Reynolds number. The mean deviations are shown in Fig. 5.12. From this figure the quoted uncertainty in ISO 5167-2:2003 appears reasonable.

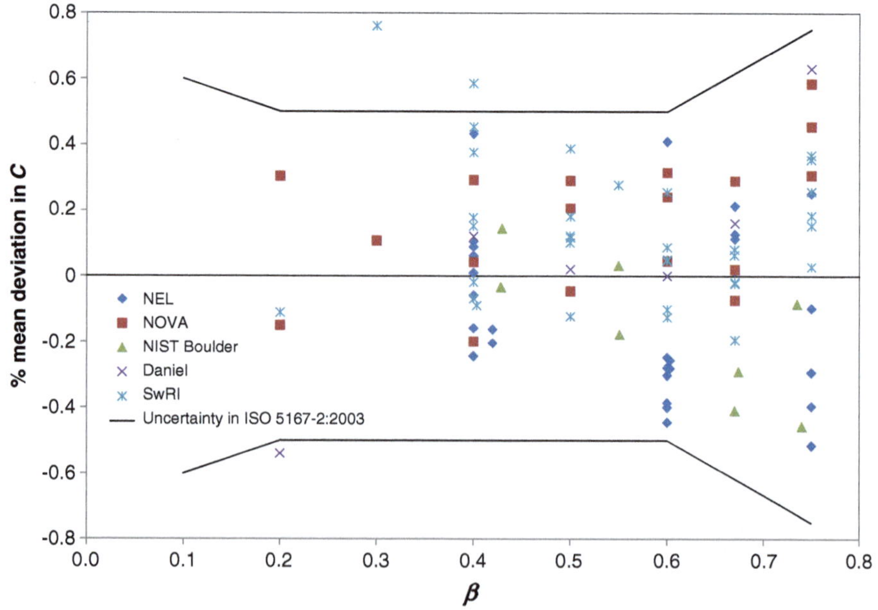

Fig. 5.12 Percentage mean deviation in C from Reader-Harris/Gallagher (1998) equation from more than 50 orifice plates

5.8 Conclusions

The derivation of the Reader-Harris/Gallagher (1998) Equation (5.22a, b), which is included in ISO 5167-2:2003, has been described. The quality of its performance, both absolutely and relative to the Reader-Harris/Gallagher (RG) Equation in API 14.3:1990 (Eqs. 5.23a, b) and the Stolz Equation in ISO 5167:1980 (Eq. 5.24), has been demonstrated.

The PR14 Equation has been described in Appendices 5.A–5.C: although it uses better tapping terms for low Reynolds number and a term for small orifice diameter rather than small pipe diameter, nevertheless for $Re_d > 10^5$, $d \geq 50$ mm, and $0.3 \leq \beta \leq 0.65$ it differs from the Reader-Harris/Gallagher (1998) Equation by less than 0.05 %. Changes in terms for low Reynolds number or small diameter have a very small effect on the equation in the range where it is most importantly used.

NOTE
Measurement of flows using orifice plates at Reynolds numbers below those covered by standards is possible, and data can be obtained particularly from Iversen (1956): he includes lists of experimenters and references and graphs of discharge coefficients using corner tappings from $Re_D = 4$ to 10,000. Johansen (1930) covers discharge coefficients at low Reynolds numbers and gives the pressure profile in the neighbourhood of the orifice for $2000 < Re_d < 20{,}000$. The maximum value of the discharge coefficient at a throat Reynolds number of the order of 500 is considered by Engel (1962). Orifice assemblies with rotatable multiple-orifice plates are considered by Voss (1974): he worked for General Motors and covers pipes of size 0.65″, 1.01″ and 1.75″ and orifices of diameter 3–31 mm.

Appendix 5.A: Better Options for Tapping Terms

When data were taken by NEL in oil in 2″ pipe in 1990 for inclusion in the European database not only discharge coefficients but also direct measurements of pressure profile were made: the pressure rise to the upstream corner from tappings at distances D, $D/2$, $D/4$ and $D/8$ upstream was measured, where D is the pipe diameter, as well as the pressure drop from the downstream corner to tappings at distances $D/8$ and $D/4$ downstream of the downstream face of the orifice plate and to the downstream $D/2$ tapping. Whereas for high Re_d (greater than approximately 80,000) the tapping terms do not depend on Re_d, the tapping terms at the Reynolds numbers obtained in oil are significantly different.

Moreover, the data collected on tapping terms by NEL in 2″ pipe (NEL 1991) both show that the reality is more complex than the analysis in Sect. 5.4.2.3 and also provide revised tapping terms which correspond much better to the tapping term collapse found in the database as a whole than the Reader-Harris/Gallagher (RG) (API) tapping terms did (Reader-Harris et al. 1992a). Figure 5.A.1 shows all

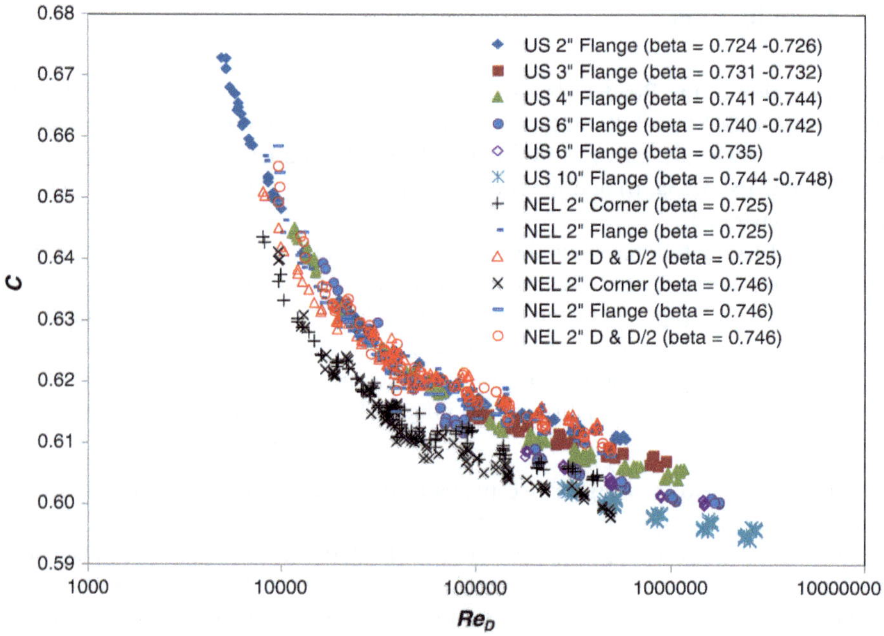

Fig. 5.A.1 American (CEESI and NIST) data (all with flange tappings) and NEL 2″ data for $\beta \approx 0.735$

the American discharge-coefficient data for $\beta \approx 0.735$ together with the NEL 2″ data. Because the tapping terms have a strong dependence on β, where there is a wide range of β the data have been divided.

It can be seen in this figure that most of the data collapse on to one another as Re_D decreases. However, the corner-tapping data are not collapsing on to the other data, or at least only to a very limited extent. The collapse of the flange tapping data on to one another can be seen more clearly in Fig. 5.7, which shows the American data for $\beta \approx 0.735$: the fact that the tapping terms are approximately constant for high Re_d can also be seen. Both Figs. 5.A.1 and 5.7 confirm the need for tapping terms which are functions of Reynolds number, but also show that the simple dependence on Re_d used in the Reader-Harris/Gallagher (1998) Equation is insufficient.

The main features of the tapping term data collected in 2″ pipe (in Figs. 83–100 of NEL (1991)) are as follows: the upstream tapping term for *D* and for *D*/2 tappings decreases with decreasing Re_d as expected from the work of Johansen and of Witte and Schröder, although the NEL data decrease a little more slowly with Re_d; the upstream tapping term for *D*/4 tappings remains approximately constant; the upstream tapping term for *D*/8 tappings increases with decreasing Re_d. The dependence of the downstream tapping terms on Re_d depends on β: for $\beta > 0.7$ they decrease in magnitude with decreasing Re_d; otherwise they are constant. At the bottom of the Re_d range the uncertainties in the data become large, especially for the

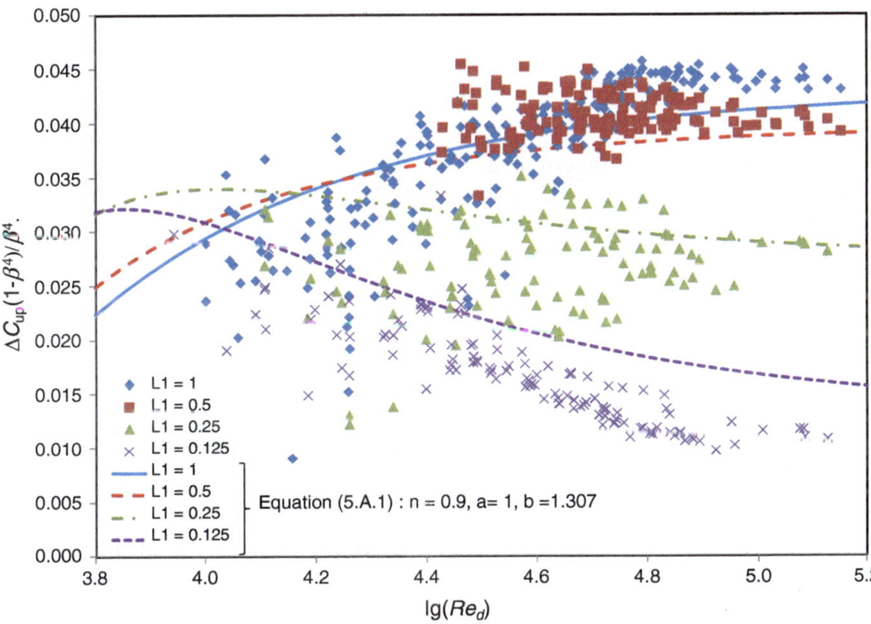

Fig. 5.A.2 Upstream tapping term for low Re_d (following ISO 80000-2:2009, $\log_{10}$ is written lg)

upstream $D/2$ tapping data because only one differential-pressure transmitter, rather than two covering different ranges, was used for this measurement.

It is important to see the pattern in the tapping term data: to do this it is necessary to do an analysis of the uncertainty of these data. It is then possible to analyse all the upstream tapping term data simultaneously, and, in particular, to verify that the dependence on $\beta^4/(1-\beta^4)$ which characterizes the data for high Re_d continues to apply for low Re_d. Figure 5.A.2 shows the change in discharge coefficient due to moving the upstream pressure tapping from the upstream corner for the four values of L_1 for which measurements were made. Where the data are multiplied by $(1-\beta^4)/\beta^4$ they fall on to a single curve for each value of L_1. Data are only plotted if $(1-\beta^4)u/(\beta^4\Delta p_c) < 0.04$, where u is the uncertainty of the pressure measurement between upstream pressure tappings at that point and Δp_c is the differential pressure across the orifice plate using corner tappings.

Various possible forms of the upstream tapping term, ΔC_{up}, for low Re_d data were tried, and the best one was found to be the following:

$$\Delta C_{up} = (0.043 + (0.090 - aA')e^{-10L_1} - (0.133 - aA')e^{-7L_1})(1 - bA')\frac{\beta^4}{1-\beta^4} \tag{5.A.1}$$

where

$$A' = \left(\frac{2100\beta}{Re_D}\right)^n$$

and a, b, and n are to be determined.

This equation is similar to Eq. 5.9a at high Reynolds number, but has a different shape for lower Reynolds number. The best fit of the form of Eq. 5.A.1 to the upstream tapping term data included in Fig. 5.A.2 was obtained, making appropriate allowance for the fact that the small product term is omitted from the final formula for the tapping terms. Allowance was also made for the fact that especially for $L_1 = 0.125$ the measured tapping terms (corrected for the product term) even for $Re_d \approx 100{,}000$ are not equal to the high Re_d values of Eq. 5.A.1; the fitted equation was therefore calculated based on data points shifted so that for each value of L_1 the mean value of the data for $Re_d > 80{,}000$ (corrected for the product term) agrees with the high Re_d version of Eq. 5.A.1, because the high Reynolds number version of Eq. 5.A.1 was fitted to many sets of data.

The best fit value for n was 0.925, but for simplicity this was rounded to 0.9, and with $n = 0.9$ the other constants were

$$a = 0.833 \quad \text{and } b = 1.307.$$

However, these values were adjusted to give a better fit to the database: the best fit to the complete database gave a larger value of a than the fit to the upstream tapping term data: a compromise value was obtained as follows: from the Figures in NEL (1991) it appears that the data for $L_1 = 1$, those for $L_1 = 0.25$ and those for $L_1 = 0.125$ meet at $Re_d \approx 13{,}000$. Since in Eq. 5.A.1 the three curves representing the three values of L_1 do not intersect at a single point, it was decided that the intersection of the curve for $L_1 = 0.167$ (corresponding to flange tappings in 6″ pipe) with the curve for $L_1 = 1$ should occur at $Re_d = 13{,}000$: this occurs for $a = 1.03$. This constant is then rounded to 1. Equation 5.A.1 with $a = 1$, $b = 1.307$, and $n = 0.9$ is then plotted in Fig. 5.A.2 for comparison with the data. This equation describes a change in the pressure profile upstream of the orifice in which, as Re_d decreases, the upstream tapping term at D decreases but the gradient of the tapping term near the corner increases.

It is unnecessarily complicated to construct a downstream tapping term which decreases in magnitude with decreasing Re_d for very large β but is constant for smaller β to fit the data in NEL (1991); since the upstream term is significant for large β, but very small for small β, this downstream Re_d effect is incorporated in the upstream term by reducing b from 1.307 to 1. The optimum upstream tapping term (incorporating a downstream effect) (for use in the PR14 equation) is therefore

$$\Delta C_{\text{up}} = (0.043 + (0.090 - A')e^{-10L_1} - (0.133 - A')e^{-7L_1})(1 - A')\frac{\beta^4}{1 - \beta^4} \qquad (5.A.2)$$

where

$$A' = \left(\frac{2100\beta}{Re_D}\right)^{0.9}.$$

With this upstream formula no change in the downstream formula from that in Eq. 5.6 is required for $Re_D > 4000$. However, from examination of the data for $Re_D < 4000$ in Sattary et al. (1992) it can be seen that over the range of data in the database for $Re_D < 4000$ the discharge coefficient using corner tappings becomes increasingly larger than that using flange or D and $D/2$ tappings as Re_D decreases; since this applies even for small β this can best be represented by the downstream tapping term being modified, although both upstream and downstream tapping terms change with Re_D. The model used was as follows:

$$\Delta C_{\mathrm{down}} = -0.031\,(M_2' - 0.8M_2'^{1.1})\left\{1 + c\max\left(\lg\left(\frac{Re_{D,\mathrm{T}}}{Re_D}\right),\ 0\right)\right\}\beta^{1.3} \quad (5.A.3)$$

where c is a constant and $Re_{D,\mathrm{T}}$ is the pipe Reynolds number at which transition to fully turbulent flow occurs. $Re_{D,\mathrm{T}}$ varies, as would be expected, from one set of data to another, but a reasonable estimate of the range of values encountered in the database is 3000–5000, and $Re_{D,\mathrm{T}} = 3700$ has been used for both the tapping term and the slope term. With this value for $Re_{D,\mathrm{T}}$ c is determined by fitting the data in Sattary et al. (1992): using the difference between flange and corner tappings only, $c = 8.20$; using the difference between D and $D/2$ and corner tappings only, $c = 7.88$; using all the data, $c = 8.04$. The agreement between the values of c obtained using flange and D and $D/2$ tappings is very good, and the optimum downstream tapping term (for use in the PR14 Equation) is as follows:

$$\Delta C_{\mathrm{down}} = -0.031\,(M_2' - 0.8M_2'^{1.1})\left\{1 + 8\max\left(\lg\left(\frac{3700}{Re_D}\right),\ 0\right)\right\}\beta^{1.3} \quad (5.A.4)$$

Appendix 5.B: Small Orifice Diameters Within the EEC/API Database

As a result of collecting the NEL 2″ data, which included measurements of edge sharpness, it was appropriate to add an additional term for small orifice diameters to the equation accepted at New Orleans. The problem is that it is extremely difficult to obtain a sufficiently sharp edge where the orifice diameter is small: Fig. 5.B.1, which gives measured edge radii (r) from the plates used in the European tests, in which the pipe size was in the range 2″–24″, shows that for orifice diameter, d, less than about 50 mm the plates rarely met the requirements of ISO 5167-1:1991 (ISO 1991). For pipe sizes other than 2″ the edge radii shown are averages of

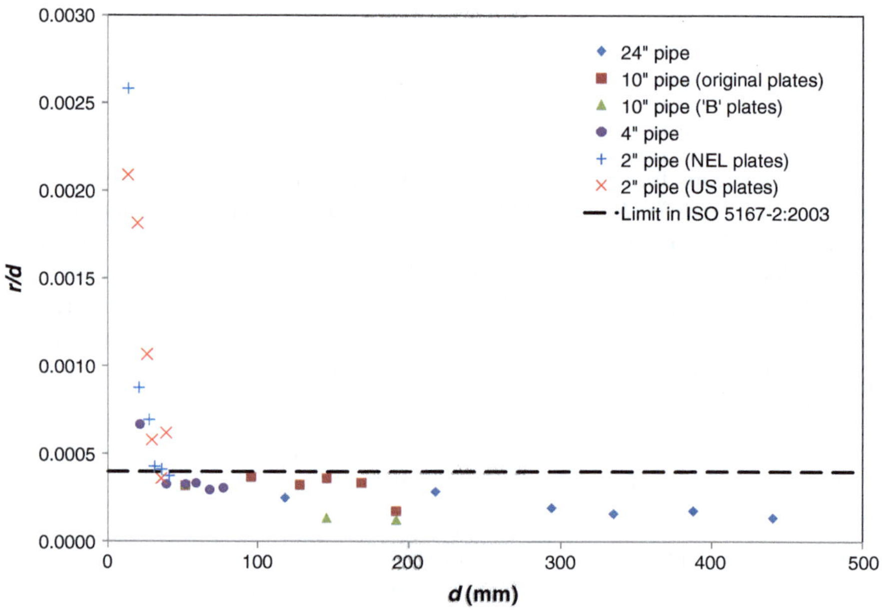

Fig. 5.B.1 r/d as a function of d (for the orifice plates used in the EEC tests)

measurements by several European laboratories; for 2″ pipes the measurements are those of NEL alone since NEL was the only European laboratory at which 2″ tests were performed. It is clear that for $d < 25$ mm large shifts in C were to be expected. When the edge radii themselves are plotted as in Fig. 5.B.2, it appears that the edge radius, r, increased as d decreased from about 50 mm, whereas to meet the standard it needed to decrease fairly rapidly.

The change in discharge coefficient due to edge roundness, ΔC_{round}, was measured by Hobbs (1989) and seen to be a function of change in edge radius, Δr, that can be expressed approximately (see Sect. 2.2.4) as

$$\Delta C_{\text{round}} = 3.33\Delta r/d. \tag{5.B.1}$$

It was assumed that the mean value of r, $\bar{r}$, for $d < d_1$, where $d_1 \approx 50$ mm, was given by

$$\bar{r} = 0.0002d_1 + B(d_1 - d) \tag{5.B.2}$$

where B is a constant, since this is linear with d and gives $\bar{r}/d$ equal to 0.0002 where $d = d_1$. Given that the discharge-coefficient equation for large d is based on $\bar{r}/d$ being approximately equal to 0.0002, the additional term for $d < d_1$ was calculated to be

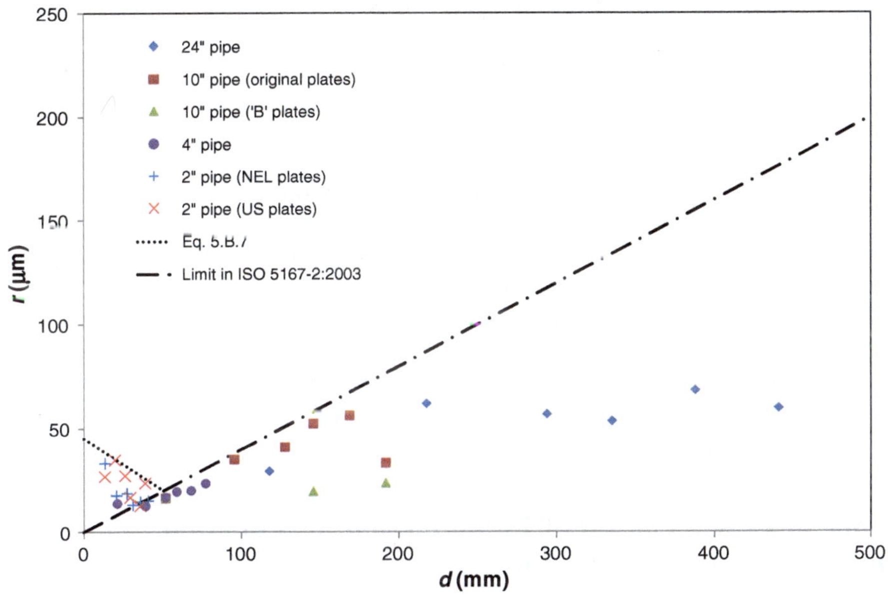

Fig. 5.B.2 Edge radius as a function of d (for the orifice plates used in the EEC tests)

$$\Delta C_{\text{round}} = 3.33\left(\frac{\bar{r}}{d} - 0.0002\right), \tag{5.B.3}$$

which on substituting from Eq. 5.B.2 becomes

$$\Delta C_{\text{round}} = 3.33(B + 0.0002)\left(\frac{d_1}{d} - 1\right). \tag{5.B.4}$$

When ΔC_{round} was determined by fitting the database the minimum standard deviation was obtained for $d_1 = 44$ mm, but over the range 40 mm $< d_1 <$ 50 mm the overall standard deviation was within 0.0002 % of the minimum. It was therefore reasonable to approximate d_1 by 50 mm since to round d_1 to 40 mm would mean that one plate would still be too rounded (i.e. it would lie outside the limit corresponding to Eq. 5.B.7). With $d_1 = 50$ mm a good approximation to the small orifice diameter term was

$$\Delta C = 0.0015\max\left(\frac{50}{d} - 1,\ 0\right), \tag{5.B.5}$$

which corresponds to $B = 0.00025$ and to

$$\bar{r} = \max(0.0225 - 0.00025d, 0.0002d). \quad (5.B.6)$$

The maximum value of r, r_{max}, is equal to $2\bar{r}$, i.e.

$$r_{max} = \max(0.045 - 0.0005d,\ 0.0004d), \quad (5.B.7)$$

and from Fig. 5.B.2 it can be seen that all the plates lie within this limit. Clearly this term gives rise to an increase in uncertainty for $d < 50$ mm.

Twenty years after this work, plates with an edge radius of around 9 μm were being made by one manufacturer using spark erosion (see Fig. 2.A.3). If such were normal practice the small orifice diameter term for the orifice equation would be smaller than the one in this Appendix.

Appendix 5.C: The PR14 Equation and an Equation in Terms of Friction Factor

5.C.1 The PR14 Equation

Given the optimum tapping terms in Eqs. 5.A.2 and 5.A.4 and the small orifice diameter term in Eq. 5.B.5, the coefficients in the C_∞ and slope terms were obtained to fit the EEC/API database in Sect. 5.3. The forms of the C_∞ and slope terms were as in Eqs. 5.11 and 5.17. Previously m_1 had been taken to be equal to 2, but a better fit was obtained with a smaller value, and following a suggestion of Stolz (1991) the same value as the exponent of β in the downstream tapping term (Eq. 5.A.4) was used. The optimum value of m_1 in terms of the lowest standard deviation of the data about the equation lay between 1.2 and 1.3. The mean Re_D at which the flow becomes fully turbulent was taken to be 3700. This gave c_1 in terms of c_2. c_2 was obtained by trying appropriate values in turn and obtaining the best overall fit: $c_2 = 4800$ gave an excellent overall fit. With m_2 and n_1 as in earlier work a least-squares fit of the complete database was performed: on rounding the constants, the C_∞ and slope terms became

$$\begin{aligned} C_\infty + C_s = {} & 0.5934 + 0.0232\beta^{1.3} - 0.201\beta^8 + 0.000515\ (10^6\beta/Re_D)^{0.7} \\ & + (0.0187 + 0.0400A')\beta^{3.5}\max\{(10^6/Re_D)^{0.3},\ 23.1 - 4800\ (Re_D/10^6)\}. \end{aligned} \quad (5.C.1)$$

The complete PR14 Equation can be brought together from Eqs. 5.A.2, 5.A.4, 5.B.5 and 5.C.1 to give

$$
\begin{aligned}
C = {} & 0.5934 + 0.0232\beta^{1.3} - 0.201\beta^8 + 0.000515\,(10^6\beta/Re_D)^{0.7} \\
& + (0.0187 + 0.0400A')\beta^{3.5}\max\{(10^6/Re_D)^{0.3},\ 23.1 - 4800\,(Re_D/10^6)\} \\
& + (0.043 + (0.090 - A')\mathrm{e}^{-10L_1} - (0.133 - A')\mathrm{e}^{-7L_1})(1 - A')\frac{\beta^4}{1-\beta^4} \\
& - 0.031\,(M_2' - 0.8M_2'^{1.1})\left\{1 + 8\max\left(\lg\left(\frac{3700}{Re_D}\right),\ 0.0\right)\right\}\beta^{1.3} \\
& + 0.0015\max\left(\frac{50}{d} - 1,\ 0\right)
\end{aligned}
\tag{5.C.2}
$$

where d is in mm,

$$A' = \left(\frac{2100\beta}{Re_D}\right)^{0.9} \text{ and } M_2' = \frac{2L_2'}{1-\beta}.$$

Although the PR14 Equation has tapping terms that fit the pressure-profile data for low Reynolds number better than those in the Reader-Harris/Gallagher (1998) Equation and a term for small orifice diameter rather than small pipe diameter, nevertheless for $Re_d > 10^5$, $d \geq 50$ mm, and $0.3 \leq \beta \leq 0.65$ it differs from the Reader-Harris/Gallagher (1998) Equation by less than 0.05 %. Changes in terms for low Reynolds number or small diameter have a very small effect on the equation in the range where it is most importantly used. Using $\beta^{1.3}$ in C_∞ rather than β^2 makes a difference for small β but it is difficult to choose the best exponent because most data for very small β also have small d and thus possible edge rounding.

5.C.2 An Equation in Terms of Friction Factor

Using the tapping terms in Eqs. 5.A.2 and 5.A.4 and the small orifice diameter term in Eq. 5.B.5 but including the effect of friction factor as in Eq. 5.15 the database was refitted with the C_∞ and slope terms of the following form:

$$
\begin{aligned}
C_\infty + C_s = {} & a_1 + a_2\beta^{1.3} + a_3\beta^8 + b_1(10^6\beta/Re_D)^{0.7} \\
& + (b_2 + b_3A')\beta^{3.5}\max\{\lambda,\ c_1 - c_2(Re_D/10^6)\}.
\end{aligned}
\tag{5.C.3}
$$

The mean Re_D at which the flow becomes fully turbulent was taken to be 3700. This gave c_1 in terms of c_2. c_2 was obtained by trying appropriate values in turn and obtaining the best overall fit: $c_2 = 4800$ gave an excellent overall fit. The C_∞ and slope terms were

$$C_\infty + C_s = 0.5945 + 0.0157\beta^{1.3} - 0.2417\beta^8 + 0.000514\,(10^6\beta/Re_D)^{0.7} + (3.134 + 4.726A')\beta^{3.5}\max\{\lambda,\ 0.1704 - 35(Re_D/10^6)\}. \tag{5.C.4}$$

This equation is theoretically desirable and indeed is used to determine roughness limits in Sect. 2.3.3.1. However, it would be difficult to use commercially.

Appendix 5.D: The Effect on the Discharge-Coefficient Equation of Changing the Expansibility-Factor Equation

In the database (Sect. 5.3) the discharge coefficients were calculated using the expansibility-factor equation in ISO 5167-1:1991 (ISO 1991); when this was changed to Eq. 6.13 the discharge coefficients should have been changed, and there would have been an effect on the calculated discharge-coefficient equation. This did not happen; however, it is shown in this appendix that to have recalculated the discharge coefficients would have had an insignificant effect on the discharge-coefficient equation.

In order to prove that recalculating the discharge coefficients would have had an insignificant effect on the discharge-coefficient equation, for each point of the database the value of discharge coefficient which would have been obtained if an alternative equation for expansibility factor had been used was calculated and the equation refitted. This work was done before the actual Eq. 6.13 was determined, but, using other proposed expansibility-factor equations, showed that the discharge-coefficient equation was insensitive to the expansibility-factor equation. If the value of discharge coefficient given in the database is termed C_I, based on the expansibility factor given by the equation in 8.3.2.2 of ISO 5167-1:1991, ε_I, then C_N, the value of discharge coefficient based on an alternative value of expansibility factor, ε_N, is given by

$$C_N\varepsilon_N = C_I\varepsilon_I, \tag{5.D.1}$$

where

$$\varepsilon_I = 1 - (0.41 + 0.35\beta^4)\frac{\Delta p}{\kappa p_1}. \tag{6.1}$$

Where the value of ε_I is given in the discharge-coefficient database it was possible to calculate the isentropic exponent κ using the values of β, Δp and p_1 and then to calculate ε_N. Where ε_I was not given in the database it was necessary, in the first instance, to estimate what value of κ might have been used on the basis of other data; if it had been shown to be the case that the discharge-coefficient equation fitted to the values of C_N differed significantly from that fitted to the values of C_I it would have been necessary to obtain better values for κ.

The only sets of gas data for which ε_I was not provided in the discharge-coefficient equation database were those from SwRI and Ruhrgas. For SwRI the downstream expansion factor, Y_2, (see API 1990) was provided but not ε_I. So for SwRI and Ruhrgas values for κ of 1.41 and 1.32, respectively, were used: 1.41 is appropriate for nitrogen; 1.32 is a typical value for natural gas.

Three alternative equations for ε_N were used: they were as follows:

$$\varepsilon_{N,1} = 1 - (0.35 + 0.38\beta^4)\frac{\Delta p}{\kappa p_1}, \tag{5.D.2}$$

$$\varepsilon_{N,2} = 1 - (0.352 + 0.433\beta^4)\frac{\Delta p}{\kappa p_1}, \tag{5.D.3}$$

and

$$\varepsilon_{N,3} = 1 - (0.357 + 0.557\beta^4)\frac{\Delta p}{\kappa p_1}. \tag{5.D.4}$$

$\varepsilon_{N,1}$ and $\varepsilon_{N,2}$ were taken from Eqs. 9 (rounded as in the conclusions of the paper) and 10 of Kinghorn (1986) and $\varepsilon_{N,3}$ was taken from Eq. 8 (the recommended equation) of Seidl (1995).

The absolute value of the coefficient of $\frac{\Delta p}{\kappa p_1}$ for the equation in ISO 5167-1:1991 (Eq. 6.1) and for the alternative equation numbers 1, 2 and 3 (Eqs. 5.D.2, 5.D.3 and 5.D.4 respectively) and the absolute value of the coefficient of $1 - \left(\frac{p_2}{p_1}\right)^{1/\kappa}$ for the equation in ISO 5167-2:2003 (Eq. 6.13) are shown in Fig. 5.D.1.

Calculating $C_{N,i}$ on the basis of $\varepsilon_{N,i}$ for $i = 1,3$ and, using the tapping terms given in Eqs. 5.10a, b, the sum of the other terms (the C_∞, C_s and small pipe diameter, ΔC_D, terms), C_Σ, was refitted, assuming it was of the following form:

$$\begin{aligned} C_\Sigma = C_\infty + C_s + \Delta C_D = a_1 + a_2\beta^2 + a_3\beta^8 + b_1(10^6\beta/Re_D)^{0.7} \\ + (b_2 + b_3 A)\beta^{3.5}\max\{(10^6/Re_D)^{0.3},\ 22.7 - 4700\,(Re_D/10^6)\} \\ + h(0.75 - \beta)\max(2.8 - D/25.4, 0.0). \end{aligned} \tag{5.D.5}$$

In Eqs. 5.D.5–5.D.9 D is in mm.

In each case the standard deviation of the data in the database about the equation, s, the number of points which are shifted by more than 0.2 %, N_s, and the largest magnitude of shift for a point in the database, S_M, were calculated. The results were as follows:

$$\begin{aligned} C_{\Sigma,N,1} = 0.59590 + 0.02638\,\beta^2 - 0.21794\,\beta^8 + 0.0005288\,(10^6\beta/Re_D)^{0.7} \\ + (0.01904 + 0.005864\,A)\beta^{3.5}\max\{(10^6/Re_D)^{0.3},\ 22.7 - 4700\,(Re_D/10^6)\} \\ + 0.01135\,(0.75 - \beta)\max(2.8 - D/25.4, 0.0). \\ s_{N,1} = 0.0016775\,;\quad N_{s,1} = 191\,;\quad S_{M,1} = 0.98\%. \end{aligned} \tag{5.D.6}$$

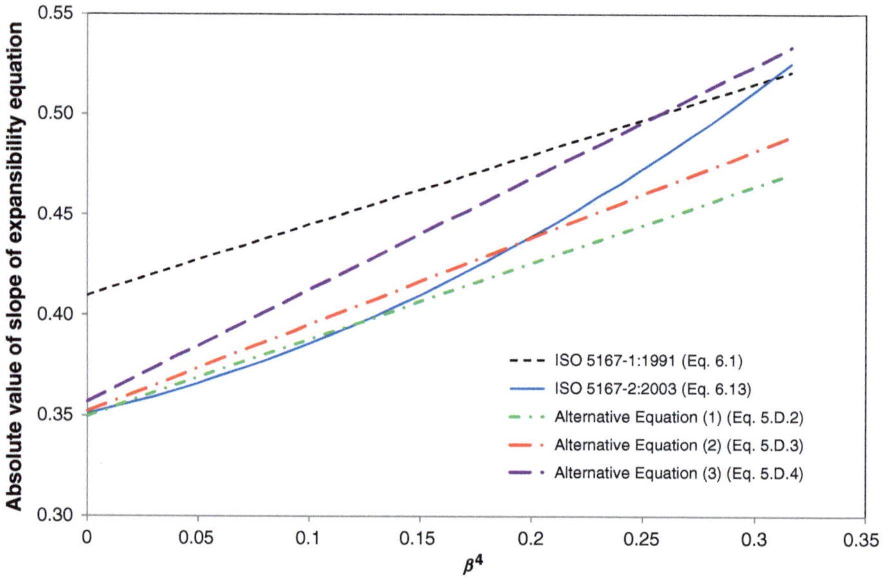

Fig. 5.D.1 Absolute value of slope of expansibility-factor equation for different equations

$$
\begin{aligned}
C_{\Sigma,N,2} &= 0.59591 + 0.02645\,\beta^2 - 0.21778\,\beta^8 + 0.0005286\,(10^6\beta/Re_D)^{0.7} \\
&\quad + (0.01895 + 0.005894\,A)\beta^{3.5}\max\{(10^6/Re_D)^{0.3},\ 22.7 - 4700\,(Re_D/10^6)\} \\
&\quad + 0.01133\,(0.75 - \beta)\max(2.8 - D/25.4, 0.0). \\
s_{N,2} &= 0.0016768\,;\quad N_{s,2} = 177\,;\quad S_{M,2} = 0.85\%.
\end{aligned}
\tag{5.D.7}
$$

$$
\begin{aligned}
C_{\Sigma,N,3} &= 0.59592 + 0.02662\,\beta^2 - 0.21740\,\beta^8 + 0.0005281\,(10^6\beta/Re_D)^{0.7} \\
&\quad + (0.01876 + 0.005965\,A)\beta^{3.5}\max\{(10^6/Re_D)^{0.3},\ 22.7 - 4700\,(Re_D/10^6)\} \\
&\quad + 0.01129\,(0.75 - \beta)\max(2.8 - D/25.4, 0.0). \\
s_{N,3} &= 0.0016784\,;\quad N_{s,3} = 139\,;\quad S_{M,3} = 0.76\%.
\end{aligned}
\tag{5.D.8}
$$

Since small differences between equations were being investigated the constants for C_Σ are also required where ε_I was used:

$$
\begin{aligned}
C_{\Sigma,I} &= 0.59615 + 0.02609\,\beta^2 - 0.21675\,\beta^8 + 0.0005216\,(10^6\beta/Re_D)^{0.7} \\
&\quad + (0.01874 + 0.006071\,A)\beta^{3.5}\max\{(10^6/Re_D)^{0.3},\ 22.7 - 4700\,(Re_D/10^6)\} \\
&\quad + 0.01101\,(0.75 - \beta)\max(2.8 - D/25.4, 0.0). \\
s_I &= 0.0016747.
\end{aligned}
\tag{5.D.9}
$$

The constants have been rounded and then rebalanced to ensure that there is no mean deviation between the equation and the database.

It can be seen that the differences in s and thus in overall quality of fit are very small. Moreover the coefficients in Eqs. 5.D.6–5.D.9 are very similar. The largest value of S_M for Eqs. 5.D.6–5.D.8 occurs for Eq. 5.D.6, as expected from Fig. 5.D.1; however, even in this case the largest magnitude of difference between the discharge-coefficient equations (i.e. between $C_{\Sigma,N,1}$ and $C_{\Sigma,I}$) is 0.04 % for any values of β, D and Re_D except at the very lowest end of the Reynolds number range, below 4000. Therefore the choice of the expansibility-factor equation has very little effect on the discharge-coefficient equation. Equation 6.13 itself generally lies between the ISO 5167 1:1991 Equation and Eq. 5.D.2, and the difference between $\frac{\Delta p}{\kappa p_1}$ and $1 - \left(\frac{p_2}{p_1}\right)^{1/\kappa}$ is small for the points in the database, most of which were collected at high pressure. Moreover, whereas changing the expansibility-factor equation would reduce the discharge coefficient in the database, at the same time the isenthalpic temperature correction (see Sect. 4.6.2) was introduced, which, if it had been applied to all the points in the database, would have increased the discharge coefficient in the database. Both effects will be larger for small β than for large β. However, although the discharge-coefficient equation is very little affected by the choice of expansibility-factor equation, the choice of expansibility-factor equation has a significant effect on some data points in the field, and it is important that the best equation is used.

Appendix 5.E: Orifice Plates in Pipes of Small Diameter or with No Upstream or with No Downstream Pipeline or with No Upstream and No Downstream Pipeline

5.E.1 Introduction

This area is covered by ISO/TR 15377:2007, and this appendix gives a justification for the appropriate sections in ISO/TR 15377 (see also Reader-Harris et al. 2006). Practical applications of Sects. 5.E.3 to 5.E.5 would be flow out of and/or into a reservoir or tank.

5.E.2 Orifice Plates in Pipes of Small Diameter

ISO/TR 15377:1998 permitted pipe diameter $D \geq 25$ mm and orifice diameter $d \geq 5.75$ mm and gave an additional uncertainty on the discharge coefficient of 1 %.

However, when the data collected in the EEC/API Orifice Project in the 1980s (see Sect. 5.3) are analysed the uncertainty increases rapidly when $d < 12.5$ mm (only API had data collected for $d < 12.5$ mm or diameter ratio $\beta < 0.2$). For the water data (Whetstone et al. 1989) for $d \approx 9$ mm the deviations from the Reader-Harris/Gallagher (1998) Equation can be as large as 1.5 %; for $d \approx 6$ mm they can be as large as around 3 %. The deviations are a very weak function of Reynolds number. It appears likely that in most cases the shifts are due to edge sharpness, but that explanation would not suit one set (4″ pipe, $\beta = 0.06$, run 2). Therefore in ISO/TR 15377:2007 it states that the uncertainty increases significantly if $d < 12.5$ mm.

The OSU data for orifice plates (Beitler 1935) in 1″ and 1.5″ pipes were compared with the Reader-Harris/Gallagher (1998) Equation as they form an excellent set of data for $D < 50$ mm. Provided that the uncertainty derived from 5.3.3.1 of ISO 5167-2:2003 is increased by 0.33, 95 % of those points for which $d > 12.5$ mm and $\beta < 0.762$ lie within the uncertainty. In ISO/TR 15377:2007 the increase in uncertainty was rounded up to 0.5 %.

Only corner tappings are permitted: flange tappings would give $M'_2 > 4$ (i.e. a tapping in the pressure recovery zone: see Fig. 5.6) for sufficiently high β. Although flange tappings could be used for sufficiently small β it is generally wise to design a system so that β may be easily changed if the flow is higher or lower than the design flow.

5.E.3 Orifice Plates with No Upstream or Downstream Pipeline

ISO/TR 15377 permits orifice plates without either an upstream pipeline or a downstream pipeline or both. Data from Sect. 5.3 (Whetstone et al. 1989; Britton et al. 1988) with $d \geq 12.5$ mm and $\beta = 0.1$ are plotted in Fig. 5.E.1.

All the oil data (Britton et al. 1988) for $d \geq 9$ mm and $\beta \leq 0.22$ are plotted in Fig. 5.E.2. On the basis of Figs. 5.E.1 and 5.E.2 it appears reasonable to suppose that the Reader-Harris/Gallagher (1998) Equation will continue to perform well as $\beta \rightarrow 0$.

NOTE In Appendix 2.A it is shown that the Reader-Harris/Gallagher (1998) Equation continues to perform well for β down to 0.016 provided that allowance is made for the roundness of the orifice plates for very small d.

It is also necessary to check over what range of β a lower throat Reynolds number limit of 3500 (chosen for ISO/TR 15377:2007) might be appropriate. From Eq. 5.16 the discharge coefficient C is the sum of two Reynolds number terms, the throat term, which can be seen from Figs. 5.E.1 and 5.E.2 to be applicable down to at least $Re_d = 3500$, and the velocity profile term, which is applicable down to the beginning of transition to laminar flow, giving a limit of $Re_D \geq 5000$ in ISO 5167-2:2003. However, if β is sufficiently small, the velocity profile term is small.

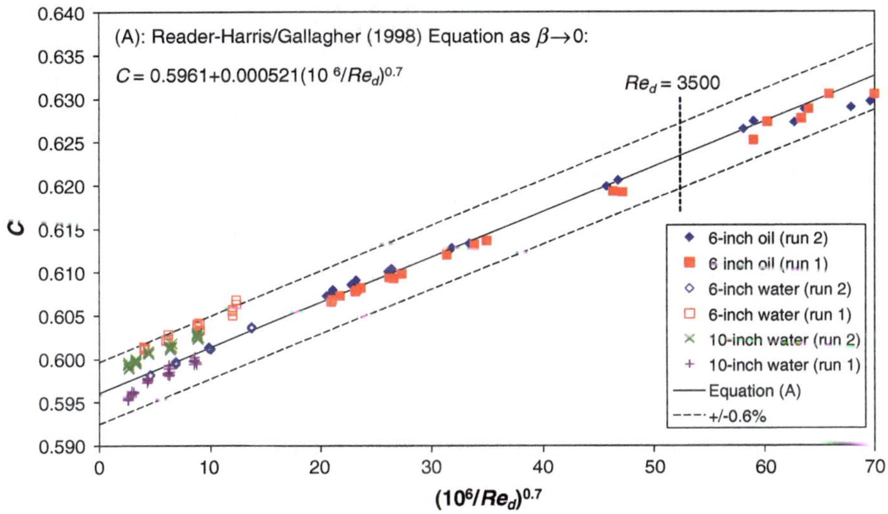

Fig. 5.E.1 Data (all API: NIST and CEESI) from EEC/API orifice project taken for $\beta = 0.1$ and $d > 12.5$ mm

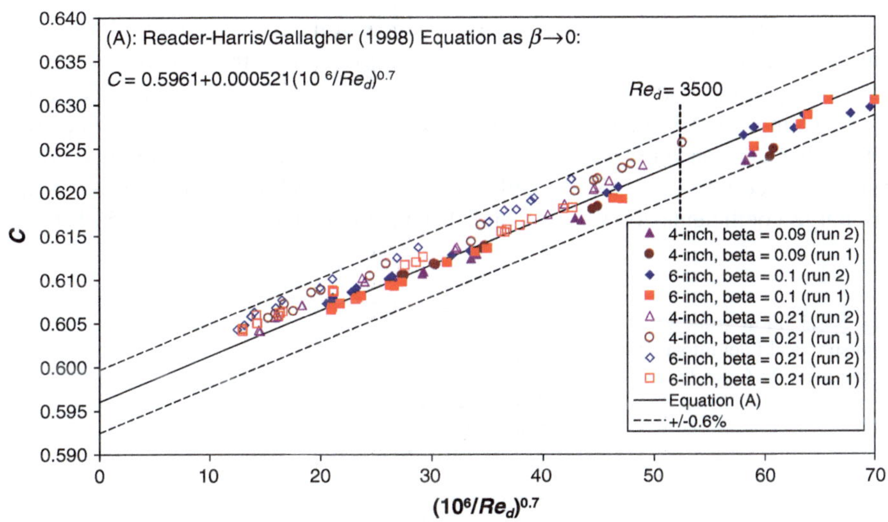

Fig. 5.E.2 Data (all CEESI) from EEC/API orifice project taken in oil for $\beta < 0.22$ and $d > 9$ mm

In Fig. 5.E.2 there appears to be no problem for $\beta = 0.21$. In the complete discharge-coefficient equation (5.21) fitted to the complete EEC/API database including points for $Re_D < 4000$ there are changes to the velocity-profile and to one other term for $Re_D < 3700$. From Eq. 5.21, for $\beta \leq 0.2$ when $Re_d \geq 3500$, the maximum error in C due to just using the equation in ISO 5167-2:2003 is 0.53 %. Provided that $\beta \leq 0.2$, $d \geq 12.5$ mm and $Re_d \geq 3500$, it seems reasonable to use the Reader-Harris/Gallagher (1998) Equation as in ISO 5167-2:2003 with an uncertainty of 1 % (for $Re_D < 5000$), as stated in NOTE 2 of 5.3.2.2.2 of ISO/TR 15377:2007. The simplified Equation (A) in both Figs. 5.E.1 and 5.E.2 can be used for $d \geq 12.5$ mm and $Re_d \geq 3500$ with an uncertainty of 1 % with a large upstream space as in 5.3.2.2.3 of ISO/TR 15377:2007.

5.E.4 Orifice Plates with No Upstream Pipeline

Section 5.E.3 showed the suitability of the Reader-Harris/Gallagher (1998) Equation for use where there is neither an upstream nor a downstream pipe. To investigate the suitability of this equation for the situation where there is no upstream pipe but there is a downstream pipe Computational Fluid Dynamics (CFD) was used. Solutions were obtained using the commercial code Fluent v6.1 with the realizable $k - \varepsilon$ turbulence model with $d = 100$ mm and $Re_d = 2.5 \times 10^5$. It was not easy to obtain converged solutions. With some grids converged solutions could only be obtained with first-order interpolation. However, with the same fine grid in the vicinity of the orifice second-order solutions were obtained for both Cases 1 and 2, shown in Figs. 5.E.3 and 5.E.4. 20 square cells were used along the orifice bore. The cells were concentrated around the orifice plate edges, expanding outwards to reduce cell-count. With 35,500 cells in Case 1 and 52,600 cells in Case 2, the discharge coefficients for Cases 1 and 2 with corner tappings were 0.59291 and

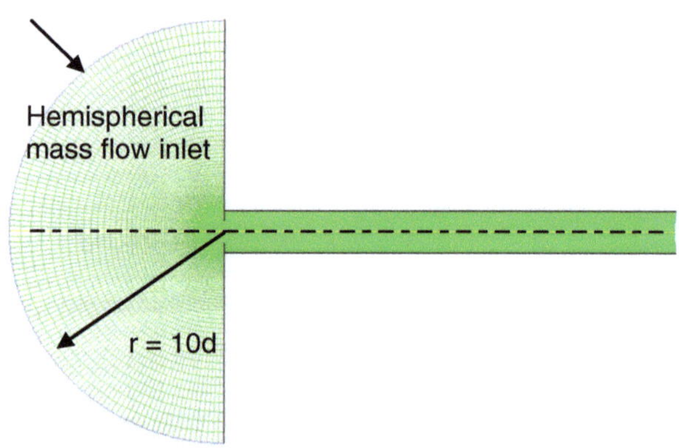

Fig. 5.E.3 Flow from a hemispherical inlet into a pipe of diameter 2d (case 1)

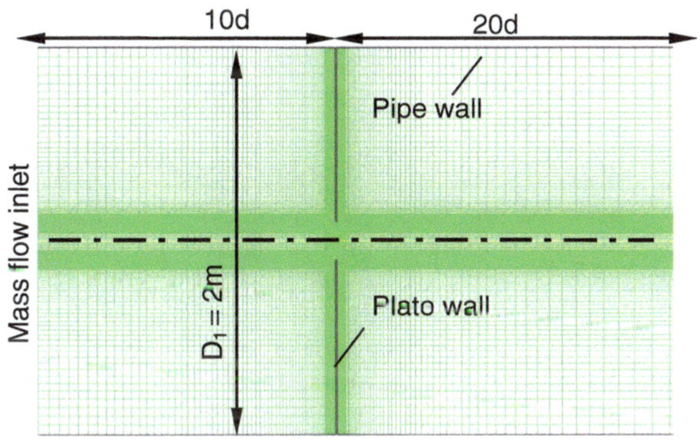

Fig. 5.E.4 Flow from large pipe ($\beta = 0.05$) into large pipe (case 2)

0.59324 respectively. So, if corner tappings are used, a flow from a large space into a downstream pipe of diameter 2*d* can be considered to have the same discharge coefficient as a flow from a large pipe into a large pipe.

So since the Equation works well where there is neither an upstream nor a downstream pipe, the CFD implies that it will work well for flow from a plenum into a downstream pipe of diameter 2*d*, and thus by obvious extension for flow into a pipe of diameter greater than 2*d*. This confirms 5.3.2.2.2 of ISO/TR 15377:2007.

5.E.5 Orifice Plates with No Downstream Pipeline

In the case of an orifice plate with no downstream pipeline, ISO/TR 15377:1998 implied that an orifice plate discharging into a large space from a pipe of finite bore will have the same discharge coefficient using corner tappings as an orifice plate installed with a downstream pipe of the same diameter as the upstream one. However, the permissible upstream diameter ratio was unclear. In 5.3.3.1 (simul) the upstream pipeline should be no greater than 2.5*d* (that is, $\beta \le 0.4$). In 5.3.3.1 (ter) it stated that $0.4 \le \beta \le 0.7$, which is very different from the previous statement.

A total of four simulations were run for this case: Case 3: pipe of diameter 2.5*d* upstream, large pipe (diameter 10*d*) downstream ($\beta = 0.4$ upstream); Case 4: pipe of diameter 2.5*d* upstream and downstream ($\beta = 0.4$ upstream); Case 5: pipe of diameter 1.33*d* upstream, large pipe (diameter 10*d*) downstream ($\beta = 0.75$ upstream); Case 6: pipe of diameter 1.33*d* upstream and downstream ($\beta = 0.75$ upstream), where $d = 100$ mm, as before. Cases 5 and 6 were run to see whether the diameter of the upstream pipe influences the solution.

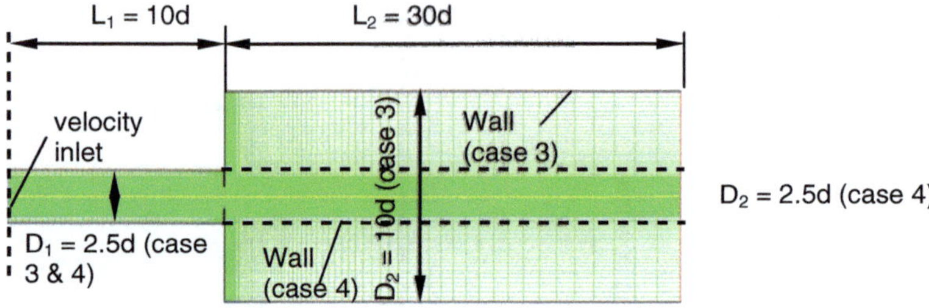

Fig. 5.E.5 Grids for a small upstream pipe issuing into a larger diameter pipe where $\beta = 0.4$ (case 3) and for a standard $\beta = 0.4$ orifice meter (case 4)

Table 5.E.1 Results for orifice with a large space or a pipe downstream

Case	Downstream	Cells	β	C_{corner}
3	Large space	22,400	0.4	0.59962
4	Pipe	17,000	0.4	0.59962
5	Large space	22,400	0.75	0.60106
6	Pipe	17,000	0.75	0.60123

As the inlet was always a pipe in these cases, instead of a mass flow inlet a velocity inlet boundary condition was used, in which a fully developed flow profile was superimposed on the inlet. The profiles of velocity magnitude and turbulence were produced by running a separate model with cyclic boundary conditions. Figure 5.E.5 shows the grid nomenclature for Cases 3 and 4, where $\beta = 0.4$; the dotted lines show the portion of the grid used to represent the case where the orifice was installed (as normal) with the same pipe diameter upstream and downstream of it. The grids for cases 5 and 6 were similar. As in Sect. 5.E.4 there were 20 square cells along the orifice bore and the cells expanded outwards from there. All of the solutions in this case converged using second order interpolation. Table 5.E.1 summarizes the results for C.

Provided that corner tappings are used the discharge coefficient is the same with a plenum downstream as it would have been with a pipe downstream of the same diameter as the upstream pipe. Since this is true for $\beta = 0.4$ and 0.75 (and trivially for $\beta = 0.1$) it is reasonable to suppose that it will be true for all β. On this basis the Reader-Harris/Gallagher (1998) Equation as in ISO 5167-2:2003 can be used as in ISO/TR 15377:2007.

Figure 5.E.6 shows the axial pressure profiles along the pipe wall, with $\beta = 0.4$. It is clear that there is no recovery of pressure when the jet issues from the orifice plate into a large diameter pipe, but that although the pressure in Case 4 varies downstream of the orifice it converges in the corner with that in Case 3.

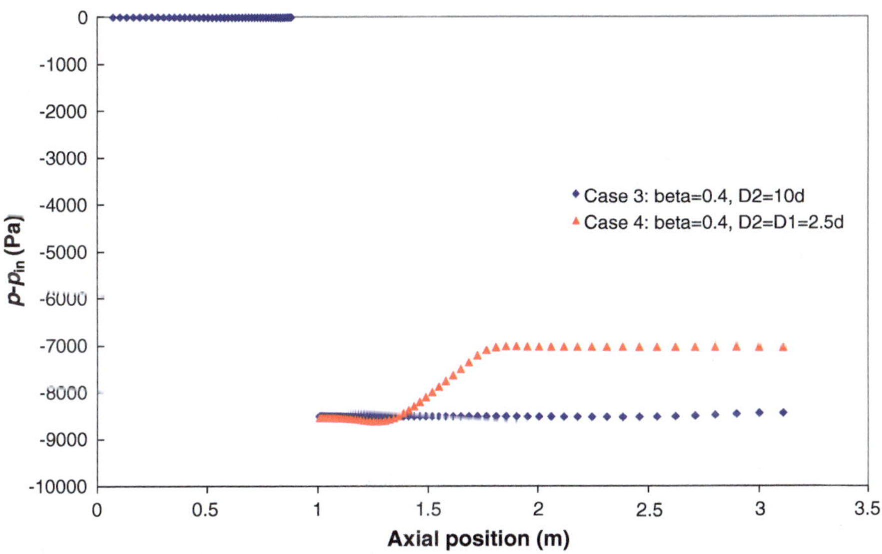

Fig. 5.E.6 Axial pressure profiles along the wall for cases 3 and 4 ($\beta = 0.4$)

Appendix 5.F: Lower Reynolds Number Limit for the Reader-Harris/Gallagher (1998) Equation

There are two lower limits on Reynolds number for each pair of tappings in 5.3.1 of ISO 5167-2:2003. Firstly, the Equation should not be used outside the range of the database. Secondly, even where there are data in the database, there is a lower limit of 5000 for Re_D (rather than 4000 as had been expected). The reason for choosing 5000 is that it is about the maximum value at which the change in the slope of the discharge coefficient as the flow becomes fully turbulent upstream of the orifice meter was seen to occur: see the 4″ and 6″ API oil data for $\beta = 0.37$ shown in Fig. 5.F.1: for each pipe size Set 1 and Set 2 were taken with different plates in the same pipe.

The change in slope may, of course, occur for Re_D well below 5000: see, for example, Fig. 5.F.2: Set 1 and Set 2 were taken with different plates in the same pipe; Set 3 was taken with the same plate as in Set 2 but after the pipe was plated.

A typical value for Re_D at which the change in slope occurs is presumably 3700, since in fitting the data this value was calculated (see Sect. 5.4.3). If β is around 0.1 or below there is no significant change in slope: this is clear by calculation from Eq. 5.21; see also Sect. 5.E.3. Because of the variation in the Reynolds number at which the change of slope occurs and the steep gradient of the discharge coefficient below that Reynolds number for medium or large β, orifice plates are rarely used for $Re_D < 5000$ except for small β (for example as in Sect. 5.E.3).

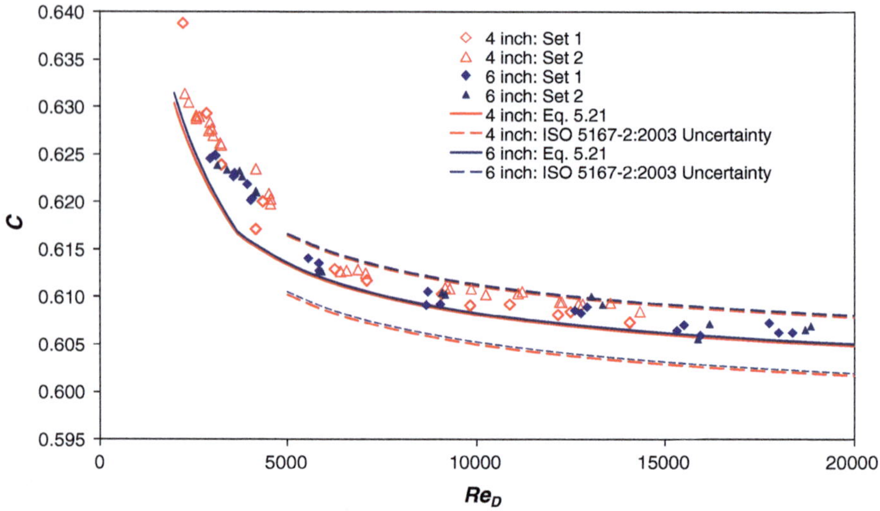

Fig. 5.F.1 The API oil data (collected by CEESI) in 4″ and 6″ pipes for $\beta = 0.37$ with flange tappings

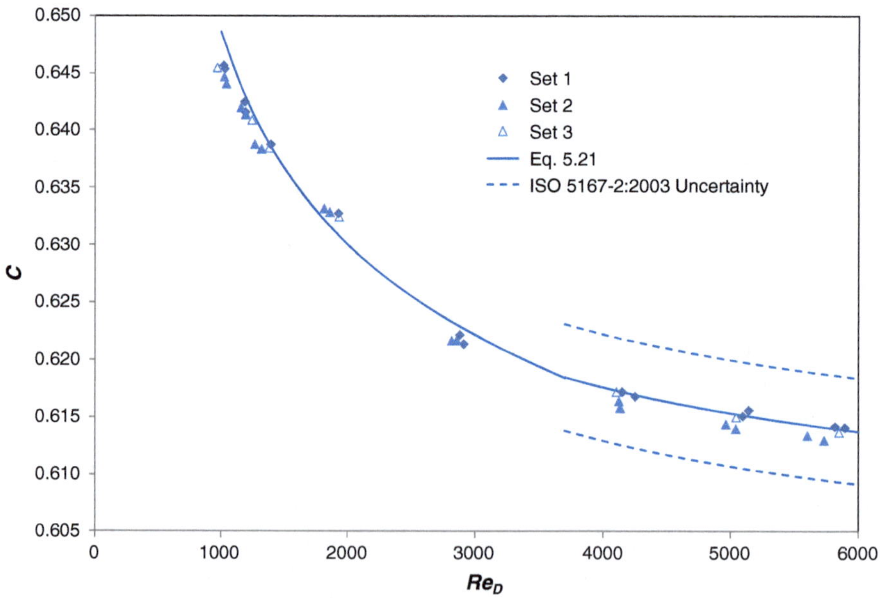

Fig. 5.F.2 The API oil data (collected by CEESI) in 2″ pipes for $\beta = 0.36$ with flange tappings

References

AGA (1954) Large diameter orifice meter tube tests. Final report of Supervising Committee. Research Project NX-4. American Gas Association

API (1985) Manual of Petroleum Measurement Standards, Chapter 14—Natural gas fluids measurement, Section 3—Orifice metering of natural gas and other related hydrocarbon fluids. ANSI/API 2530, 2nd edn

API (1988) API Coefficients of discharge for concentric, square-edged, flange-tapped orifice meters: equation data set—supporting documentation for floppy diskettes. American Petroleum Institute, Washington, DC

API (1990) Manual of petroleum measurement standards, Chapter 14—Natural gas fluids measurement, Section 3—Concentric, square-edged orifice meters, Part 1—General equations and uncertainty guidelines. MPMS Chapter 14, Section 3, Part 1, 3rd edn. Washington DC: American Petroleum Institute

API (2012) Orifice metering of natural gas and other related hydrocarbon fluids—concentric, square-edged orifice meters, Part 1—General equations and uncertainty guidelines. Manual of petroleum measurement standards, Chapter 14.3.1. September 2012: Errata 2013. American Petroleum Institute, Washington DC

Bean HS, Benesh ME, Buckingham E (1931) Experiments on the metering of large volumes of air. Bur Stand J Res 7:93–145

Beitler S R (1935) The flow of water through orifices. A study in 1-in., 1½-in., 2-in., 3-in., 6-in., 10-in, and 14-in. lines. Engineering Experiment Station bulletin No 89. Ohio State University, Columbus

Britton CL, Caldwell S, Seidl W (1988) Measurements of coefficients of discharge for concentric, flange-tapped, square-edged orifice meters in white mineral oil over a low Reynolds number range. American Petroleum Institute, Washington DC

Darrow RR (ed) (1954) Investigation of orifice meter installation requirement, Interim Research Report No 2. American Gas Association, Washington, DC

Engel FVA (1953) Fluid-flow measuring devices. Eng 176: 6–10, 35–37

Engel FVA (1961) Durchflußmessung in Rohrleitungen. Bericht über ein Symposium in East Kilbride (Schottland). Brennst-Wärme-Kraft: 13: 125–133

Engel FVA (1962) New interpretations of the discharge characteristics of measuring orifices—an approach to improved accuracy. ASME paper 62-WA-237. American Society of Mechanical Engineers, New York

Gasunie (1988) 600 mm orifice plate project. Vol XII. Discharge coefficients. TP/T 88.R.2052. Gasunie, Groningen

George DL, Morrow TB (2001) Orifice meter calibration for backwards-facing orifice plates. GRI Report No 01/0074 on SwRI Project No 18-8890. Gas Research Institute, Chicago

Herning F (1955) Untersuchung des Druckverlaufs an Normblenden grosser Rohrweiten. Brennst-Wärme-Kraft 7:303–307

Herning F, Bellenberg H (1960) Neue Versuche mit Normblenden. Brennst-Wärme-Kraft 12:89–96

Hobbs JM (1985) Experimental data for the determination of basic 100 mm orifice meter discharge coefficients (European programme). Report EUR 10027. Commission of the European Communities, Brussels, Belgium

Hobbs JM (1987) The EEC orifice plate project, part II: critical evaluation of data obtained. Progress report no PR5: EUEC/17. National Engineering Laboratory, East Kilbride, Glasgow. (Available on microfiche as Report EUR 13754. Commission of the European Communities, Brussels, Belgium, 1991)

Hobbs JM (1989) Determination of the effects of orifice geometry upon discharge coefficients. Report No ORIF/01. National Engineering Laboratory, East Kilbride, Glasgow

Hobbs JM, Sattary JA (1986) Experimental data for the determination of 100 mm orifice meter discharge coefficients under different installation conditions (European programme). Report EUR 10074. Commission of the European Communities, Brussels, Belgium, 1986

Hobbs JM, Sattary JA, Maxwell A D (1987a) Experimental data for the determination of basic 250 mm orifice meter discharge coefficients (European programme). Report EUR 10979. Commission of the European Communities, Brussels, Belgium, 1987

Hobbs JM, Sattary JA, Maxwell AD (1987b) Experimental data for the determination of 250 mm orifice meter discharge coefficients under different installation conditions (European programme). Report EUR 10980. Commission of the European Communities, Brussels, Belgium

Husain ZD, Teyssandier RG (1986) The effects of plate thickness and bevel angle in a 150 mm line size orifice meter. In: Proceedings Flow Measurement in the Mid 80s. National Engineering Laboratory, East Kilbride, Glasgow

ISO (1967) Measurement of fluid flow by means of orifice plates and nozzles. ISO/R 541:1967. International Organization for Standardization, Geneva

ISO (1980) Measurement of fluid flow by means of orifice plates, nozzles and Venturi tubes inserted in circular cross-section conduits running full. ISO 5167:1980. International Organization for Standardization, Geneva

ISO (1991) Measurement of fluid flow by means of pressure differential devices—Part 1: Orifice plates, nozzles and Venturi tubes inserted in circular cross-section conduits running full. ISO 5167-1:1991. International Organization for Standardization, Geneva

ISO (1998) Measurement of fluid flow by means of pressure differential devices—Part 1: Orifice plates, nozzles and Venturi tubes inserted in circular cross-section conduits running full. AMENDMENT 1. ISO 5167-1:1991/Amd. 1:1998. Geneva: International Organization for Standardization

ISO (2003) Measurement of fluid flow by means of pressure differential devices inserted in circular cross-section conduits running full—Part 2: Orifice plates. International Organization for Standardization, Geneva, ISO 5167-2:2003

Iversen HW (1956) Orifice coefficients for Reynolds numbers from 4 to 50,000. Trans ASME 78 (2):359–364

Johansen FC (1930) Flow through pipe orifices at low Reynolds numbers. Aeron Res Committee, Report and Memo No 1252. HM Stationery Office, London

Keyser DR, Friedman JR (2005) Introduction to improved orifice flow measurement methods in ASME PTC 19.5. Proc. PWR 2005. ASME Power. PWR2005-50072. Chicago, Illinois

Kinghorn FC (1986) The expansibility correction for orifice plates: EEC data. In: Proceedings Flow Measurement in the mid 80s, paper 5.2. National Engineering Laboratory, East Kilbride, Glasgow

Martin CNB (1986) Effects of upstream bends and valves on orifice plate pressure distributions and discharge coefficients. NEL Report No 702. National Engineering Laboratory, East Kilbride, Glasgow

Morrow TB (2004) Orifice meter expansion factor tests in 4-inch and 6-inch meter tubes. GRI Report GRI-04/0042. Gas Research Institute, Chicago

Morrow TB (2005) Metering research facility program: additional studies of orifice meter installation effects and expansion factor. GRI Report GRI-04/0246 on SwRI Project No 18.10315. Gas Research Institute, Chicago

Morrow TB, Park JT (1992) Baseline conditions for orifice meter calibration. GRI Report GRI-92/0097. Gas Research Institute, Chicago (as amended by Errata, 1993)

NEL (1988) Certificate of calibration on determination of the longitudinal pressure profile through six thin orifice plates mounted in the EEC 100 mm assembly using AGA tappings throughout for the European Economic Community DGXII Research. Science and Education, National Engineering Laboratory, East Kilbride, Glasgow

NEL (1991) 2-inch orifice plate investigation for Commission of the European Communities Directorate General for Science, Research and Development Community Bureau of Reference. Report EEC004. National Engineering Laboratory, East Kilbride, Glasgow

Pfeffer W (1968) Untersuchungen an Blenden mit Eckanbohrungen. Brennst-Wärme-Kraft 20:108–115

Rapier C (1979) The extrapolation of orifice calibrations to high Reynolds numbers. In: Proceedings Fluid Mechanics Silver Jubilee Conference paper 4.3. National Engineering Laboratory, East Kilbride, Glasgow

Reader-Harris MJ (1990) The tapping terms in the orifice plate discharge coefficient formula. Progress Report No PR8: EUEC/17. National Engineering Laboratory, East Kilbride, Glasgow. (Available on microfiche as Report EUR 13759. Commission of the European Communities, Brussels, Belgium, 1991)

Reader-Harris MJ (2002) The effect on orifice plates of diameter steps in upstream pipework and of swirl. Report No 2002/75 on Project No FDDP02. NEL, East Kilbride, Glasgow

Reader-Harris MJ, Barton N, Hodges D (2010) The effect of contaminated orifice plates on the discharge coefficient. In: Proceedings 15th FLOMEKO, Taipei

Reader-Harris MJ, Brunton WC, Nicholson IG, Rushworth R (2003) Ageing effects on orifice metering. In: Proceedings 21st North Sea Flow Meas Workshop, Norway

Reader-Harris MJ, Gibson J, Hodges D, Rushworth R (2006) The revision of ISO/TR 15377 (orifice plates, nozzles and Venturi tubes beyond the scope of ISO 5167). In: Proceedings 6th International Symposium Fluid Flow Meas, Querétaro, Mexico

Reader-Harris MJ, Hodges D, Rushworth R (2008a) The effect of drain holes in orifice plates on the discharge coefficient. In: Proceedings of 26th International North Sea Flow Meas Workshop, St Andrews. NEL, East Kilbride, Glasgow

Reader-Harris MJ, Keegans W (1986) Comparison of computation and LDV measurement of flow through orifice and perforated plates, and computation of the effect of rough pipework on orifice plates. In: Proceedings of International Symposium Fluid Flow Meas, Washington DC, 653–666

Reader-Harris MJ, Sattary JA (1990) The orifice plate discharge coefficient equation. Flow Meas Instrum 1:67–76

Reader-Harris MJ, Sattary JA (1996) The orifice plate discharge coefficient equation—the equation for ISO 5167-1. In: Proceedings of 14th North Sea Flow Meas Workshop, Peebles: paper 24. National Engineering Laboratory, East Kilbride, Glasgow

Reader-Harris MJ, Sattary JA, Spearman EP (1992a) The orifice plate discharge coefficient equation. Progress Report No PR14: EUEC/17 (EEC005). National Engineering Laboratory Executive Agency, East Kilbride, Glasgow

Reader-Harris MJ, Sattary JA, Spearman EP (1992b) The orifice plate discharge coefficient equation—further work. In: Proceedings of 10th North Sea Flow Meas Workshop, Peebles: paper 1.1. National Engineering Laboratory Executive Agency, East Kilbride, Glasgow

Reader-Harris MJ, Sattary JA, Spearman EP (1995) The orifice plate discharge coefficient equation—further work. Flow Meas Instrum 6:101–114

Reader-Harris MJ, Stobie G, Kelley T (2008b) The discharge coefficient of orifice plates with diameters from 1/16″ (1.6 mm) to 3/8″ (9.5 mm). In: Proceedings of the Production and Upstream Flow Meas Workshop, Houston, Texas

Sattary JA, Spearman EP (1992) Experimental data for the determination of basic 600 mm orifice meter discharge coefficients—European programme. Progress Report No PR11: EUEC/17 (EEC005). National Engineering Laboratory Executive Agency, East Kilbride, Glasgow

Sattary JA, Spearman EP, Reader-Harris MJ (1992) Experimental data for the determination of basic 50 mm orifice meter discharge coefficients—European programme. Progress Report No PR12: EUEC/17 (EEC005). National Engineering Laboratory Executive Agency, East Kilbride, Glasgow

Seidl W (1995) The orifice expansion correction for a 50 mm line size at various diameter ratios. In: Proceedings of 3rd International Symposium Fluid Flow Meas, San Antonio, Texas

Sens M (1977) Etude expérimentale des variations de la pression au voisinage d'un diaphragme normalisé selon ISO/DIS/5167. ISO/TC 30/SC 2 (France-11) 85, March

Spearman EP, Sattary JA, Reader-Harris, MJ (1992) The EEC orifice plate project: index to the data tables. Progress Report No PR13: EUEC/17 (EEC005). National Engineering Laboratory Executive Agency, East Kilbride, Glasgow

Spencer EA (1993) Bibliography of the EEC Orifice Plate Project. Report EUR 14885 EN. Commission of the European Communities, Brussels

Stolz J (1978) A universal equation for the calculation of discharge coefficients of orifice plates. In: Dijstelbergen HH, Spencer EA (eds) Flow Measurement of Fluids, North Holland Publishing Company, Amsterdam, p 519–534

Stolz J (1988) Private communication from J Stolz regarding data collected in Stuttgart

Stolz J (1991) Orifice plate discharge coefficients correlation—a new concept in the formulation of tap terms. Communication of 16 October 1991

Studzinski W, Karnik U, LaNasa P, Morrow T, Goodson D, Husain Z, Gallagher J (2001a) White paper on orifice meter installation configurations with and without flow conditioners. White paper prepared for API 14.3 Part 2. GRI Report GRI 99/0262, Gas Research Institute, Chicago

Studzinski W, Weiss M, Geerligs J, Attia J (2001b) Effect of reducers, expanders, a gate valve and two bends in perpendicular planes on orifice meter performance. In: Proceedings of Flow Meas 2001 International Conference Peebles: paper 3.1. National Engineering Laboratory, East Kilbride, Glasgow

Teyssandier RG, Husain ZD (1987) Experimental investigation of an orifice meter pressure gradient. Trans ASME J Fluids Eng 109:144–148

Teyssandier RG, Miller RW (1979) A new coefficient equation for concentric sharp-edged orifice: Part I—Fundamental considerations. ASME Paper 79-WA/FM-3. American Society of Mechanical Engineers, New York

Voss LR (1974) The design of small multiple-orifice meters for 1 percent air flow measurements. In: Dowdell R (ed) Flow its measurement and control in science and industry, vol 1(2). Instrument Society of America, Pittsburgh, : pp 507–515.(Conference was in May 1971)

Whetstone JR, Cleveland WG, Bateman BR, Sindt CF (1988) Measurements of coefficients of discharge for concentric, flange-tapped, square-edged orifice meters in natural gas over a Reynolds number range of 25,000–16,000,000. NIST Technical Note TN-1270. National Institute of Standards and Technology, Washington DC

Whetstone JR, Cleveland WG, Baumgarten GP, Woo S (1989) Measurements of coefficients of discharge for concentric, flange-tapped, square-edged orifice meters in water over a Reynolds number range of 1000–2,700,000. NIST Technical Note TN-1264. NIST, Washington DC

Witte R (1931) Die Strömung durch Düsen und Blenden. Forschung auf dem Gebiet des Ingenieurwesens 2: 245 and 291 (two issues)

Chapter 6
Orifice Expansibility Factor

Abstract The equation for the expansibility factor is required to measure gas flows. Whereas for Venturi tubes the expansibility-factor equation comes directly from theory, that for orifice plates is derived from experiment. This chapter considers the expansibility-factor equation for orifice plates, its history, the new data collected at the same time as the discharge-coefficient equation was revised, the analysis, the equation in ISO 5167-2:ISO 2003, a theoretical model and subsequent work in the US to derive the equation in API 14.3.1:2012. The expansibility-factor equation in ISO 5167-2:2003 is significantly different from that in ISO 5167-1:1991, especially for small β. At $p_2/p_1 = 0.95$, $\beta = 0.2$, and $\kappa = 1.3$ it differs from that in ISO 5167-1:1991 by 0.22 %. Moreover, the uncertainty of the equation in ISO 5167-2:2003 is lower than that of the equation in ISO 5167-1:1991. The work described here provided the necessary case for change in ISO 5167. There is good agreement between the ISO 5167-2:2003 equation and Buckingham's theoretical model. Moreover, for $p_2/p_1 \geq 0.95$ and $\kappa = 1.3$ the equation in API 14.3.1:2012 differs in magnitude from that in ISO 5167-2:2003 by less than 0.044 % for $0.2 \leq \beta \leq 0.66$.

6.1 Introduction

The equation for the expansibility (or expansion) factor is required to measure gas flows (see Eq. 1.15):

$$q_m = \frac{C\varepsilon}{\sqrt{1-\beta^4}}\frac{\pi d^2}{4}\sqrt{2\rho_1 \Delta p}. \qquad (1.15)$$

Whereas for Venturi tubes the expansibility-factor equation comes directly from theory (see Eq. 1.13), that for orifice plates is derived from experiment. This chapter considers the expansibility-factor equation for orifice plates, its history, the new data collected at the same time as the discharge-coefficient equation was revised, the

M. Reader-Harris, *Orifice Plates and Venturi Tubes*,
Experimental Fluid Mechanics, DOI 10.1007/978-3-319-16880-7_6

analysis, the equation in ISO 5167-2:2003 (see Eq. 6.13), a theoretical model and subsequent work in the US.

The effect of the expansibility-factor equation on the measured flowrate increases when the ratio of differential pressure to static pressure increases. As orifice plates are usually designed to give a similar differential pressure no matter what the static pressure is in the application (the differential pressure is a little lower at ambient pressure than at high pressure), the ratio of differential pressure to static pressure tends to be higher when gases are at ambient and low pressures than when they are at high pressure.

6.2 History and Theory

The equation for the orifice expansibility factor used in ANSI/API 2530-1985 (API 1985), ISO 5167:1980 (ISO 1980), API MPMS 14.3.1:1990 (API 1990) and ISO 5167-1:1991 (ISO 1991) was derived by Buckingham (1932) largely on the basis of data collected at tests in Los Angeles in 1929. Using these data Buckingham derived the equation:

$$\varepsilon = 1 - (0.41 + 0.35\beta^4)\frac{\Delta p}{\kappa p_1}, \tag{6.1}$$

where ε is the expansibility factor, Δp is the differential pressure across the orifice plate, p_1 is the static pressure at the upstream tapping and κ is the isentropic exponent.

The expansibility factor is given the symbol ε_1 in ISO 5167:1980 and ISO 5167-1:1991 and Y_1 in API MPMS 14.3.1:1990 and API MPMS 14.3.1:2012 (API 2012).

It is significant that in analysing the data Buckingham neglected the effect of Reynolds number on the grounds that above a throat Reynolds number, Re_d, of 2×10^5 the discharge coefficient was considered to be constant. It is now known that the discharge coefficient continues to change with Re_d above this value (see Chap. 5). The work of Buckingham and his co-worker Bean has nevertheless set a pattern for subsequent workers in this area. Murdock and Foltz (1953) tested the Buckingham Equation and found that it fitted their steam data very well. Thibessard (1960) used an equation of the form

$$\varepsilon = 1 - (a + b\beta^4)\left(1 - \left(\frac{p_2}{p_1}\right)^{1/\kappa}\right)^c \tag{6.2}$$

where β is the diameter ratio, a, b and c are constants and p_2 is the static pressure at the downstream pressure tapping. He obtained $c = 0.935$. If c is set equal to 1 this equation can be expanded for small $\frac{\Delta p}{\kappa p_1}$ to give

$$\varepsilon \approx 1 - (a + b\beta^4)\left(\frac{\Delta p}{\kappa p_1} - \frac{1}{2}(\kappa - 1)\left(\frac{\Delta p}{\kappa p_1}\right)^2\right). \tag{6.3}$$

NOTE If the expansibility-factor equation for Venturi tubes (Eq. 1.13) is expanded for small $\frac{\Delta p}{\kappa p_1}$ it becomes

$$\varepsilon \approx 1 - \left(\frac{3}{4} + \frac{\beta^4}{1 - \beta^4}\right)\frac{\Delta p}{\kappa p_1}.$$

This approximation is within 0.1 % of Eq. 1.13 for $p_2/p_1 \geq 0.95$ and $\kappa \geq 1.25$.

6.3 The Database

As part of the EEC Orifice Project data were collected on expansibility factors. At NEL (Reid 1981) using the 4″ orifice run data were collected for three diameter ratios, 0.2, 0.57 and 0.75 in air with 1.4 bara < p_1 < 8 bara. Reference sonic nozzles were used to give the reference mass flowrate, and they were installed upstream of the orifice plate. A valve was installed downstream of the orifice plate, which allowed the static pressure at the orifice plate to be varied. The air was exhausted to atmosphere downstream of the valve. When using a sonic, or critical flow, nozzle, as long as it remains choked, the mass flow is a function of the nozzle upstream pressure and is independent of the downstream pressure. By varying the valve position the static pressure at the orifice plate could be varied while the mass flowrate remained constant. Provided that the nozzle remained choked, the Reynolds number of that set of data was held approximately constant. Over any set of data used in the subsequent analysis the Reynolds number did not differ from its mean value by more than 2 %. So, when ε was determined from Eq. 1.15, C could be assumed constant.

Further data for $\beta = 0.2$ were collected at NEL (1988) for the EEC. This report also suggested that in the NEL data a correction should have been made from stagnation to static temperature at least for large β (see Sect. 4.6.1). This correction was then applied to the NEL data with a temperature probe recovery factor of 0.7. An isenthalpic correction to the temperature from the downstream measurement to the upstream pressure tapping was also made, following on from the work, for example, of Gasunie (1988): this correction now appears in 5.4.4.1 of ISO 5167-1: 2003; see Sect. 4.6.2. The effect of these corrections was small.

Gaz de France (Benzoni and Kervevan 1983) collected data using the 4″ orifice run for a diameter ratio of 0.66 in natural gas at $Re_D = 1.2 \times 10^6$. Reference sonic nozzles were used to measure the mass flowrate.

Gasunie (see Hobbs 1985) did not collect data on expansibility factor directly, but within the data collected by them on the 4″ orifice run for the discharge coefficient-database it was possible to identify sets of data taken over both a significant range of static pressure and a small range of Reynolds number. The data used were for diameter ratios 0.2, 0.57 and 0.66.

Similarly CEAT (see Hobbs et al. 1987) did not collect data on expansibility factor directly, but within the data collected by them on the 10″ orifice run for the discharge-coefficient database it was possible to identify sets of data taken over both a significant range of static pressure and a small range of Reynolds number. The data used were for a diameter ratio of 0.2.

In addition to the European work CEESI (Seidl 1995) in a 2″ orifice run collected data for six diameter ratios, 0.242, 0.363, 0.484, 0.5445, 0.6655 and 0.726, in air with 1.185 bara $< p_1 <$ 21.3 bara. Reference sonic nozzles were used to give the mass flowrate.

6.4 Analysis

Analysis of the expansibility-factor data in Reid (1981), Benzoni and Kervevan (1983) and Hobbs (1985) was carried out by Kinghorn (1986). He concluded that the expansibility-factor equation should be revised. His conclusions, although clear, did not lead to a revised expansibility-factor equation in ISO 5167-1:1991. They did lead to NEL collecting the additional data in NEL (1988). When CEESI's data (Seidl 1995) confirmed Kinghorn's conclusions it was clear that the orifice expansibility-factor equation in ISO 5167-1:1980 should be changed. The requirement was that the best equation should now be used.

In order to analyse the NEL (and similar) data a common method had been to calculate

$$\varepsilon|_{calc} = \frac{(C\varepsilon)_{\text{measured}}}{C_{\text{water}}} \tag{6.4}$$

and then to use the method of least-squares to determine the constants in

$$\varepsilon|_{calc} = 1 - a - bf\left(\frac{\Delta p}{p_1}, \kappa\right) \tag{6.5}$$

and then to fit the slope terms, b.

The problem with this method was that there is always some bias between a gas flow laboratory and a water flow laboratory. So a better estimate of ε is given by

$$\varepsilon|_{calc,2} = \frac{\varepsilon|_{calc}}{1-a} = 1 - \frac{b}{1-a} f\left(\frac{\Delta p}{p_1}, \kappa\right) \tag{6.6}$$

which is equivalent to fitting $(C\varepsilon)_{\text{measured}}$ as $C_{\text{incompressible}}(1 - b^{*}f)$ without assuming a value for $C_{\text{incompressible}}$. Then $b^{*} = b/(1 - a)$.

In the case of the Gaz de France data the value of C obtained when ε is as close as possible to 1 had been taken as the reference value (equivalent to C_{water} in Eq. 6.4); this reference value was a single measurement, and so in the analysis used here all the data including the reference value were fitted and values of $b/(1-a)$ calculated. The data from CEESI had already been analysed in a similar manner to that recommended here and so in most cases a was calculated to be 0.

For the data from Gasunie and CEAT the Reynolds number was not constant and so it was necessary to apply corrections to the measured values of $C\varepsilon$ so that all the values are effectively taken at one Reynolds number. To do this the dependence of C on Reynolds number given in the Reader-Harris/Gallagher (1998) Equation in ISO 5167-2:2003 (ISO 2003) was assumed (with the term in A taken as negligible):

$$\varepsilon|_{calc} = \frac{(C\varepsilon)_{\text{measured}}}{\left(C_{\infty} + 0.000521\left(\frac{10^6\beta}{Re_D}\right)^{0.7} + 0.0188\,\beta^{3.5}\left(\frac{10^6}{Re_D}\right)^{0.3}\right)} \tag{6.7}$$

where an assumed value for C_{∞}, the value of the discharge coefficient in incompressible flow at infinite Reynolds number, is used. Use of Eq. 6.6 effectively removes the effect of the choice of C_{∞}.

For each set of data calculations were performed for two functions, f:

$$f\left(\frac{\Delta p}{p_1}, \kappa\right) = \frac{\Delta p}{\kappa p_1} \tag{6.8}$$

and

$$f\left(\frac{\Delta p}{p_1}, \kappa\right) = 1 - \left(\frac{p_2}{p_1}\right)^{1/\kappa} \tag{6.9}$$

The first equation is simpler but the latter is based on the best physical understanding (see Eq. 6.2 (with $c = 1$) and Eq. 6.3 for the relationship between Eqs. 6.8 and 6.9). As stated above, the values of b^{*} ($=b/(1 - a)$), the coefficient of f, were calculated, and are given in Table 6.1. In performing the calculations the standard deviation of the data about the fitted line, s, was obtained:

$$s = \sqrt{\frac{1}{n-2}\sum_{i=1}^{n}\left\{\frac{\varepsilon|_{calc}}{1-a} - \left(1 - \frac{b}{1-a}f\right)\right\}^2} \tag{6.10}$$

Table 6.1 Experimental data: the coefficient of f in Eq. 6.6

Laboratory	Diameter ratio	Tappings	f as in Eq. 6.8		f as in Eq. 6.9	
			b^*	s	b^*	s
NEL (Reid 1981)	0.2006	Corner	0.3643	0.00077	0.3503	0.00051
NEL (Reid 1981)	0.2006	Flange	0.3539	0.00102	0.3399	0.00091
NEL (Reid 1981)	0.2006	D & D/2	0.3491	0.00157	0.3353	0.00124
NEL (Reid 1981)	0.5700	Corner	0.4252	0.00058	0.4084	0.00030
NEL (Reid 1981)	0.5700	Flange	0.4170	0.00080	0.4004	0.00048
NEL (Reid 1981)	0.5700	D & D/2	0.4015	0.00135	0.3849	0.00097
NEL (Reid 1981)	0.7501	Corner	0.5433	0.00075	0.5228	0.00050
NEL (Reid 1981)	0.7501	Flange	0.5444	0.00070	0.5236	0.00026
NEL (Reid 1981)	0.7501	D & D/2	0.5458	0.00112	0.5269	0.00142
NEL (1988)	0.2006	Corner	0.3938	0.00329	0.3873	0.00329
NEL (1988)	0.2006	Flange	0.3881	0.00090	0.3812	0.00091
NEL (1988)	0.2006	D & D/2	0.3769	0.00086	0.3704	0.00085
Gaz de France	0.6601	Corner	0.4534	0.00209	0.4283	0.00144
Gaz de France	0.6601	Flange	0.4553	0.00180	0.4301	0.00078
Gaz de France	0.6601	D & D/2	0.4663	0.00189	0.4408	0.00107
CEESI	0.2420	Flange	0.3791	0.00062	0.3654	0.00041
CEESI	0.3630	Flange	0.3569	0.00319	0.3405	0.00280
CEESI	0.4840	Flange	0.3835	0.00107	0.3696	0.00083
CEESI	0.5445	Flange	0.4025	0.00113	0.3862	0.00080
CEESI	0.6655	Flange	0.4406	0.00091	0.4265	0.00080
CEESI	0.7260	Flange	0.5310	0.00239	0.5132	0.00264
Gasunie	0.2006	Corner	0.3615	0.00067	0.3564	0.00066
Gasunie	0.2006	Flange	0.3415	0.00108	0.3367	0.00108
Gasunie	0.2006	D & D/2	0.3573	0.00073	0.3523	0.00073
Gasunie	0.5700	Corner	0.3983	0.00035	0.3924	0.00037
Gasunie	0.5700	Flange	0.3962	0.00054	0.3903	0.00056
Gasunie	0.5700	D & D/2	0.3998	0.00034	0.3939	0.00035
Gasunie	0.6601	Corner	0.4383	0.00047	0.4320	0.00044
Gasunie	0.6601	Flange	0.4204	0.00043	0.4143	0.00041
Gasunie	0.6601	D & D/2	0.4237	0.00006	0.4177	0.00003
CEAT	0.2007	Corner	0.3376	0.00207	0.3342	0.00207
CEAT	0.2007	Flange	0.3407	0.00191	0.3371	0.00192
CEAT	0.2007	D & D/2	0.3290	0.00182	0.3256	0.00182

where n is the number of points in the data set. From these data it was clear that where there is a wide range of values of p_2/p_1, that is in the NEL (Reid 1981), Gaz de France and CEESI data, using Eq. 6.9 gave a smaller standard deviation of the data about the obtained fit, and so Eq. 6.9 was used to reduce bias in the final equation. More complex forms of equation than Eqs. 6.8 and 6.9 were tried but only

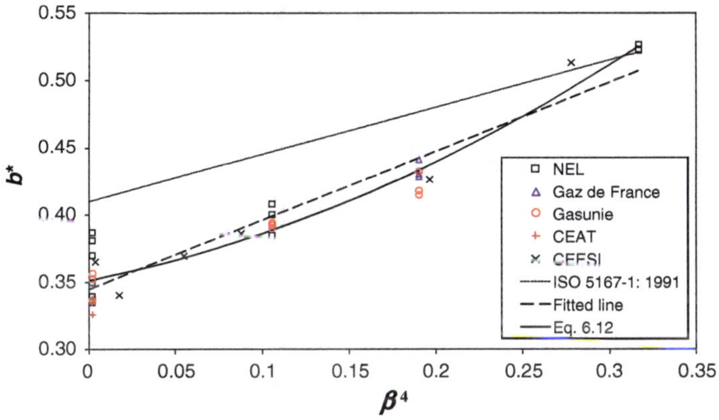

Fig. 6.1 Experimental data: coefficient of f in Eq. 6.6 (f as in Eq. 6.9)

gave marginal improvements in quality of fit. The values of b^* (for use with f in Eq. 6.9) were dependent on β and are shown in Fig. 6.1. For $\beta \le 0.66$ there was a linear dependence on β^4. At higher diameter ratios the measured values lie above a fitted line and so, to avoid bias at small β due to the points for large β, b^* was fitted as follows:

$$b^* = a_1 + a_2\beta^4 + a_3\beta^8. \tag{6.11}$$

Different functions of β did not give a worthwhile improvement in fit. On fitting the data the following equation was obtained:

$$b^* = 0.351 + 0.256\beta^4 + 0.93\beta^8. \tag{6.12}$$

The standard deviation of the data about this equation was 0.0148.
Hence,

$$\varepsilon = 1 - (0.351 + 0.256\beta^4 + 0.93\beta^8)\left\{1 - \left(\frac{p_2}{p_1}\right)^{1/\kappa}\right\}. \tag{6.13}$$

This equation was published in Reader-Harris (1998) and accepted as the expansibility-factor equation in ISO 5167-2:2003.

The Gaz de France values in Table 6.1 were calculated assuming $\kappa = 1.31$; the effect of using the true value would be very small.

It is clear that within the database in Table 6.1 there are sets of data with higher uncertainty. The uncertainty of each of the values of b was evaluated, and if those data with uncertainty in b greater than 0.03 were removed and the equation refitted the value of ε would change by about $0.43\Delta p/(\kappa p_1)$ % at most. There was no need to restrict the database and to change Eq. 6.13.

The uncertainty of a value of ε from Eq. 6.13 is not less than that due to the fit to derive Eq. 6.12 (which, as stated above, has a standard deviation of 0.0148). This uncertainty is

$$0.0296\left\{1-\left(\frac{p_2}{p_1}\right)^{1/\kappa}\right\}. \tag{6.14}$$

In percentage terms this becomes at most

$$2.96\left(1-\left(\frac{p_2}{p_1}\right)^{1/\kappa}\right)\left\{1-0.5250\left(1-\left(\frac{p_2}{p_1}\right)^{1/\kappa}\right)\right\}^{-1}$$

For $p_2/p_1 \geq 0.75$ and $\kappa \geq 1.14$ this is less than

$$3\frac{\Delta p}{p_1}, \tag{6.15}$$

and for $p_2/p_1 \geq 0.75$ and for all κ it does not exceed

$$3.413\frac{\Delta p}{\kappa p_1}. \tag{6.16}$$

This uncertainty is higher than it might have been because the points with high uncertainty were included. If the points with $s > 0.0015$ were excluded this uncertainty would be reduced to $2.79\frac{\Delta p}{\kappa p_1}$. Moreover, some of the spread in the data may come from the differences between laboratories. On the other hand the uncertainty in the values of b^* themselves should be included. Moreover, the uncertainty of the value of ε should also include the effect of errors due to the data not exactly following Eq. 6.9.

Of the data sets available the largest value of $\frac{\Delta p}{\kappa p_1}$ is found in the Gaz de France data. The result of fitting all their data taken with flange tappings, including those outside the range of ISO 5167-2:2003, is shown in Fig. 6.2. No point deviates from the fitted line by more than 0.001.

It is also important to note that the orifice expansibility-factor equation has to apply to plates other than those used for the determination of the equation (see Sect. 6.6).

No formal calculation method was used but in ISO 5167-2:2003 the relative uncertainty of ε is given as

$$3.5\frac{\Delta p}{\kappa p_1}\%. \tag{6.17}$$

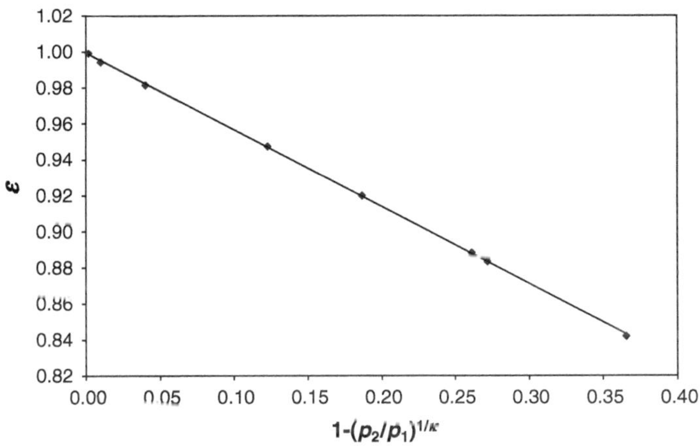

Fig. 6.2 Expansibility-factor data for $\beta = 0.66$ from Gaz de France with flange tappings

6.5 Theoretical Model

It was desirable to check that the modifications to the existing equation were not inconsistent with theory. Buckingham (1931) used Bernoulli's Equation between an upstream location and the location of maximum convergence, i.e. the *vena contracta*, and the momentum theorem for both an incompressible flow and an isentropic flow. He then postulated that if the upstream density, the upstream pressure and the mass flow were the same in both cases the force exerted by the fluid on the orifice plate would be the same in both cases. The consequence of this assumption is to give (following some manipulation) the following equation for the ratio of the contraction coefficient in isentropic flow, $C_{c,ise}$ to that in incompressible flow, $C_{c,inc}$, where the contraction coefficient is the ratio of the area of the *vena contracta* to that of the orifice:

$$\frac{C_{c,ise}}{C_{c,inc}} = \frac{\phi\, C_{c,inc}}{B}\left(1 - \sqrt{1 - \frac{B(1-\tau)}{\phi^2 C_{c,inc}^2}}\right) \tag{6.18}$$

where

$$B = \tau^{\frac{1}{\kappa}}\left\{\left(2C_{c,inc} + 2C_{c,inc}^2\beta^4 - 1\right)\phi - C_{c,inc}^2\beta^4\tau^{\frac{1}{\kappa}}(1-\tau)\right\} \tag{6.19}$$

$$\phi = \frac{\kappa}{\kappa - 1}\tau^{\frac{1}{\kappa}}\left(1 - \tau^{\frac{\kappa-1}{\kappa}}\right) \tag{6.20}$$

and

$$\tau = \frac{p_2}{p_1}. \tag{6.21}$$

It is then necessary to equate the mass flowrate from Bernoulli's Equation for an isentropic flow with that from the equation for flowrate in ISO 5167-1:2003. This gives

$$C\varepsilon = C_{c,ise}\sqrt{\frac{\kappa\tau^{\frac{2}{\kappa}}(1-\tau^{\frac{\kappa-1}{\kappa}})(1-\beta^4)}{(\kappa-1)(1-\tau)(1-C_{c,ise}^2\beta^4\tau^{\frac{2}{\kappa}})}} \tag{6.22}$$

A similar process for incompressible flow gives

$$C = C_{c,inc}\sqrt{\frac{1-\beta^4}{1-C_{c,inc}^2\beta^4}} \tag{6.23}$$

and combining Eqs. 6.22 and 6.23 gives

$$\varepsilon = \frac{C_{c,ise}}{C_{c,inc}}\sqrt{\frac{\kappa\tau^{\frac{2}{\kappa}}(1-\tau^{\frac{\kappa-1}{\kappa}})(1-C_{c,inc}^2\beta^4)}{(\kappa-1)(1-\tau)(1-C_{c,ise}^2\beta^4\tau^{\frac{2}{\kappa}})}} \tag{6.24}$$

C is now determined from the Reader-Harris/Gallagher Equation in ISO 5167-2:2003 (ISO 2003) using pressure tappings D upstream and at the minimum pressure location downstream (although the *vena contracta* may not occur at exactly the same axial location as the minimum pressure point (see NOTE to 1.10) the wall pressure changes very slowly with axial distance near *vena contracta*). Then $C_{c,inc}$ can be obtained by solving Eq. 6.23, $C_{c,ise}$ from Eq. 6.18 and ε from Eq. 6.24. This was done for a range of values of β, κ, and τ. Calculated values of ε for values of τ from 1 to 0.75 at intervals of 0.01 were fitted by an equation of the following form:

$$\varepsilon = 1 - bf\left(\frac{\Delta p}{p_1}, \kappa\right) \tag{6.25}$$

where f is as in Eqs. 6.8 and 6.9. In this way the coefficients in Table 6.2 were obtained. In determining C a value of pipe Reynolds number of $Re_D = 1.5 \times 10^6\,\beta^2$ was used: this was just above the Reynolds number for the NEL and CEESI data but below the value for the other data. The standard deviation, s, is given by:

Table 6.2 Theoretical modelling: the coefficient of *f* as in Eq. 6.25

Diameter ratio	β^4	κ	*f* as in Eq. 6.8		*f* as in Eq. 6.9	
			b	*s*	*b*	*s*
0.1000	0.0001	1.200	0.3822	0.00038	0.3757	0.00016
0.1000	0.0001	1.300	0.3834	0.00039	0.3743	0.00010
0.1000	0.0001	1.400	0.3844	0.00039	0.3732	0.00006
0.1000	0.0001	1.500	0.3853	0.00038	0.3722	0.00003
0.5700	0.1055	1.200	0.4286	0.00028	0.4213	0.00004
0.5700	0.1055	1.300	0.4302	0.00031	0.4201	0.00002
0.5700	0.1055	1.400	0.4316	0.00032	0.4190	0.00004
0.5700	0.1055	1.500	0.4328	0.00033	0.4181	0.00007
0.6777	0.2110	1.200	0.4773	0.00016	0.4692	0.00012
0.6777	0.2110	1.300	0.4795	0.00021	0.4682	0.00015
0.6777	0.2110	1.400	0.4814	0.00024	0.4673	0.00017
0.6777	0.2110	1.500	0.4830	0.00027	0.4666	0.00018
0.7500	0.3164	1.200	0.5323	0.00002	0.5232	0.00032
0.7500	0.3164	1.300	0.5351	0.00007	0.5225	0.00033
0.7500	0.3164	1.400	0.5376	0.00013	0.5219	0.00033
0.7500	0.3164	1.500	0.5397	0.00018	0.5213	0.00032

$$s = \sqrt{\frac{1}{n-1}\sum_{i=1}^{n}(\varepsilon - (1 - bf))^2} \tag{6.26}$$

In the calculations as described here no irrecoverable loss in pressure was assumed: assuming no loss at a lower Reynolds number gives the same contraction coefficient as assuming a loss at a higher Reynolds number. It was clear that a better fit was obtained by using an equation of the form of Eq. 6.9 rather than Eq. 6.8. Moreover, the range of coefficients as a function of κ was generally much smaller. Indeed, with Eq. 6.9 the value of *b* did not change by more than 0.01 from its value at $\kappa = 1.4$ for $1.0001 < \kappa < 4$. Fitting the data gave the following equation:

$$\varepsilon = 1 - (0.374 + 0.405\beta^4 + 0.193\beta^8)\left\{1 - \left(\frac{p_2}{p_1}\right)^{1/\kappa}\right\} \tag{6.27}$$

The introduction of a term in β^8 improved the fit, and it does have a positive coefficient. Agreement with the experimental data is good; the difference between Eqs. 6.13 and 6.27 is approximately equal to the uncertainty of Eq. 6.13 at maximum. This work tends to confirm the form of Eq. 6.13.

6.6 Subsequent Work

ISO/TC 30/SC 2 decided to include Eq. 6.13 in ISO 5167-2:2003; API decided to collect additional expansibility-factor data in a wider range of pipe sizes. The additional data taken using orifice plates whose edges were sharp and which were installed with a long upstream length are shown in Fig. 6.3: they have been taken from Hodges (2006). The new CEESI data were taken in 2″, 3″, 4″, 6″ and 10″ pipes; the SwRI data in this figure were taken in a 4″ pipe. The SwRI tests are described in Morrow (2004, 2005). The SwRI data were calculated assuming a value of 1.3 for κ; George (2008) states that the using the true value of κ would have a negligible effect on the plotted data. Data taken with a flow conditioner $7D$ or $10D$ upstream of an orifice plate are given in Appendix 6.A. The key test for the expansibility-factor equation in ISO 5167 (as for the discharge-coefficient equation) is that it works for orifice plates to which it was not originally fitted: it can be seen that all the new points lie within the uncertainty band for the equation in ISO 5167-2:2003.

API omitted the European data, fitted a revised equation and included it in API 14.3.1:2012 (API 2012): it is given by

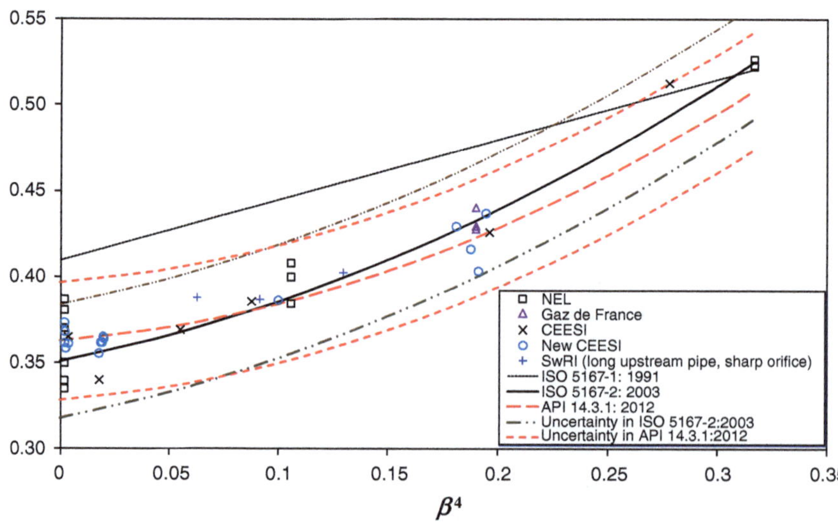

Fig. 6.3 Expansibility-factor equations and flange-tapping data: coefficient of f in Eq. 6.6 (f as in Eq. 6.9) *NOTE 1* The uncertainty in both ISO 5167-2:2003 and in API 14.3.1:2012 is a percentage figure and so the absolute values depend slightly on $\Delta p/(\kappa p_1)$; the uncertainty plotted for API 14.3.1:2012 also depends on κ; accordingly the uncertainty curves shown in this figure are approximate. *NOTE 2* Only data for which Reynolds number corrections were not required are shown in this figure

$$\varepsilon = 1 - (0.3625 + 0.1027\beta^4 + 1.132\beta^8)\left\{1 - \left(\frac{p_2}{p_1}\right)^{1/\kappa}\right\}. \tag{6.28}$$

In API 14.3.1:2012 the relative uncertainty of ε is given as

$$2.6\frac{\Delta p}{p_1}\%. \tag{6.29}$$

The r.m.s. deviations of the data in Fig. 6.3 about the ISO and API equations are almost identical.

6.7 Conclusions

A revised expansibility-factor equation was derived in the 1990s (Reader-Harris 1998) and is given in Eq. 6.13. This equation was significantly different from the expansibility-factor equation in ISO 5167-1:1991, especially for small β. At $p_2/p_1 = 0.95$, $\beta = 0.2$, and $\kappa = 1.3$ Eq. 6.13 differs from that in ISO 5167-1:1991 by 0.22 %. The work described here provided the necessary case for change in ISO 5167.

Moreover, the uncertainty of Eq. 6.13 is lower than the uncertainty of the equation in ISO 5167-1:1991: the uncertainty of Eq. 6.13 is given in Eq. 6.17. There is good agreement between the ISO 5167-2:2003 Equation and Buckingham's theoretical model.

The data collected to derive the new equation in API 14.3.1:2012 provide confirmation of the equation in ISO 5167-2:2003. For $p_2/p_1 \geq 0.95$ and $\kappa = 1.3$ the new API equation differs in magnitude from that in ISO 5167-2:2003 by less than 0.044 % for $0.2 \leq \beta \leq 0.66$.

Appendix 6.A: Data Taken with a Flow Conditioner 7*D* or 10*D* from the Orifice Plate

The SwRI data taken with a flow conditioner upstream of the orifice plate are shown in Fig. 6.A.1 The 6″ data were taken in a 21*D* tube with a CPA 50E Flow Conditioner 10*D* upstream of the orifice plate; the 3″ data were taken in a 17*D* tube with a CPA 50E Flow Conditioner 7*D* upstream of the orifice plate. On the basis of the data in Morrow (1997) the NOVA (now CPA) 50E did not pass the flow conditioner test in 7.4.1.7 of ISO 5167-1:2003 for $\beta = 0.75$ when the flow conditioner was 7*D* upstream of the orifice plate.

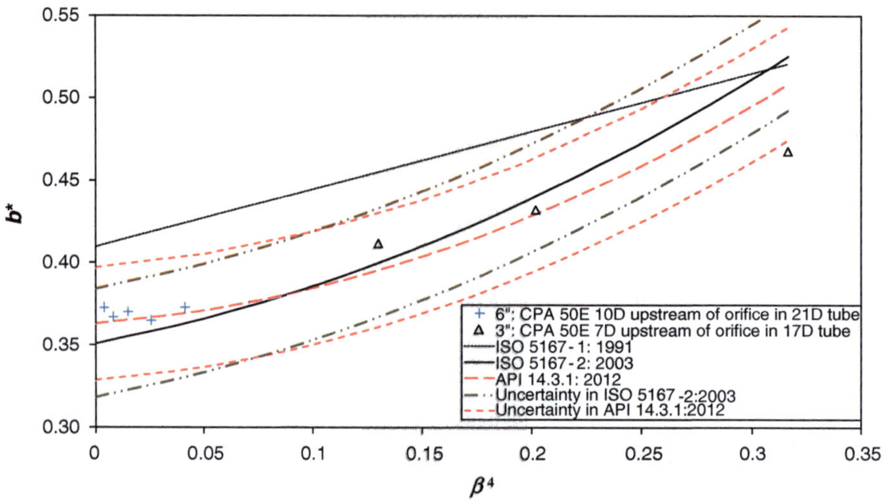

Fig. 6.A.1 Expansibility-factor equations and flange-tapping data for SwRI tests downstream of the CPA 50E flow conditioner: coefficient of f in Eq. 6.6 (f as in Eq. 6.9) *NOTE* The uncertainty in both ISO 5167-2:2003 and in API 14.3.1:2012 is a percentage figure and so the absolute values depend slightly on $\Delta p/(\kappa p_1)$; the uncertainty plotted for API 14.3.1:2012 also depends on κ; accordingly the uncertainty curves shown in this figure are approximate

Where the flow conditioner passed the flow conditioner test in 7.4.1 of ISO 5167-1:2003 the points in Fig. 6.A.1 lie within the uncertainty band for the equations in ISO 5167-2:2003 and API 14.3.1:2012.

References

API (1985) Manual of petroleum measurement standards, Chap. 14—Natural gas fluids measurement, Sect. 3—Orifice metering of natural gas and other related hydrocarbon fluids. ANSI/API 2530, 2nd edn

API (1990) Manual of petroleum measurement standards, Chap. 14—Natural gas fluids measurement, Sect. 3—Concentric, square-edged orifice meters, part 1—General equations and uncertainty guidelines. MPMS Chap. 14, Sect. 3, Part 1, 3rd edn. American Petroleum Institute, Washington DC

API (2012) Orifice metering of natural gas and other related hydrocarbon fluids—Concentric, square-edged orifice meters, part 1—General equations and uncertainty guidelines. Manual of petroleum measurement standards, Chap. 14.3.1. Sept 2012: Errata 2013. American Petroleum Institute, Washington DC

Benzoni A, Kervevan P (1983) Coefficients de décharge des diaphragmes détermination en gaz naturel dans les conditions de référence (CEE—Phase 2). Gaz de France, Alfortville, France

Buckingham E (1931) Note on contraction coefficients of jets of gas. Bur Stand J Res 6 (Research paper no. 303)

Buckingham E (1932) Notes on the orifice meter: the expansion factor for gases. Bur Stand J Res 9 (Research paper no. 459)

Gasunie (1988) 600 mm orifice plate project. Part X: temperature profile measurements. Report TP/T 88.R.2050. NV Nederlandse Gasunie, Groningen

George DL (2008) Revised analysis of orifice meter expansion factor data. Catalog no. L52299 of Pipeline Research Council International Inc, Technical Toolboxes Inc, Texas, Houston

Hobbs JM (1985) Experimental data for the determination of basic 100 mm orifice meter discharge coefficients (European programme). Report EUR 10027. Commission of the European Communities, Brussels, Belgium

Hobbs JM, Sattary JA, Maxwell AD (1987) Experimental data for the determination of basic 250 mm orifice meter discharge coefficients (European programme). Report EUR 10979. Commission of the European Communities, Brussels, Belgium

Hodges C (2006) Paper summarizing API investigative testing for existing square edge flange tap orifice expansion factor equations. In: Proceedings of the 6th International Symposium Fluid Flow Meas, Querétaro, Mexico

ISO (1980) Measurement of fluid flow by means of orifice plates, nozzles and Venturi tubes inserted in circular cross-section conduits running full. ISO 5167:1980. International Organization for Standardization, Geneva

ISO (1991) Measurement of fluid flow by means of pressure differential devices—Part 1: Orifice plates, nozzles and Venturi tubes inserted in circular cross-section conduits running full. ISO 5167-1:1991. International Organization for Standardization, Geneva

ISO (2003) Measurement of fluid flow by means of pressure differential devices inserted in circular cross-section conduits running full—Part 2: Orifice plates. ISO 5167-2:2003. International Organization for Standardization, Geneva

Kinghorn FC (1986) The expansibility correction for orifice plates: EEC data. In: Proceedings of Flow Measurement in the mid 80s, paper 5.2. National Engineering Laboratory, East Kilbride, Glasgow

Morrow TB (1997) Orifice meter installation effects: development of a flow conditioner performance test. Report prepared by SwRI as GRI report no. GRI-97/0207, Gas Research Institute, Chicago

Morrow TB (2004) Orifice meter expansion factor tests in 4-inch and 6-inch meter tubes. GRI Report GRI-04/0042. Gas Research Institute, Chicago

Morrow TB (2005) Metering research facility program: additional studies of orifice meter installation effects and expansion factor. GRI Report GRI-04/0246 on SwRI project no. 18.10315. Gas Research Institute, Chicago

Murdock JW, Foltz CJ (1953) Experimental evaluation of expansion factors for steam. Trans ASME 75(5):953–959

NEL (1988) Commissioning tests with 0.2 β ratio plate in EEC expansibility rig in air with Re_d of 35 000 and re-examination of existing EEC data on expansibility. Report no. EUEC/21 for EEC Community Bureau of Reference (BCR). National Engineering Laboratory, East Kilbride, Glasgow

Reader-Harris MJ (1998) The equation for the expansibility factor for orifice plates. In: Proceedings of FLOMEKO 98, Lund, Sweden

Reid J (1981) Tests to determine orifice plate expansibility factors. Progress report no. PR6: EUEC/03 for EEC Community Bureau of Reference (BCR). National Engineering Laboratory, East Kilbride, Glasgow

Seidl W (1995) The orifice expansion correction for a 50 mm line size at various diameter ratios. In: Proceedings of the 3rd International Symposium Fluid Flow Meas, San Antonio

Thibessard G (1960) Über die Expansionszahl bei der Durchflußmessung mit Normblenden. Brennst-Wärme-Kraft 12(3):97–101

Chapter 7
Venturi Tube Discharge Coefficient in High-Pressure Gas

Abstract In this chapter the performance of Venturi tubes in high-pressure gas is described: the discharge coefficient tends to increase with Reynolds number largely because of static-hole error, the effect that pressure tappings of finite size do not measure the pressure which would have been measured using an infinitely small hole. There are also effects of throat velocity, including humps and dips in the calibration curve, sometimes an audible tone. The discharge coefficient depends on the shape of the Venturi tube: of the different shapes tested the best results were obtained with a Venturi tube identical to the standard Venturi tube with a machined convergent except that the convergent angle was 10.5° (instead of 21°).

7.1 Introduction

Until the mid 1990s it was generally assumed that the discharge coefficient of a Venturi tube is constant provided that the pipe Reynolds number is greater than around 2×10^5. However, work in the 1990s showed the problems with this view. The work of Jamieson et al. (1996), Van Weers et al. (1998) and Metz et al. (1991) showed that the performance of Venturi tubes in high-pressure gas is very different from that in water. A significant programme of work to examine this problem was carried out at NEL. Very careful manufacture of the Venturi tubes was undertaken to reduce the risk that the results were affected by poor manufacture. This chapter describes experimental work carried out on new Venturi tubes in water and dry gas and gives a partial explanation of the behaviour of the discharge coefficient in high-pressure gas. The first section describes work carried out on Venturi tubes of standard convergent angle. Attempts at improvement by revising the shape are described in Appendix 7.A: reducing the convergent angle helped, but more interesting shapes gave no significant benefit. Finally more work was undertaken with the optimum design using a reduced convergent angle: this work gave rise to a section in ISO/TR 15377:2007 and is summarized in the main text.

It should be noted that the main design specifications for classical Venturi tubes are given in Chap. 3; installation effects are covered in Chap. 8.

M. Reader-Harris, *Orifice Plates and Venturi Tubes*,
Experimental Fluid Mechanics, DOI 10.1007/978-3-319-16880-7_7

7.2 Experimental Work: Standard Shape

7.2.1 Description of the Venturi Tubes

Fifteen classical Venturi tubes with machined convergent sections were manufactured by Jordan Kent Metering Systems/Seiko and Crane Perflow for calibration in water and in gas: the former manufactured Venturi tubes with diameter ratio $\beta = 0.4$, 0.6 and 0.75 for nominal diameter 2″, 4″ and 8″, together with $\beta = 0.5$, 0.65 and 0.7 for nominal diameter 4″ only; the latter manufactured Venturi tubes with $\beta = 0.4$, 0.6 and 0.75 for nominal diameter 4″ only. They were manufactured to drawings with tight tolerances designed to ensure that where possible the results were not affected by uncontrolled variables. Each Venturi tube was manufactured out of solid metal so that there would be no steps due to welding within the Venturi tube. They were made of stainless steel and were suitable for use at pressures up to 70 with ANSI Class 600 flanges fitted. They were designed not only to meet the requirements of ISO 5167-1:1991 (ISO 1991) but to follow its recommendations. The Standard recommends the use of a divergent angle between 7° and 8°: therefore, 7½° was specified for the Venturi tubes used in this project.

So that the results would not be corrupted by the introduction of steps at joins in the pipework, an upstream length of 8*D* (6*D* for 8″ pipe) and a downstream length of 4*D*, where *D* is the diameter of the entrance cylinder, were manufactured with machined bores; this ensured that in no case was there a step in diameter greater than 0.0035*D* at the upstream flange of the Venturi tube. The lengths of pipework were manufactured by boring out Schedule 80 pipe to the bore of a Schedule 40 pipe. The lengths of pipe and the Venturi tubes were dowelled to ensure concentricity; O rings were used to ensure that there would not be recesses or protruding gaskets. The distance from the upstream pressure tappings to the first upstream flange was 1.5, 1.1 and 0.7*D* for 2″, 4″ and 8″ pipes respectively.

In addition to the shorter lengths of pipework already described, an additional 21*D* (23*D* for 8″ pipe) length was manufactured by welding a 19*D* (21*D* for 8″ pipe) length of Schedule 40 pipe to a 2*D* length of pipe machined to the bore of the other pipes, smoothing off any step at the weld. This length of pipe was installed with the machined length adjacent to the machined pipe already described, so that there was at least 8*D* of machined pipework, whose bore matched that of the Venturi tube very accurately, immediately upstream of the Venturi tube. In total there was 29*D* of pipe of the same schedule with no recesses, protruding gaskets or significant steps upstream of the Venturi tube. Upstream of this assembly there was generally further pipe of the same nominal diameter preceded by a flow conditioner.

The Standard recommends that the radii of curvature at the intersections of the entrance cylinder and the convergent section, the convergent section and the throat, and the throat and the divergent section be equal to zero, although significantly larger values are permitted (0.25*D*, 0.25*d* and 0.25*d* respectively). The drawings requested a maximum radius of curvature of 1 mm. In order to measure the radius of curvature at the intersection between the convergent section and the throat measurements of profile were made through the convergent and the throat with one trace per Venturi tube (this was not possible for some of the Venturi tubes). The average of the measured radii of curvature at the intersection between the convergent section and the throat was, in fact, 18 mm (this sounds like a very large error, but a radius of curvature of 18 mm actually corresponds to a deviation from the intended profile of 76 μm at maximum). The average ratio of radius of curvature to throat diameter at this intersection was 0.34 (the average ratio would be 0.26 if the '4 inch*' set in Fig. 7.2 were omitted, as its ratio was 1.17). Half of them lay outside the standard, but on the basis of Appendix 3.A it is expected that this will have a negligible effect: a radius of 3.625*d* (more than 10 times the average radius in the experiment) gives a maximum computed shift in C of −0.3 %.

The typical surface finish was $R_a \approx 1.7$ μm, where R_a is the arithmetical mean deviation of the roughness profile. All the Venturi tubes had $10^{-5} < R_a/d < 10^{-4}$.

Except for the Venturi tube with a throat diameter of 20 mm the pressure tappings were 4 mm in diameter; the throat pressure tappings were of constant diameter for a depth of 94 mm and the upstream tappings for a depth of 53 mm. The Venturi tube with a throat diameter of 20 mm had pressure tappings of diameter 2.6 mm; the throat pressure tappings were of constant diameter for a depth of 62 mm and the upstream tappings for a depth of 37 mm. Tappings were of constant diameter because the work described by Jamieson et al. suggested that such a design might be beneficial; it may also reduce the number of variables. The tappings were connected in triple-T arrangements (see Fig. 4.8).

For most of the Venturi tubes the convergent angles were determined both from three measurements of diameter for each convergent cone and from the wall profile measurements from which intersection radii were obtained. The measured value of the convergent angle never differed from the nominal value by more than 0.14°.

Except for one measured diameter in the 20 mm throat (2″, $\beta = 0.4$) the measured throat diameters were within 0.1 % of the mean value of the throat diameters at the pressure tappings. The measured diameters of the entrance cylinders were within 0.05 % of the mean value of the entrance cylinder diameters at the pressure tappings.

7.2.2 Calibration in Water

The Venturi tubes were calibrated in water in 1997–1999 in the UKAS-accredited (ISO 17025) National Standard facility for water flow measurement at NEL. For each Venturi tube the data in water lay on a straight line and with a small scatter. The slopes were small: when fitted against Re_D the majority had a positive sign, but, since several had a negative sign, it seemed appropriate to represent the discharge coefficient of each Venturi tube by its mean value (over the range of the data the average increase in discharge coefficient with Reynolds number was 0.0005). The results are shown in Fig. 7.1. Where two Venturi tubes were made from the same drawings the maximum difference in mean discharge coefficient was 0.57 %. The

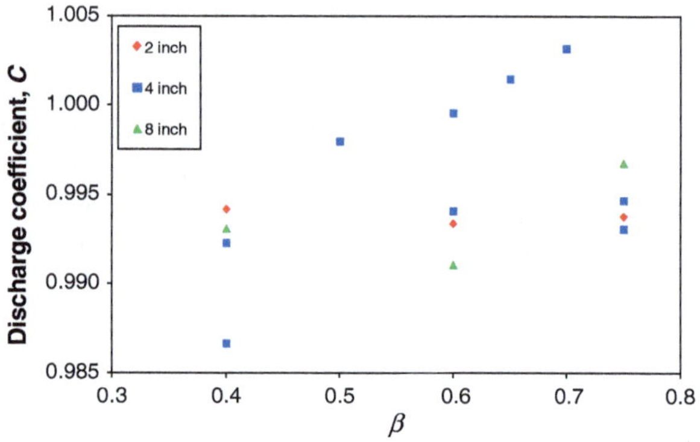

Fig. 7.1 Mean discharge coefficients in water

Reynolds number range over which the data were used to calculate the mean was that over which C was approximately constant. In one case there was a significant hump in the data for $1.2 \times 10^5 < Re_D < 2 \times 10^5$ (see Fig. 3.7); the throat Reynolds number below which C decreased rapidly varied but was typically about 2.5×10^5.

7.2.3 Calibration in Gas

The Venturi tubes were calibrated in gas in 1998–1999 in the UKAS-accredited (ISO 17025) National Standard facility for high-pressure gas flow measurement at NEL at two static pressures, 20 and 60 (or 70) barg, and the data are presented in Figs. 7.2, 7.3, 7.4 and 7.5. All the gas data were collected in air except the data for 4″ for $\beta = 0.4$ from Jordan Kent Metering Systems/Seiko, which were collected in nitrogen. Since there are two 4″ Venturi tubes for $\beta = 0.4$, 0.6 and 0.75 the ones from Jordan Kent Metering Systems/Seiko are described as ‘4 inch’ and those from Crane Perflow are described as ‘4 inch*’. In Fig. 7.4 ‘Yokogawa DP’ and ‘Mensor DP’ refer to the devices used to measure the differential pressure; they were used for two separate calibrations.

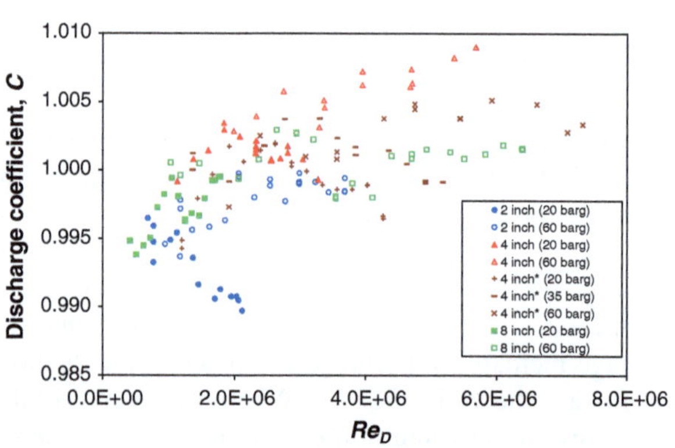

Fig. 7.2 Calibration in gas against Reynolds number: $\beta = 0.4$

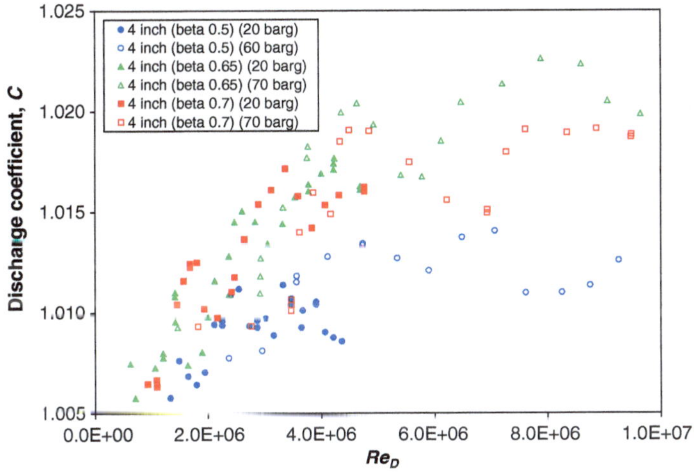

Fig. 7.3 Calibration in gas against Reynolds number: $\beta = 0.5$, 0.65 and 0.7

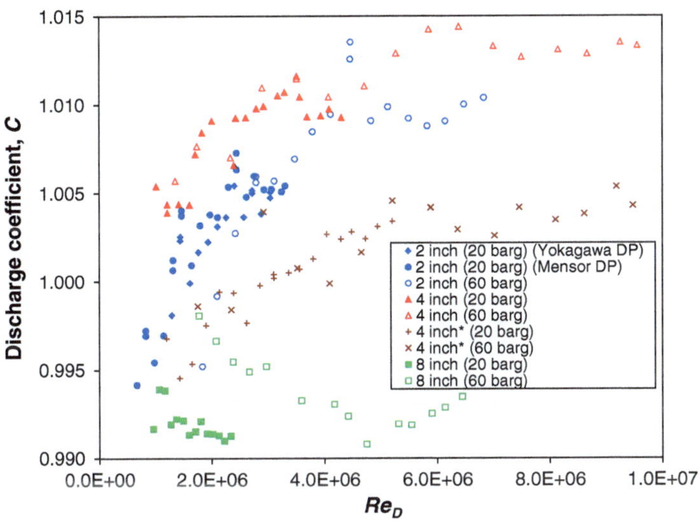

Fig. 7.4 Calibration in gas against Reynolds number: $\beta = 0.6$

The data collected in gas were more scattered than those taken in water. There were peaks and troughs in the data sets. In general better agreement was obtained between the two sets of data obtained at different static pressures when the data were fitted against Reynolds number; however, it is clear that the locations of the peaks and troughs depended on the throat velocity. As an example the $\beta = 0.65$ data from Fig. 7.3 are presented as a function of throat velocity in Fig. 7.6.

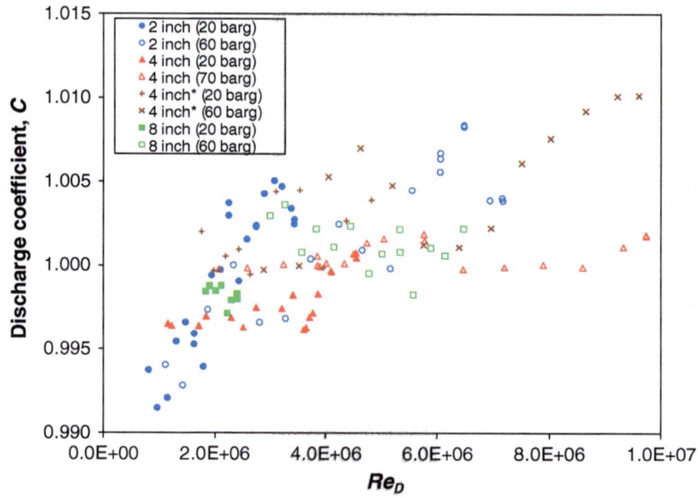

Fig. 7.5 Calibration in gas against Reynolds number: $\beta = 0.75$

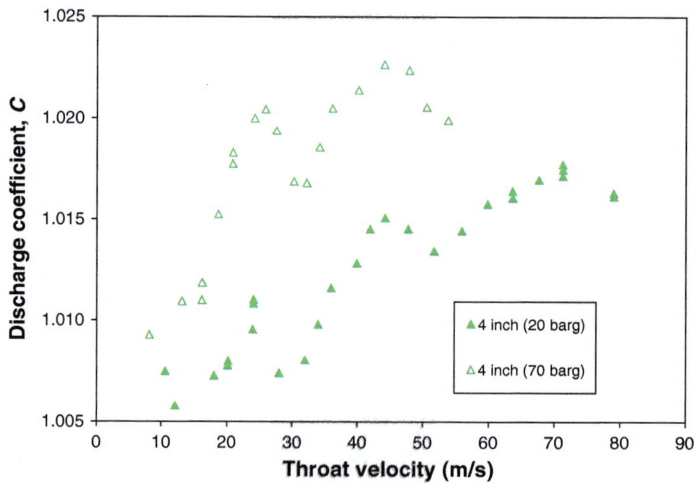

Fig. 7.6 Calibration in gas against throat velocity: $\beta = 0.65$

In all data fitting described in this chapter the points with $\Delta p/p_1 > 0.08$ were excluded since these points displayed a reduction in discharge coefficient from what would have been expected from other data from the same Venturi tube. It is assumed that this effect is due to expansibility effects which are not incorporated in the expansibility-factor equation. The decrease in discharge coefficient was, however, less than the uncertainty of ε as given by Sect. 5.8 of ISO 5167-4:2003 (ISO 2003) (Eq. 1.14 in this book).

Sometimes there were audible tones at particular velocities, perhaps in the same way that blowing across the top of a bottle gives an audible tone. It is unknown whether the throat velocity effects depend on the facility in which the Venturi tube is installed as well as on the Venturi tube itself. Sometimes there was a peak in discharge coefficient at both a particular velocity and twice that velocity.

7.3 Interpretation and Analysis of Data

7.3.1 Static-Hole Error

A major cause for the change in discharge coefficient from that obtained in water is static-hole error. Static-hole error is the effect that pressure tappings of finite size do not provide the pressure which would have been measured using an infinitely small hole. The effect of static-hole error is that the measured pressure using a pressure tapping is higher than the static pressure would have been if the tapping had been infinitely small. This effect is considered in many papers (e.g. Franklin and Wallace 1970; Gibson et al. 1999). If the increase in measured pressure is denoted by e_p and the wall shear stress by τ_w, then

$$\frac{e_p}{\tau_w} = f(Re_{tap}), \tag{7.1}$$

where Re_{tap} is the tapping-hole Reynolds number defined by

$$Re_{tap} = \frac{u_\tau d_{tap}}{\nu}, \tag{7.2}$$

d_{tap} is the tapping diameter, ν is the kinematic viscosity, u_τ is the friction velocity, $\sqrt{\frac{\tau_w}{\rho}}$, and ρ is the density. Because the velocity and therefore the wall shear stress are much higher in the throat than in the entrance cylinder, the static-hole error leads to a reduction in the measured differential pressure and an increase in the measured value of C. To calculate the static-hole error it is necessary to have an estimate of the relationship between τ_w and $\bar{u}$, the mean velocity at the tapping plane. Following Schlichting (1960) (and generalizing Eq. 1.20) this is expressed in terms of the friction factor, λ, where

$$\tau_w = \frac{1}{8}\lambda\rho\bar{u}^2. \tag{7.3}$$

Then (assuming incompressible flow) the predicted value of the total reduction in differential pressure, $e_{p,\mathrm{total}}$, is given by

$$e_{p,\text{total}} = \frac{1}{8}\bar{u}_{th}^2 \rho \left(\lambda_{th} f(Re_{tap,th}) - \lambda_{up} \beta^4 f(Re_{tap,up}) \right), \tag{7.4}$$

where the subscripts '*th*' and '*up*' refer to conditions at the throat tapping location and the upstream cylinder tapping location, respectively. This corresponds to an increase in discharge coefficient of approximately

$$\frac{\lambda_{th} f(Re_{tap,th}) - \lambda_{up} \beta^4 f(Re_{tap,up})}{8(1 - \beta^4)}. \tag{7.5}$$

7.3.2 Measurements of Static-Hole Error at High Tapping-Hole Reynolds Number

There are many sets of measurements of static-hole error, both at high tapping-hole Reynolds number (typical of high-pressure gas) and at low tapping-hole Reynolds number (as generally found in water or air at ambient pressure). Those taken at high tapping-hole Reynolds number are shown in Fig. 7.7: see Flack (1978), Moulden et al. (1977), Plentovich and Gloss (1986), Rainbird (1967) and McKeon and Smits (2002) for the experimental data. Jeff Gibson of NEL did significant computational work in this area: for the CFD points see Gibson et al. (1999). Those curves with equation numbers are derived from work described elsewhere in this book. Flack, Moulden et al. and Plentovich and Gloss worked in transonic flow.

Flack states that Mach number had little effect on e_p/τ_w (his Mach number was around 1) and that his results were independent of d_{tap}/δ^* (where δ^* is the

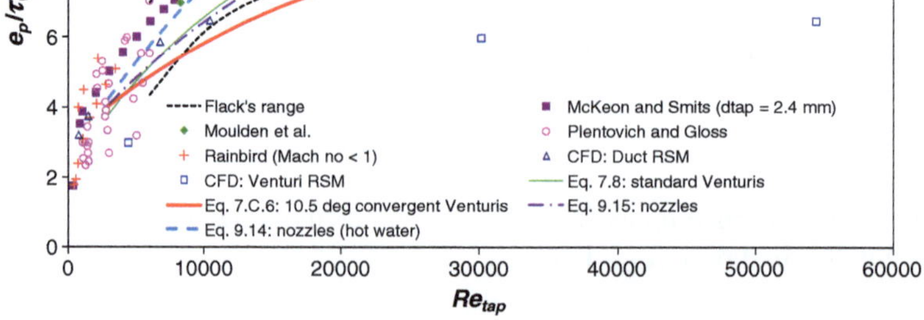

Fig. 7.7 Static-hole error at high tapping-hole Reynolds number

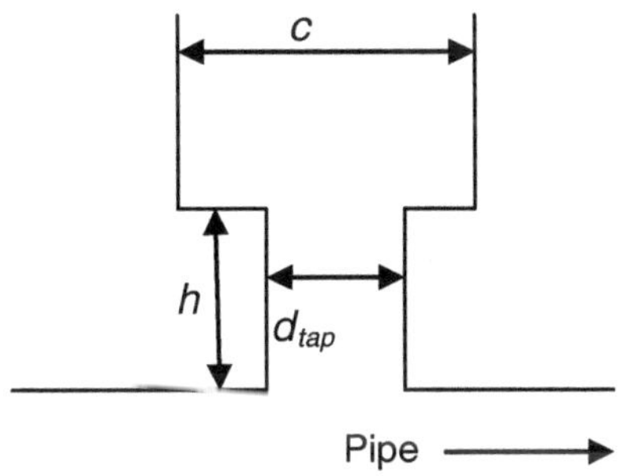

Fig. 7.8 Pressure tapping

boundary-layer displacement thickness) and of pressure gradient over the range tested; his pressure tappings had constant depth, h (see Fig. 7.8), and entered a chamber whose diameter, c, was such that $c = 2\ h$; h/d_{tap} varied from 0.77 to 4.33, with the higher values at the lower values of Re_{tap}.

The data of Moulden et al. were taken at a Mach number around 0.8 with $h = 2.8\ d_{tap}$ and $c < d_{tap}$.

The data of Plentovich and Gloss were taken at a Mach number of 0.4–0.72: $c > d_{tap}$; h is not stated, but it appears from Plentovich and Gloss (1982) that $h = 3\ d_{tap}$.

Rainbird's data shown here are those taken in subsonic flow: the Mach number was 0.56; $h \approx 3\ d_{tap}$; except for the two largest holes $c \approx 2\ d_{tap}$; the sum of the depth of the chamber and the depth of the tapping (h) was always 38 mm and the impulse line was 1.6 mm in diameter; d_{tap}/δ^* varied from 0.3 to around 5, the higher values occurring at higher Re_{tap}.

McKeon and Smits (2002) carried out work in the Princeton Superpipe with d_{tap} in the range 0.254–2.381 mm in a pipe of diameter, D, equal to 129.36 mm at Reynolds numbers up to 3.5×10^7 at pressures up to 190 bara. For $Re_{tap} < 500$ the static-hole error depended only on Re_{tap}, but for larger Re_{tap} the static-hole error showed a dependence not only on Re_{tap} but also on d_{tap}/D even though d_{tap}/D in their case was in the range 0.002–0.0184, smaller than the value of d_{tap}/d for most Venturi tubes: the static-hole error for larger tappings was smaller than that obtained using a smaller diameter hole at the same Re_{tap}. h/d_{tap} was effectively infinite.

7.3.3 Measurements of Static-Hole Error at Low Tapping-Hole Reynolds Number

The measurements of static-hole error obtained at the tapping-hole Reynolds numbers generally obtained in water are given in Fig. 7.9.

Shaw (1960) obtained a common curve for data with $1.5 \le h/d_{tap} \le 6$ for $d_{tap} \le D/16$, where D is the internal pipe diameter; $c/d_{tap} = 2$. With small holes

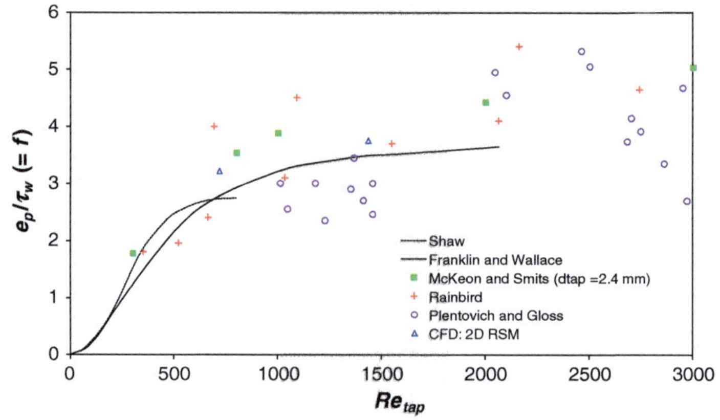

Fig. 7.9 Static-hole error at low tapping-hole Reynolds number

(0.3–1.6 mm) in a 190 mm nozzle with h/d_{tap} = 4 and c/d_{tap} = 2 Stratford and Ascough (1962) obtained excellent agreement with Shaw. Shaw's impulse line was of constant diameter equal to c and of length 240 d_{tap}; Stratford and Ascough had an impulse line of diameter 1.6 mm; so in some cases there was an expansion from the chamber into the impulse line and in others a contraction. It appears that a change in diameter between the chamber and the impulse line had a negligible effect.

Franklin and Wallace collected data in a subsonic wind-tunnel of width, w, equal to 150 mm with Mach number below 0.5: h/d_{tap} > 15; c/d_{tap} ≈ 2.3; $0.003 \le d_{tap}/w \le 0.042$. They had $0.15 < d_{tap}/\delta^* < 4$, but found no effect of d_{tap}/δ^*: they suggest that Rainbird's results differ from others' because of Mach number not d_{tap}/δ^*.

For McKeon and Smits, Plentovich and Gloss and Rainbird see Sect. 7.3.2.

The static-hole error at the tapping-hole Reynolds numbers obtained in water is quite well defined; so static-hole error in water is not too large a problem: the discharge coefficient may be around 0.6 % higher with real tappings than it would be with infinitely small tappings, but ISO 5167-4:2003 is based on real tappings and so there is no problem in use. Static-hole error is the reason why the discharge coefficient even in water can exceed 1. At high tapping-hole Reynolds numbers the static-hole error has a wider spread than at low tapping-hole Reynolds numbers.

7.3.4 The Effect of Tapping Depth on Static-Hole Error

The effect of the depth of a tapping has been measured: Roshko (1955) considered cavities of different depths up to the width of the cavity and measured the pressure

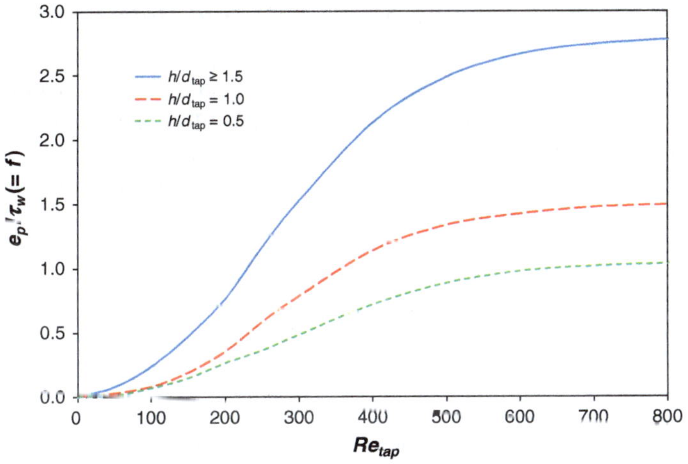

Fig. 7.10 Static-hole error as a function of h/d_{tap} (from Shaw (1960))

on the bottom of the cavity. Although Ray (1956) stated that (in the notation of Fig. 7.8) c/d_{tap} is unimportant unless h/d_{tap} is very small, there are differences between the results of Shaw (1960), who had $c/d_{tap} = 2$, and those of Livesey et al. (1962), who had $c/d_{tap} = 14$.

If $c = 2d_{tap}$, then from Shaw (1960) f is as shown in Fig. 7.10 (based on Figs. 7–9 of Shaw's paper). If $h/d_{tap} \geq 1.5$ the static hole error is unaffected by h/d_{tap}; if h/d_{tap} is less than 1.5 the static-hole error differs from that for $h/d_{tap} \geq 1.5$. The same curves are obtained for a range of values of d_{tap}, provided that, as can be seen from Shaw's paper, $d_{tap} \leq 3.2$ mm (0.125″) (his data were taken in a 2″ pipe).

Livesey et al. considered tappings of depth (h) between 0.3 and $7.4d_{tap}$, which beyond the tapping of stated depth entered a chamber whose diameter, c, was $14d_{tap}$: they stated that holes with $h/d_{tap} < 5$ were sensitive to variations of tapping depth. It is, however, rare in practice for tappings to enter a chamber of diameter $14d_{tap}$; for a wall tapping to have $c/d_{tap} = 2$ or 3 is more common.

From consideration of Livesey et al. and Shaw it appears better to avoid large values of c/d_{tap}.

The data in Sect. 7.2.3 had $h/d_{tap} = 23.5$ at the throat tappings, where static-hole error is more important, and $h/d_{tap} = 13.2$ at the upstream tappings (14.2 for the 2″ $\beta = 0.4$ upstream tappings); the chamber was ¼ BSP, the aim was that the data should be unaffected by tapping depth.

The standard requires $h/d_{tap} \geq 2.5$. However, for higher tapping-hole Reynolds numbers the discharge coefficient changes with h/d_{tap} even for $h/d_{tap} \geq 2.5$ and $c/d_{tap} = 2.25$: see Appendix 7.B for NEL tests.

In Appendix 7.C it is shown (see Fig. 7.C.2) that even with $h/d_{tap} = 18.5$ at the throat tappings surprising results can be obtained where a tapping of diameter 2 mm enters a ¼ BSP fitting.

7.3.5 The Effect of Tapping Shape on Static-Hole Error

The static hole error is a function not only of the tapping diameter but of its shape. Significant effort has been undertaken to examine the effect of rounding a tapping or inserting it at an angle (Rayle 1959; Repik and Kuzenkov 1990).

Tappings for flow measurement are almost always perpendicular to the pipe wall and should be as sharp as possible.

7.3.6 The Effect of a Burr or a Protruding Tapping on Static-Hole Error

The effect of a burr on the tapping hole was examined by Shaw (1960): since burrs of height $d_{tap}/127$ may be significant, Venturi tubes with small d_{tap}/d could be affected by the roughness of the wall of the Venturi tube even if it is within the limit given by ISO 5167-4:2003.

The effect of protruding tappings may be found in Zogg and Thomann (1971). The effect of both protruding tappings and those in recesses may be found in Allen and Hooper (1932).

7.3.7 Analysis of the Gas Data in Sect. 7.2.3

In a standard Venturi tube in both tapping planes Lindley (1966) made measurements of τ_w from which the friction factor, λ, (see Eq. 1.20) can be deduced. In the entrance cylinder the curve for λ appeared to become asymptotic to 0.012 as Reynolds number increased. For $10^6 < Re_D < 10^7$ and $k/D = 5 \times 10^{-5}$ λ will always be within 10 % of 0.012 in a straight pipe according to the Moody Diagram (see Schlichting); so it seems an appropriate value to use for λ_{up}. In the throat Lindley's measurements of λ were approximately 18 % higher than would be obtained in a straight pipe of the same relative roughness and Reynolds number; so in determining the static-hole error a figure of 0.015 was used for λ_{th} (for $10^6 < Re$ and $k/d = 10^{-4}$, 1.18λ will always be within 6 % of 0.015 according to the Moody Diagram). Therefore, values of 0.012 and 0.015 were used for λ_{up} and λ_{th}, respectively, in Eq. 7.3, so that Re_{tap} could be obtained from Eq. 7.2, and in Eq. 7.5.

In deriving an equation one method was to consider the change in C from that found in water. In a calibration in water the discharge coefficient in water is approximately constant with Reynolds number, and a typical value of e_p/τ_w at the throat tapping was around 3.8 (see Fig. 7.9); so f was written as $f^* + 3.8$; it was then taken that $f^* = 0$ for $Re_{tap} < 3000$, i.e. over the typical range of $Re_{tap,th}$ at which most water data were taken. f reduces rapidly for $Re_{tap} < 800$, but in the water data

$Re_{tap,th}$ was rarely less than 800; $Re_{tap,up}$ was often less than 800, but where $Re_{tap,up}$ is small the second term in Eq. 7.5 was generally much less significant than the first). Assuming that

$$f^* = a(e^{-3000\,n} - e^{-nRe_{tap}}) \quad \text{for}\, Re_{tap} > 3000 \tag{7.6}$$

the measured values of $C - C_{\text{water}}$ were fitted, where C_{water} is the mean value for the particular Venturi tube as given in Fig. 7.1, and the best fit of f^* to the data was

$$f^* = \begin{cases} 7.165 - 8.839\,e^{-0.00007\,Re_{tap}} & \text{for } Re_{tap} > 3000 \\ 0 & \text{for } Re_{tap} \leq 3000. \end{cases} \tag{7.7}$$

This fit has an uncertainty based on two standard deviations of 0.0074. The true static-hole error (rather than the difference between the static-hole error in high-pressure gas and that in water) is

$$f = 10.965 - 8.839\,e^{-0.00007\,Re_{tap}} \quad \text{for}\, Re_{tap} > 3000 \tag{7.8}$$

and is shown in Fig. 7.7. This is consistent with the other data. This is encouraging in that there is a possible physical explanation for the values measured in gas. The complete equation is obtained by fitting the data in Fig. 7.1 and adding the additional term for high Reynolds number to give

$$C = 0.9878 + 0.0123\beta + \frac{0.015 f^*(Re_{tap,th}) - 0.012\beta^4 f^*(Re_{tap,up})}{8(1-\beta^4)} \tag{7.9}$$

with f^* given by Eq. 7.7. When values of C from Eqs. 7.7 and 7.9 were compared with the complete database of values of C in gas, twice the r.m.s. deviation was 1.24 %.

Results at high Reynolds numbers, however, depend on other parameters besides Re_{tap}: this can be seen in the experimental results in Sect. 7.2.3, which are very varied at high Reynolds numbers. The cause of the peaks and troughs in the data in Sect. 7.2.3 is not entirely clear: they are probably the result of unsteady effects (e.g. acoustic effects), whereas the basic static-hole error is a steady effect.

7.3.8 Conclusions to Sect. 7.3

Static-hole error has been examined and results presented at both high and low Reynolds number. The experimental data from this chapter and from Chap. 9 have made a significant contribution to the knowledge about static-hole error at high Reynolds number.

To fit the data in Sect. 7.2.3 much effort went into determining the best equation for Venturi tubes using a static-hole-error model, but, although the static-hole-error model gives understanding it did not provide an equation that commanded general acceptance. Accordingly in (informative) Annex B of ISO 5167-4:2003 the discharge coefficient for a classical Venturi tube with a machined convergent is given as

$$1.010 \pm 3\,\% \quad \text{for}\; 2 \times 10^6 < Re_d < 10^8. \tag{7.10}$$

It may be that this range has too high a mean: a population of 28 Venturi tubes used to construct and to test the new wet-gas equation derived in Chap. 11 have an average discharge coefficient of 0.9971 and a spread (2 standard deviations) of ± 2.7 %. Geach and Jamieson (2005) present a graph with a similar mean and a spread of ± 2 %. The discharge coefficient given by Eq. 7.9 only reaches 1.010 when β is equal to about 0.65 and above; at $Re_d = 10^7$, $D = 100$ mm and $d_{tap} = 4$ mm, C varies from 1.0056 at $\beta = 0.4$ to 1.0096 at $\beta = 0.75$.

It is worth noting that the discharge coefficient increases with Re_{tap}; a very large Venturi tube in water could have a throat Reynolds number of 10^7, but the same value of Re_{tap} and the same discharge coefficient as a Venturi tube whose throat diameter and throat Reynolds number are ten times smaller.

Tappings must be perpendicular to the wall, sharp and without burrs.

Tappings of diameter d_{tap} should have a depth, h, at least equal to 2.5 d_{tap}, as required by 5.4.5 of ISO 5167-4:2003. If it is possible to make tappings that are longer than 2.5 d_{tap} that is desirable. For higher tapping-hole Reynolds numbers the discharge coefficient changes with h/d_{tap} even for $h/d_{tap} \geq 2.5$

One important point is that the standards depend on using the tappings as described: if much smaller tappings are used than those prescribed in ISO 5167-4:2003 the discharge coefficient will be significantly different from that in ISO 5167-4:2003; so there will be a corresponding error in the measured flowrate if the discharge coefficient is taken from ISO 5167-4:2003.

It is possible to calibrate a Venturi tube and thus to reduce the uncertainty: it is desirable that calibration should be undertaken at at least two static pressures so that Reynolds number and throat velocity effects can be distinguished.

NOTE It is worth noting that static-hole error is rarely important with orifice plates: Venturi tubes are so seriously affected because of the tappings in the throat, where the velocity is high; the velocity in the pipe is smaller than that in the throat by a factor of approximately β^2, and the upstream tapping term (here meaning that due to static-hole error) is thus smaller than the throat tapping term by a factor of approximately β^4. The ratio of the orifice upstream or downstream tapping term to the differential pressure is smaller than the ratio of the Venturi throat tapping term to the differential pressure by a factor of at most $0.36\beta^4$, assuming the orifice plate discharge coefficient is 0.6, and in most cases by a factor less or much less than $0.36\beta^4$ since (except with a D tapping and an upstream flange tapping in a small pipe) the upstream and downstream pressure tappings are in fairly low velocity regions of the pipe. These points can be seen by rewriting Eq. 1.15 for incompressible flow as

$$\Delta p = \frac{1}{2}\rho\left(\frac{\bar{u}_{th}}{C}\right)^2\left(1-\beta^4\right) = \frac{1}{2}\rho\left(\frac{\bar{u}_{up}}{\beta^2 C}\right)^2\left(1-\beta^4\right).$$

7.4 Improved Shape

7.4.1 General

It is possible to change the shape of the Venturi tube and to reduce the uncertainty. The standard Venturi tube has a large pressure rise along the wall in the throat downstream of the upstream corner (see Fig. 3.3). This can be reduced by reducing the convergent angle. Work was also undertaken investigating different shapes with longer convergents, ones without sharp corners or even curved Venturi tubes. This work is described in Appendix 7.A and in Reader-Harris et al. (2001, 2002).

The conclusion of the work in Appendix 7.A is that the optimum shape of a Venturi tube is as described in ISO 5167-4:2003 in every respect except that the convergent angle should be 10.5°.

7.4.2 Venturi Tube with Convergent Angle 10.5°

Figure 7.11 shows all the data for Venturi tubes with a convergent angle of 10.5° and sharp corners with 4 mm triple-T tappings. Also shown are the data for a meter which has a 10.5° convergent but with rounded corners of the radius required by an 'as cast' Venturi tube. Computational Fluid Dynamics (CFD) was used to calculate the effect of such rounded corners on a Venturi tube with a 10.5° convergent without tappings: for $\beta = 0.6$ and $Re_D = 10^6$ the rounded corners caused a shift in discharge coefficient from that with sharp corners of—0.076 %.

The data are presented in terms of the Venturi-throat-tapping Reynolds number

$$Re^* = \frac{d_{tap}}{d} Re_d, \tag{7.11}$$

where d_{tap} is the diameter of the throat tapping and Re_d is the throat Reynolds number, which (see Sect. 1.4) is given by

$$Re_d = \frac{Re_D}{\beta}. \tag{7.12}$$

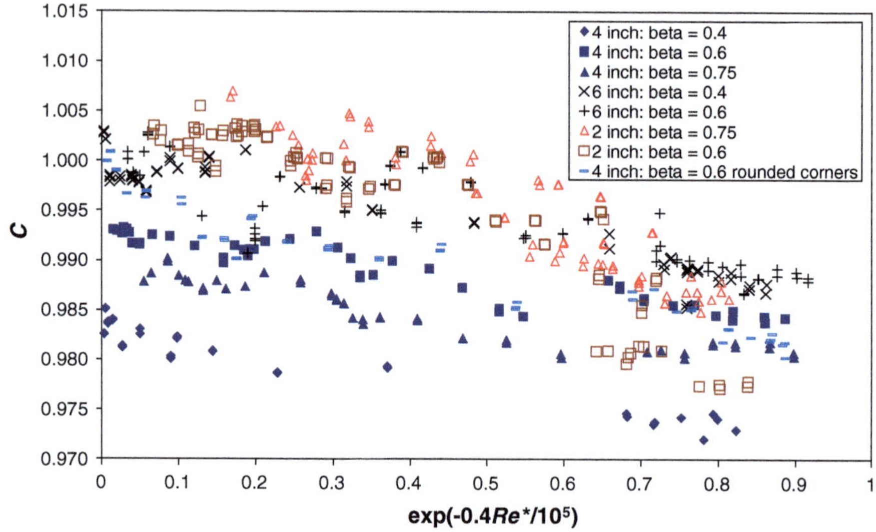

Fig. 7.11 All the gas and water discharge-coefficient data for Venturi tubes with convergent angle 10.5°

Re^* is used rather than Re_{tap} (see Eq. 7.2), because a measurement of friction factor in the throat is not available; however, Re^* does not have as much physical significance as Re_{tap}.

It is clear that the 4″ $\beta = 0.4$ data are very different from the data in the other sets. However, the surface roughness of the Venturi tubes differs significantly from one Venturi tube to another. The mean values of roughness (taking the average of all measurements) and the relative roughness are given in Table 7.1. It is not always possible to measure the roughness in the throat (where it has most effect on the discharge coefficient).

CFD (Reader-Harris et al. 2005) was used to determine the effect of roughness on discharge coefficients with infinitely small tappings: from Eq. 7.3 the friction factor in the throat, λ_{th}, is defined as

Table 7.1 Roughness of Venturi tubes with convergent angle 10.5°

Pipe NB	β	R_a (μm)	$10^5\ R_a/d$
2″	0.6	0.3	0.8
2″	0.75	0.5	1.2
4″	0.4	4.8	12
4″	0.6	3.6	5.9
4″	0.75	3.5	4.6
6″	0.4	0.8	1.3
6″	0.6	0.2	0.2
4″ (rounded corners)	0.6	1.7	2.8

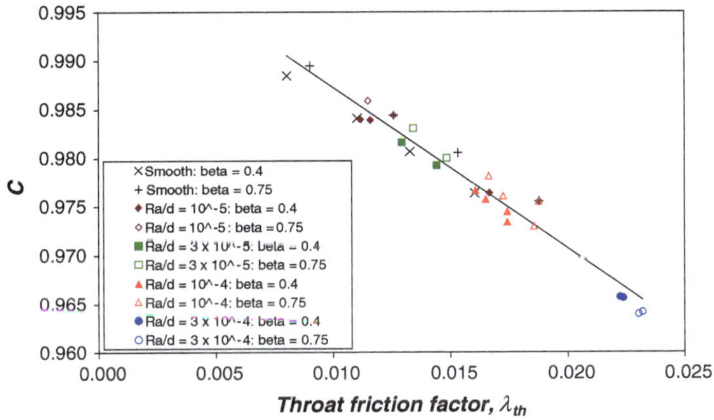

Fig. 7.12 Computed discharge coefficient for Venturi tubes with convergent angle 10.5° as a function of computed friction factor at the throat pressure tappings

$$\lambda_{th} = \frac{8\tau_w}{\rho \bar{u}_{th}^2}, \tag{7.13}$$

where τ_w is the wall shear stress at the throat pressure tappings and $\bar{u}_{th}$ is the mean throat velocity; then it is possible to plot the computed discharge coefficients against λ_{th}. These are shown in Fig. 7.12. The computed results are fitted as

$$C = 1.0039 - 1.66\lambda_{th} \tag{7.14}$$

with a standard deviation of 0.15 %. This is a remarkably small figure. In practice, of course, τ_w and therefore λ_{th} are unknown; one way of estimating λ_{th} is by calculation from the value obtained by assuming that the throat is a long straight pipe. As stated in Sect. 7.3.7 experimental measurements by Lindley (1966) in a Venturi tube of standard convergent angle gave $\lambda_{th} \approx 1.18\lambda_{sp}$, where λ_{sp} is what λ would have been in a long straight pipe.

7.4.3 The Discharge-Coefficient Equation for Venturi Tubes with Convergent Angle 10.5°

Given the dependence of discharge coefficient on friction factor seen in the CFD it is possible to fit all the data in Fig. 7.11 in a way that is consistent with the physics of the flow. This takes account of the dependence on both friction factor and static-hole error (since real Venturi tubes have pressure tappings) by including a term in *Re**. Slightly increasing the coefficient of the friction-factor term from that obtained in the CFD to fit the experimental data gave the following equation:

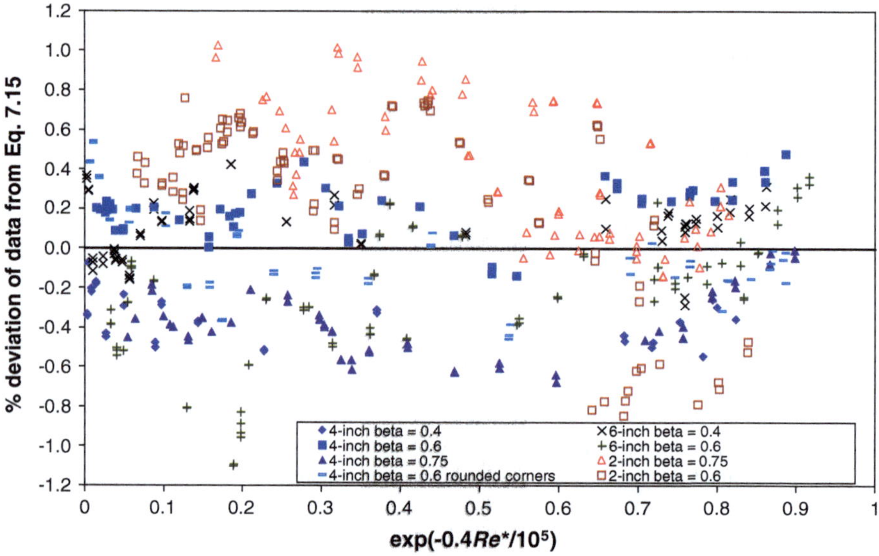

Fig. 7.13 Percentage deviations from Eq. 7.15 for all the gas and water data for Venturi tubes with convergent angle 10.5°

$$C = 1.0252 - 2.5\lambda_{sp} - 0.008\, e^{-0.4Re^*/10^5}, \tag{7.15}$$

where λ_{sp} was obtained from the Colebrook-White Equation (Schlichting 1960) in the following form:

$$\frac{1}{\sqrt{\lambda_{sp}}} = 1.74 - 2\,\mathrm{lg}\left(\frac{2\pi R_a}{d} + \frac{18.7}{Re_d\sqrt{\lambda_{sp}}}\right). \tag{7.16}$$

(see Sect. 1.5; following ISO 80000-2:2009, $\log_{10}$ is written lg). The scatter of the data about Eq. 7.15 (two standard deviations) is 0.80 %, but only 0.65 % if the 2″ data are excluded. The deviations of the measured discharge coefficient data from Eq. 7.15 are shown in Fig. 7.13.

Equation 7.15 is given in ISO/TR 15377:2007 (ISO 2007) with an uncertainty of 1 %. This includes the uncertainty of the reference.

Additional work (4″ Venturi tubes with convergent angle 10.5°, $\beta = 0.6$ and 0.75) comparing data taken in argon with those taken in nitrogen is given in Reader-Harris and Hodges (2006).

An alternative equation calculated so that the static-hole error may be obtained explicitly is given in Appendix 7.C.

The advantage of using this shape of Venturi tube is that it gives a lower uncertainty than standard Venturi tubes. Moreover, it gives more consistent data for installation effects and thus for straight lengths (see Reader-Harris et al. (2004) and Sect. 8.2.4).

7.5 Conclusions

The performance of Venturi tubes in high-pressure gas has been described. The discharge coefficient tends to increase with Reynolds number because of static-hole error. There are also effects of throat velocity: humps and dips in the calibration curve, sometimes an audible tone.

The discharge coefficient depends on the shape of the Venturi tube: of the different shapes tested the best results were obtained with a Venturi tube identical to the standard Venturi tube with a machined convergent except that the convergent angle was 10.5° (instead of 21°).

Pressure tappings should be circular and sharp-edged. Electrical Discharge Machining (spark erosion) is an excellent way of achieving this.

Uncertainty increases with the ratio of tapping diameter to throat diameter (see Sect. 7.A.7), but problems have been found with some data sets taken with 2 mm tappings probably due to the sudden expansion of the 2 mm tappings into a ¼ BSP fitting some distance from the throat itself; moreover, any tapping smaller than 4 mm is too likely to block in the field. Until an explanation is available it seems best to have $d_{tap} \geq 4$ mm, to avoid change in tapping diameter over as long a distance as possible and, where possible, to have $d_{tap}/d \leq 0.07$.

As a minimum a tapping of diameter d_{tap} should be at least 2.5 d_{tap} long. Nevertheless, in some cases (see Sect. 7.A.7) it appears that even using throat tappings of depth 18.5 d_{tap} followed by a sudden expansion from a tapping diameter of 2 mm into a ¼ BSP fitting gives unexpected results.

It is possible to calibrate a Venturi tube and thus to reduce the uncertainty: calibration should be undertaken at at least two static pressures so that Reynolds number and throat velocity effects can be distinguished.

Appendix 7.A: Shape of Venturi Tubes: Tests at NEL

7.A.1 Design

With the intention of reducing the uncertainty in flow measurement using Venturi tubes, Venturi tubes of different shapes (ones with longer convergents, ones without sharp corners or even ones with curved convergents) were tested (Reader-Harris et al. 2001, 2002).

The first design was of a Venturi tube with a machined convergent of angle 10.5° with sharp corners. Data for $\beta = 0.4$, 0.6 and 0.75 were collected first, but are presented below where appropriate. 4 mm pressure tappings connected in triple-T arrangements were used.

It had been suggested at meetings of ISO/TC 30/SC 2 (see Sect. 12.3.2) that better results would be achieved if the sharp corners in a Venturi tube with a machined convergent were rounded so that it were machined with the profile of an 'as cast' convergent. One Venturi tube of this type was lent by Atelier Pochet, of Ransart, Belgium, and is identified as AP1 and referred to subsequently as "rounded corners". Moreover, the first tests had shown that a convergent with an included angle of 10.5° gave particularly good results. So it was decided that the following three Venturi tubes should be manufactured:

- A Venturi tube with a machined convergent of angle 10.5° with rounded corners whose radii of curvature are those required by ISO 5167-1:1991 (ISO 1991) for a Venturi tube with an 'as cast' convergent. This Venturi tube is identified as 29478 and is referred to subsequently as "rounded corners, long".
- A Venturi tube with the distance between upstream and throat tappings equal to that for a Venturi tube with convergent angle 10.5° but with a machined convergent whose wall profile has continuous second derivatives. This Venturi tube is identified as 29479 and is referred to subsequently as "curved, long".
- A Venturi tube with the distance between upstream and throat tappings equal to that for a standard classical Venturi tube but with a machined convergent whose wall profile has continuous second derivatives. This Venturi tube is identified as 29480 and is referred to subsequently as "curved".

AP1 and 29478–29480 were 4″ Schedule 40 with $\beta = 0.6$: the desired values of the diameter of the entrance cylinder, D, and the throat diameter, d, were 102.26 and 61.36 mm respectively.

Much literature was reviewed and many calculations undertaken to determine the detailed form of Venturi tubes 29479 and 29480. Profiles by Witoshinsky and by Spencer (1956) were considered, but since wall profiles with continuous second derivatives were available it was felt that they would be desirable. The details of the wall profiles are given in Reader-Harris and Hodges (2001).

In the case of 29478–29480 the throat is parallel for $d/3$ upstream of the throat tapping and the upstream cylinder is parallel for at least $D/3$ downstream of the upstream tapping. In practice the throat is close to parallel over a longer distance. These parallel sections are significantly longer than those required by ISO 5167-1:1991. These three Venturi tubes were manufactured by ISA Controls Ltd, Shildon, Co. Durham. Drawings of these Venturi tubes are given in Fig. 7.A.1. The profiles which would have been obtained if sharp corners had been used are shown with dotted lines for comparison. In the case of the long convergent the profile with sharp corners is almost indistinguishable on the scale of the graph from that with rounded corners. Figure 7.A.1 also shows the shape of the Venturi tube AP1 manufactured by Atelier Pochet.

The Venturi tubes were manufactured to drawings with tight tolerances designed to ensure that where possible the results were not affected by uncontrolled variables. Details are given in Reader-Harris et al. (2001, 2002). They were made of stainless steel and were suitable for use at a design pressure of 70 barg with ANSI Class 600 flanges. A divergent angle of 7½° was specified for 29478–29480. AP1 had a divergent angle of 7°; it also had a divergent truncated by 26 % of its length.

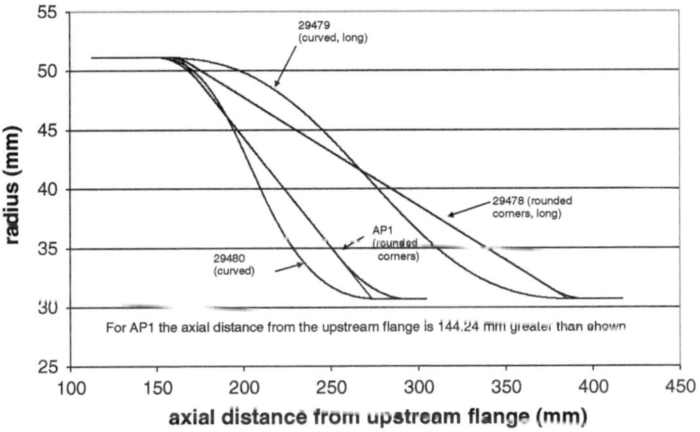

Fig. 7.A.1 Profile from the upstream tapping to the throat tapping for Venturi tubes AP1 and 29478–29480

In the case of AP1 the Venturi tube was to be used elsewhere and the diameter values supplied by the manufacturer were used. The Venturi tube was subsequently measured at NEL, and if the NEL measurements were used the discharge coefficients for AP1 would increase by 0.12 %.

The Standard then in force (ISO 5167-1:1991) required that the surface finish of the entrance cylinder, the convergent section and the throat be such that R_a/d is less than 10^{-5}, where R_a is the arithmetical mean deviation of the roughness profile. The measured values generally exceeded the permitted values by a factor of approximately 2.4; the typical surface finish was $R_a \approx 1.5$ μm. All the Venturi tubes had $10^{-5} < R_a/d < 10^{-4}$. The project wished to use Venturi tubes with surface roughnesses typical of those used in the field. All the Venturi tubes met the roughness requirements of ISO 5167-4:2003.

It was decided that 29478–29480 should have four 4 mm tappings in each tapping plane so that data collected in triple-T arrangements would be comparable with all the data described in Sect. 7.2. It had been observed that the data obtained in gas with 8 mm tappings in previous work had larger humps than the data obtained with 4 mm tappings. So 29478–29480 were each designed to have 1, 2, 6 and 8 mm single tappings in addition to the 4 mm tappings, with the exception of 29480, which did not have 1 mm tappings. For the 4, 6 and 8 mm tappings, the throat pressure tappings were of constant diameter for a depth of 94 mm and the upstream tappings for a depth of 53 mm. As actually manufactured the 1 and 2 mm tappings were much shorter; over the distance between the interior of the Venturi tube and the ¼ BSP fitting there could be up to two changes in tapping diameter. AP1 had four 4 mm tappings upstream and four 2 mm tappings in the throat. In both cases the tappings were connected by annular chambers. The 4 mm tappings were of depth 10 mm and the 2 mm tappings were described on the drawing as being of minimum depth 8 mm.

7.A.2 Calibration in Water

The Venturi tubes were calibrated first in water at NEL. In order to investigate whether the performance of tappings connected in triple-T arrangements differed from that of pairs of single tappings, data were collected with a single pair of 4 mm tappings as well as with triple-T tappings in the case of 29478 and 29479. For each Venturi tube the data in water lay on a straight line as a function of pipe Reynolds number, Re_D, and with a small scatter, provided that Re_D was above a critical value. The pipe Reynolds number below which C was not on the straight line varied, but was typically about 3×10^5. In the case of 29480 there was a significant hump in the data at $Re_D \approx 2.5 \times 10^5$.

The differences between data taken with single tappings of different diameters were surprisingly large. Some of the variation is due to the fact that the tappings are of different diameters, but there is also significant scatter due to variation between tappings of the same diameter (see Appendix 7.B for examples of two Venturi tubes each originally with four pairs of single tappings of the same diameter and depth). Therefore if a Venturi tube is used uncalibrated the use of single tappings instead of triple-T tappings (or other multiple pressure tappings) results in increased uncertainty, although there may be other practical reasons for using single tappings (see also Sect. 4.3.2). This problem was reduced in later work by using spark erosion to insert the tappings into the tube (see Sect. 7.A.7).

7.A.3 Calibration in Gas

The Venturi tubes were calibrated in nitrogen at NEL at two static pressures, 20 and 60 barg. Although discharge coefficients are presented largely in terms of Reynolds number there is an effect of throat velocity in many of the data sets, where peaks and troughs of discharge coefficient occur for both static pressures at the same throat velocity.

7.A.4 Analysis

Since the effect of static hole error at the upstream tapping was much smaller than at the throat tapping it is possible simply to correlate the data with the throat tapping Reynolds number; the simplest presentation of this is to define the Venturi-throat-tapping Reynolds number as in Eq. 7.11:

$$Re^* = \frac{d_{tap}}{d} Re_d, \qquad (7.11)$$

where d_{tap} is the diameter of the throat tapping and Re_d is the throat Reynolds number, which is equal to Re_D/β. Re^* is used rather than Re_{tap}, because a measurement of friction factor in the throat is not available; however, Re^* does not have as much physical significance as Re_{tap}.

Then for each set of data it was found that

$$C - C_{\text{water}} = a - b\, e^{-0.4(Re*/10^5)}, \tag{7.A.1}$$

where a and b are coefficients to be determined. For each data set it was possible to determine both the coefficients in Eq. 7.A.1 and the scatter about the resulting equation. In Figs. 7.A.2, 7.A.3, 7.A.4, 7.A.5 and 7.A.6 the data from Venturi tubes of revised shape taken with 4 mm tappings connected in triple-T arrangements are presented. Data for two standard Venturi tubes (included in Sect. 7.2: their numbers are 28909 and 28909C) and for Venturi tube 98491, which was standard except for convergent angle 31.5°, have been included as Figs. 7.A.7, 7.A.8 and 7.A.9. In each case data for $Re_D < 2 \times 10^5$ were excluded as the discharge coefficient tends to decrease as the Reynolds number decreases below this point. Data for $Re^* < 20000$ have also been excluded as the static hole error decreases rapidly as Re^* decreases

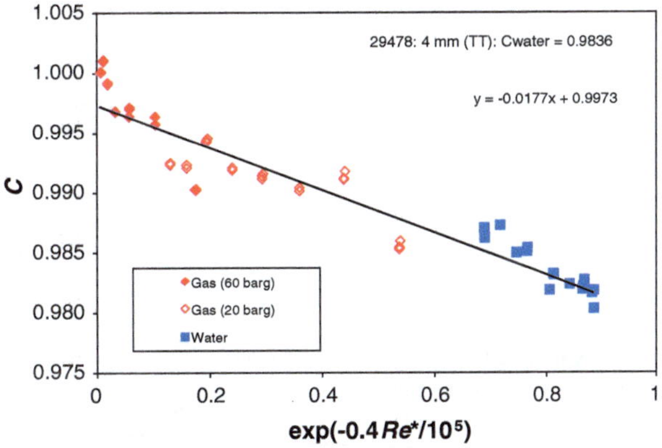

Fig. 7.A.2 Discharge coefficient for Venturi tube 29478 (rounded corners, long) [4 mm (triple-T) tappings]

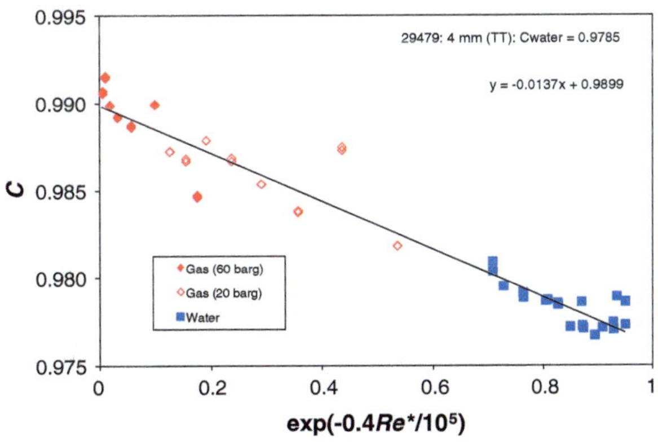

Fig. 7.A.3 Discharge coefficient for Venturi tube 29479 (curved, long) [4 mm (triple-T) tappings]

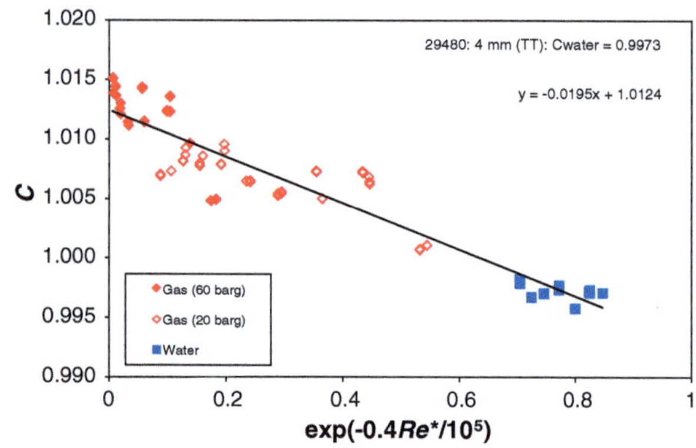

Fig. 7.A.4 Discharge coefficient for Venturi tube 29480 (curved) [4 mm (triple-T) tappings]

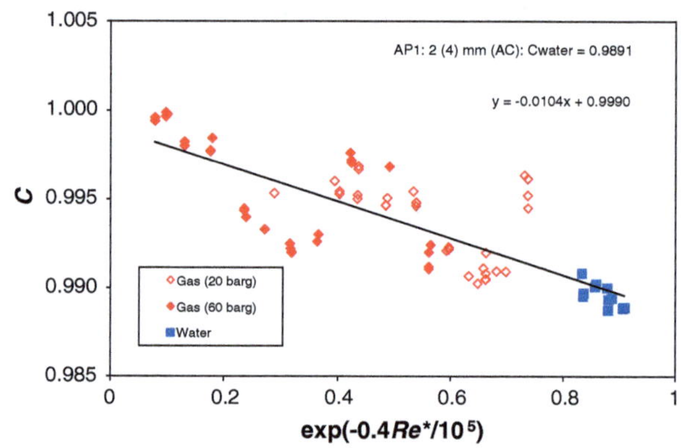

Fig. 7.A.5 Discharge coefficient for Venturi tube AP1 (rounded corners)

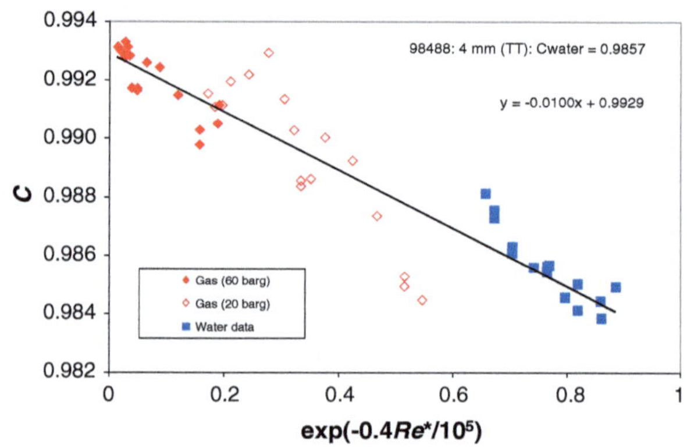

Fig. 7.A.6 Discharge coefficient for Venturi tube 98488 (10.5° convergent angle) [4 mm (triple-T) tappings]

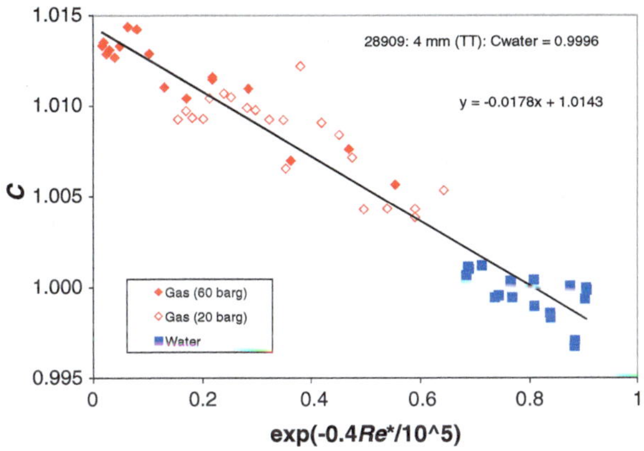

Fig. 7.A.7 Discharge coefficient for Venturi tube 28909 (standard) [4 mm (triple-T) tappings]

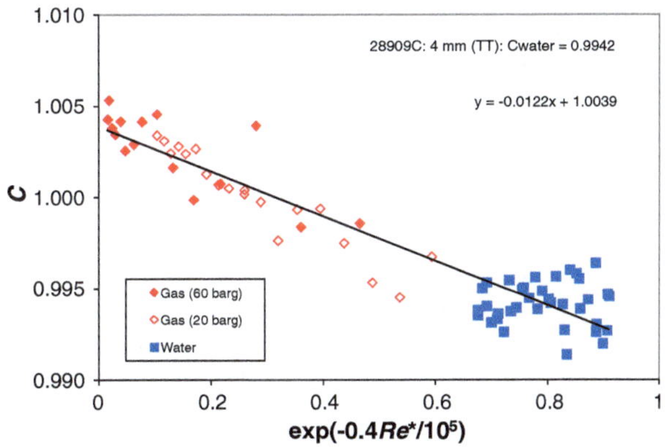

Fig. 7.A.8 Discharge coefficient for Venturi tube 28909C (standard) [4 mm (triple-T) tappings]

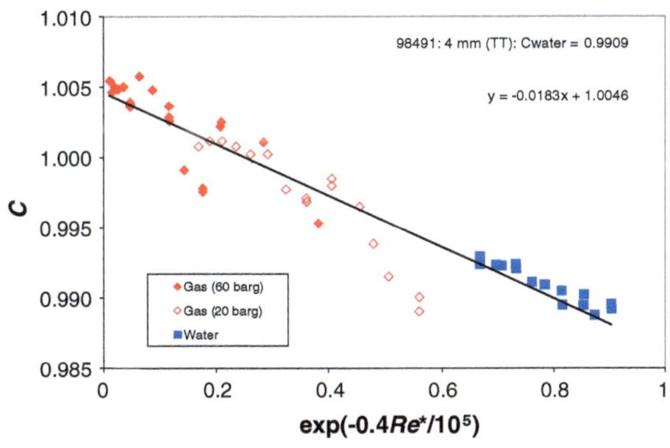

Fig. 7.A.9 Discharge coefficient for Venturi tube 98491 (31.5° convergent angle) [4 mm (triple-T) tappings]

below this point. Where data in water start to depart from a linear fit to the data as the Reynolds number decreases, for instance to form a hump, all the data below an appropriate Reynolds number were excluded.

From this work the good agreement between water and gas data is clear. The line fits from Figs. 7.A.2, 7.A.3, 7.A.4, 7.A.5, 7.A.6, 7.A.7, 7.A.8 and 7.A.9 and other data that were collected but are not presented here graphically are included in Table 7.A.1, which shows the line fits in terms both of C and of $C - C_{\text{water}}$ and the standard deviation of the data about the line fits.

Table 7.A.1 Equations fitted to water and gas data collected with Venturi tubes of different shapes

N.B. (inch)	β	Venturi shape	Serial no.	Type of tappings	C	$C - C_{\text{water}}$	Standard deviation
4	0.6	Rounded corners, long	29478	1 mm	$1.0153 - 0.0206x$	$0.0209 - 0.0206x$	0.00129
4	0.6	Rounded corners, long	29478	2 mm	$1.0005 - 0.0082x$	$0.0072 - 0.0082x$	0.00105
4	**0.6**	**Rounded corners, long**	**29478**	**4 mm (triple-T)**	$\mathbf{0.9973 - 0.0177}x$	$\mathbf{0.0137 - 0.0177}x$	**0.00191**
4	0.6	Rounded corners, long	29478	6 mm	$1.0338 - 0.0192x$	$0.0131 - 0.0192x$	0.00416
4	0.6	Rounded corners, long	29478	8 mm	$1.0037 - 0.0225x$	$0.0144 - 0.0225x$	0.00478
4	0.6	Rounded corners, long	29478	4 mm ('B')	$1.0233 - 0.0206x$	$0.0161 - 0.0206x$	0.00183
4	0.6	Curved, long	29479	1 mm	$0.9893 - 0.0062x$	$0.0051 - 0.0062x$	0.00056
4	0.6	Curved, long	29479	2 mm	$0.9923 - 0.0109x$	$0.0101 - 0.0109x$	0.00112
4	**0.6**	**Curved, long**	**29479**	**4 mm (triple-T)**	$\mathbf{0.9899 - 0.0137}x$	$\mathbf{0.0114 - 0.0137}x$	**0.00123**
4	0.6	Curved, long	29479	6 mm	$1.0102 - 0.0243x$	$0.0179 - 0.0243x$	0.00215
4	0.6	Curved, long	29479	8 mm	$0.9674 - 0.0172x$	$0.0096 - 0.0172x$	0.00362
4	0.6	Curved, long	29479	4 mm ('A')	$1.0004 - 0.0189x$	$0.0154 - 0.0189x$	0.00169
4	0.6	Curved	29480	2 mm	$1.0096 - 0.0144x$	$0.0135 - 0.0144x$	0.00172
4	**0.6**	**Curved**	**29480**	**4 mm (triple-T)**	$\mathbf{1.0124 - 0.0195}x$	$\mathbf{0.0151 - 0.0195}x$	**0.00196**
4	0.6	Curved	29480	6 mm	$1.0167 - 0.0225x$	$0.0165 - 0.0225x$	0.00262
4	0.6	Curved	29480	8 mm	$1.0130 - 0.0280x$	$0.0170 - 0.0280x$	0.00373
4	0.6	Rounded corners	AP1	2 mm throat, 4 mm upstream: annular chambers	$0.9990 - 0.0104x$	$0.0099 - 0.0104x$	0.00193
4	**0.6**	**10.5° convergent**	**98488**	**4 mm (triple-T)**	$\mathbf{0.9929 - 0.0100}x$	$\mathbf{0.0072 - 0.0100}x$	**0.00111**
4	**0.6**	**Standard**	**28909**	**4 mm (triple-T)**	$\mathbf{1.0143 - 0.0178}x$	$\mathbf{0.0147 - 0.0178}x$	**0.00140**
4	**0.6**	**Standard**	**28909C**	**4 mm (triple-T)**	$\mathbf{1.0039 - 0.0122}x$	$\mathbf{0.0097 - 0.0122}x$	**0.00138**
4	**0.6**	**31.5° convergent**	**98491**	**4 mm (triple-T)**	$\mathbf{1.0046 - 0.0183}x$	$\mathbf{0.0137 - 0.0183}x$	**0.00167**

x represents $\exp(-0.4Re^*/10^5)$; data taken with 4 mm (triple-T) tappings are in bold

Given that the standard deviation depends on the tapping diameter it was desirable to see how for 4″ $\beta = 0.6$ Venturi tubes with 4 mm triple-T tappings the standard deviation depends on the shape of the Venturi tube. On the basis of Table 7.A.2 it appeared that the best choice for subsequent work was the Venturi tube with sharp corners which is standard except for a convergent angle of 10.5°. Since two other Venturi tubes of this shape had already been manufactured and tested, numbers 98487 and 98489 of diameter ratio 0.4 and 0.75 respectively, their water and gas data are shown in Figs. 7.A.10 and 7.A.11 for comparison. The equations for 4″ Venturi tubes with convergent angle 10.5° and 4 mm triple-T tappings are presented in Table 7.A.2.

On the basis of Table 7.A.2 it was clear that the relationship between water and gas data as expressed by the equations for $C - C_{water}$ is remarkably similar for the three Venturi tubes. The standard deviation is very low. The next stage was to manufacture and test a larger population of such Venturi tubes.

One other important area is the effect of tapping diameter on Venturi tube discharge coefficients. It is clear from Table 7.A.1 that for three Venturi tubes the standard deviation increased for tapping diameters greater than 4 mm. It was obviously important to establish whether the same is true for standard Venturi tubes. Within earlier work (Reader-Harris et al. 1999) two Venturi tubes, 28908 and

Table 7.A.2 Equations fitted to water and gas data for 4″ Venturi tubes with convergent angle 10.5° and 4 mm triple-T tappings

N.B. (inch)	β	Venturi shape	Serial no.	Type of tappings	C	$C - C_{water}$	Standard deviation
4	0.4	10.5° convergent	98487	4 mm (triple-T)	$0.9831 - 0.0124x$	$0.0092 - 0.0124x$	0.00111
4	0.6	10.5° convergent	98488	4 mm (triple-T)	$0.9929 - 0.0100x$	$0.0072 - 0.0100x$	0.00111
4	0.75	10.5° convergent	98489	4 mm (triple-T)	$0.9889 - 0.0104x$	$0.0077 - 0.0104x$	0.00124

x represents $\exp(-0.4Re^*/10^5)$

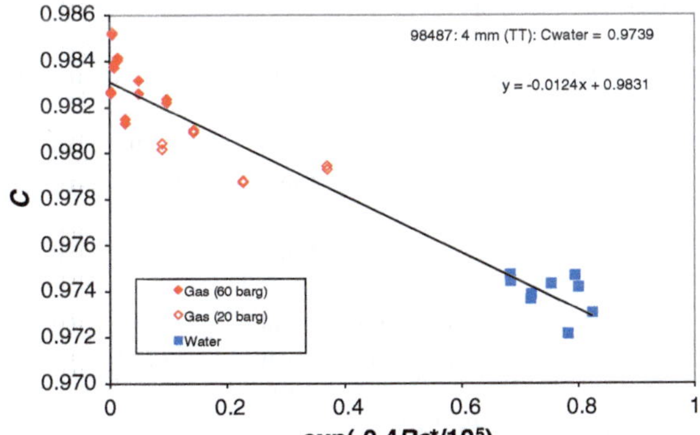

Fig. 7.A.10 Discharge coefficient for Venturi tube 98487 [4 mm (triple-T) tappings, 10.5° convergent angle, $\beta = 0.4$]

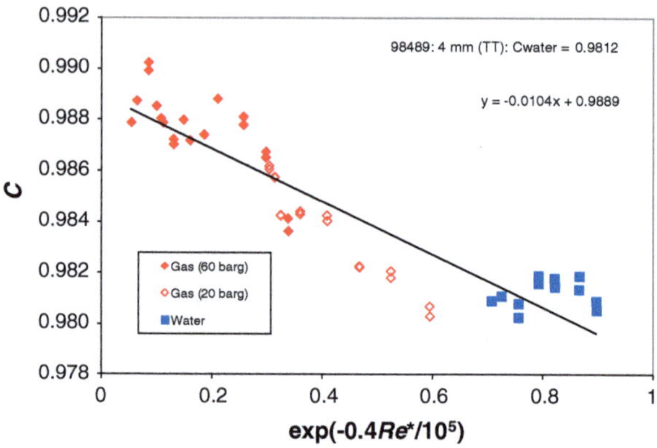

Fig. 7.A.11 Discharge coefficient for Venturi tube 98489 [4 mm (triple-T) tappings, 10.5° convergent angle, $\beta = 0.75$]

28911, of diameter ratio 0.5 and 0.7 respectively, had tapping diameters modified. Given that the tappings for smaller tapping diameters were achieved using inserts those data have not been reanalysed here. However, the 8 mm tappings were obtained by drilling, and those data together with the 4 mm tapping data have been reanalysed and presented in Table 7.A.3, although they are not plotted here.

On the basis of Tables 7.A.1 and 7.A.3 it is possible to see the effect of the diameter of pressure tappings on the standard deviation. Data from Venturi tubes with more than one diameter of pressure tapping are given in Fig. 7.A.12. The 4 mm single tapping and triple-T tapping data are slightly offset from 4 mm so that they can be distinguished. It is interesting to note that the standard deviation increased rapidly as the tapping diameter increased from 4 to 6 mm. Whether 4 mm triple-T tappings actually perform significantly better as regards standard deviation than 4 mm single tappings is not clear. The main advantage of triple-T tappings is that when a Venturi tube is used uncalibrated the discharge coefficient is likely to be

Table 7.A.3 Equations fitted to water and gas data for 4″ Venturi tubes with $\beta = 0.5$ and 0.7

N.B. (inch)	β	Venturi shape	Serial no.	Type of tappings	C	$C - C_{water}$	Standard deviation
4	0.5	Standard	28908	4 mm (triple-T)	$1.0130 - 0.0195x$	$0.0156 - 0.0195x$	0.00116
4	0.5	Standard	28908	8 mm	$1.0098 - 0.0326x$	$0.0195 - 0.0326x$	0.00409
4	0.5	Standard	28908	4 mm ('A')	$1.0181 - 0.0264x$	$0.0203 - 0.0264x$	0.00242
4	0.7	Standard	28911	4 mm (triple-T)	$1.0207 - 0.0206x$	$0.0173 - 0.0206x$	0.00200
4	0.7	Standard	28911	8 mm	$1.0089 - 0.0176x$	$0.0111 - 0.0176x$	0.00376
4	0.7	Standard	28911	4 mm ('A')	1.0202 − 0.0199x	0.0155 − 0.0199x	0.00194

x represents $\exp(-0.4Re^*/10^5)$

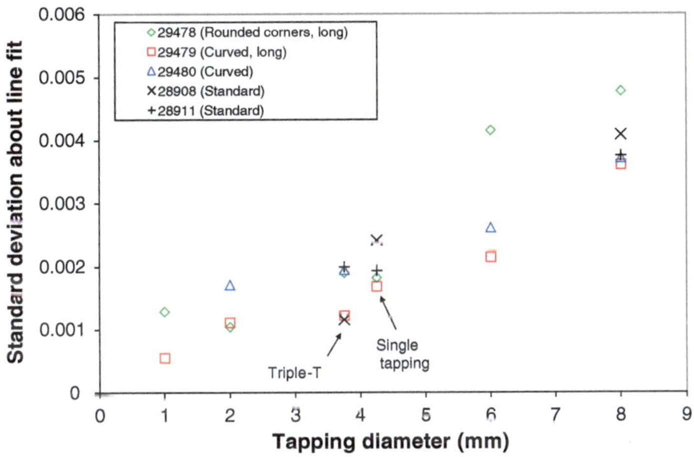

Fig. 7.A.12 The standard deviation of the water and gas discharge-coefficient data about the line fit versus $\exp(-0.4Re^*/10^5)$, as a function of tapping diameter (d was between 51 and 72 mm)

closer to the expected value for that type of Venturi tube than if a single tapping were used. They must also require shorter upstream lengths downstream of some fittings than would be required with the worst choice of circumferential location for single tappings.

The slopes of the different fits versus $\exp(-0.4Re^*/10^5)$ are plotted in Fig. 7.A.13 as a function of the tapping diameter for those Venturi tubes with more than one diameter of pressure tapping. It is noticeable that the slopes increase in absolute value with tapping diameter and that the spread of slope values is smaller for tapping diameters in the range from 2 to 6 mm than it is for tapping diameters of 1 or 8 mm.

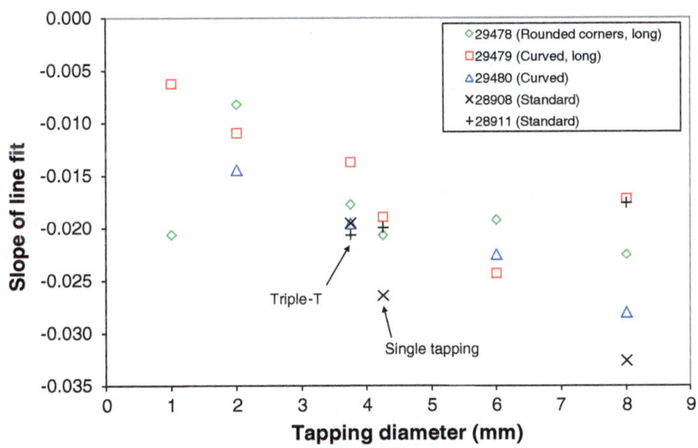

Fig. 7.A.13 The slope of the water and gas discharge-coefficient data about the line fit versus $\exp(-0.4Re^*/10^5)$, as a function of tapping diameter (d was between 51 and 72 mm)

7.A.5 Conclusions on Shape from the 4″ Venturi Tubes

The standard deviation of the data about the best-fit line was determined for each set of data with 4″ diameter and $\beta = 0.6$ but different convergent profiles. The best convergent profile was determined to be one with a 10.5° included angle with sharp corners on the basis that it gives the lowest standard deviation of the data about the fitted line. Moreover, two Venturi tubes with this particular profile and the same diameter but $\beta = 0.4$ and 0.75 were calibrated in water and gas; when they were evaluated in the same way the standard deviations of the data about best-fit lines were low. Moreover, for all three diameter ratios the slope of the line fit was very similar.

The three Venturi tubes manufactured to look for the best profile had single pairs of tappings of diameter 1 mm (except for one Venturi tube), 2, 6, and 8 mm and four pairs of tappings of diameter 4 mm. Data could be collected with a single pair of 4 mm tappings or using all four pairs connected in triple-T arrangements. All the work to determine the best profile used 4 mm tappings in triple-T arrangements. The same process of evaluation used to determine the best profile showed that 8 mm tappings and, to a lesser extent, 6 mm tappings gave worse performance than smaller diameters. 4 mm appeared to be the best choice: uncertainty increased with tapping diameter, but anything smaller than 4 mm was too likely to block.

NOTE All these Venturi tubes had similar throat diameters; so it is not clear from these data that in general performance is diminished by increasing d_{tap}; best performance might, in fact, depend on avoiding too high a value of d_{tap}/d.

7.A.6 Manufacture of Additional Venturi Tubes with 10.5° Convergent Angle and Sharp Corners

Following the work described in Sect. 7.A.5 it was decided that additional Venturi tubes of diameter 2″ and 6″ with diameter ratios of 0.4, 0.6 and 0.75 should be manufactured with a 10.5° convergent angle and sharp corners where the convergent meets the throat and the upstream cylinder. They should have four pairs of tappings of diameter 4 mm and a single pair of tappings of diameter 2 mm.

As with the standard Venturi tubes the revised Venturi tubes were manufactured to drawings with tight tolerances designed to ensure that where possible the results were not affected by uncontrolled variables. A description is given in Reader-Harris et al. (2005).

For the 4 mm tappings the throat pressure tappings were of constant diameter for a depth of 94 mm and the upstream tappings for a depth of 53 mm; for the 2 mm tappings the throat pressure tappings were of constant diameter for a depth of 37 mm and the upstream tappings for a depth of 27 mm. To give consistent tapping quality at least the final portion of each tapping (near the pipe internal wall) was produced using Electrical Discharge Machining (spark erosion).

7.A.7 Calibration of Additional Venturi Tubes in Water and in Gas

The Venturi tubes were calibrated in gas at two static pressures and in water. The 6″ gas data were collected in nitrogen in the NEL high-pressure recirculating loop. The 2″ gas data were collected in air in the NEL gravimetric facility. In order to check the quality of the individual tappings, data were collected in water with each pair of tappings separately. The data from each pair of tappings were then compared with those from each other pair and with the triple-T data. The agreement between the data from different pairs of single tappings was excellent, with a typical spread of line fits to the data of 0.2 %. In the case of the 6″ Venturi tubes the triple-T data were very close to the mean of the single-tapping data, differing by only 0.07 % at most in terms of line fits. In the case of the 2″ Venturi tubes the triple-T data were in one case higher and in the other case lower than the mean of the single-tapping data, differing by about 0.3–0.4 %. For each Venturi tube the data in water lay on a straight line as a function of pipe Reynolds number, Re_D, and with a small scatter, provided that Re_D was above a critical value.

For the sets of data it is possible to determine both the coefficients in Eq. 7.A.1 and the scatter about the resulting equation. In Figs. 7.A.14, 7.A.15, 7.A.16 and 7.A.17 triple-T data for 2″ and 6″ Venturi tubes are presented; the equations are given in Table 7.A.4. Where data in water start to depart from a linear fit to the data as the Reynolds number decreases, all the data below an appropriate Reynolds number (usually $Re_d \approx 5 \times 10^5$) have been excluded.

The 2″ data appear significantly different from the 4″ and 6″ data. This is probably due to the fact that d_{tap}/d is not negligible:

- As already stated above, in the case of the 2″ Venturi tubes (for which d_{tap}/d was 0.10 and 0.13) the triple-T data were in one case higher and in the other case lower than the mean of the single-tapping data, differing by about 0.3–0.4 %, whereas for the 6″ Venturi tubes (for which d_{tap}/d was 0.043 and 0.065) the

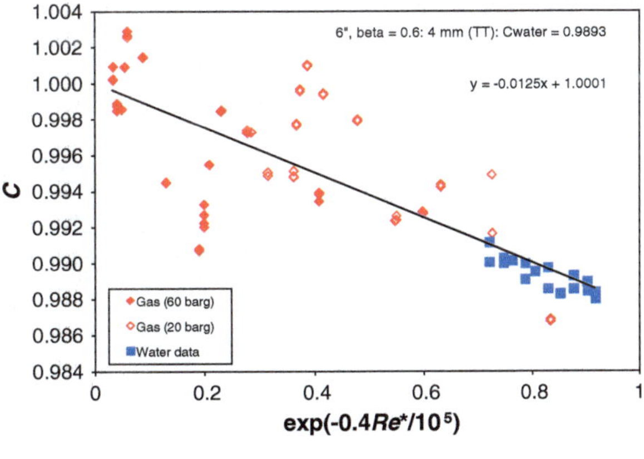

Fig. 7.A.14 Discharge coefficient for Venturi tube 29624 with convergent angle 10.5°: 6″, $\beta = 0.6$, 4 mm triple-T tappings

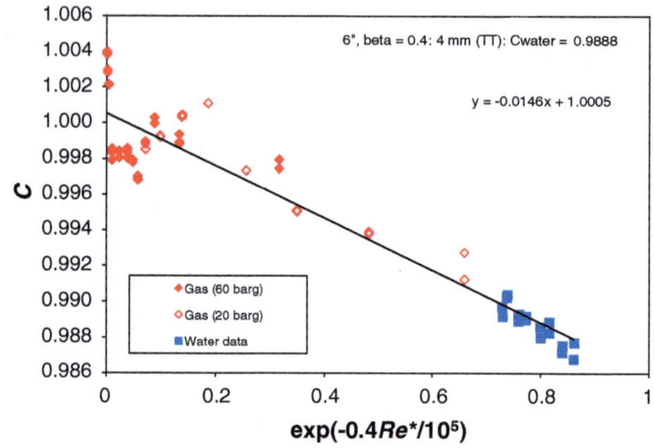

Fig. 7.A.15 Discharge coefficient for Venturi tube 29623 with convergent angle 10.5°: 6″, $\beta = 0.4$, 4 mm triple-T tappings

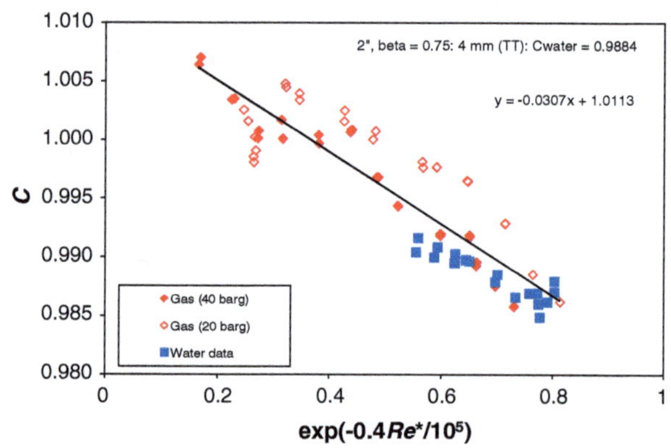

Fig. 7.A.16 Discharge coefficient for Venturi tube 29622 with convergent angle 10.5°: 2″, $\beta = 0.75$, 4 mm triple-T tappings

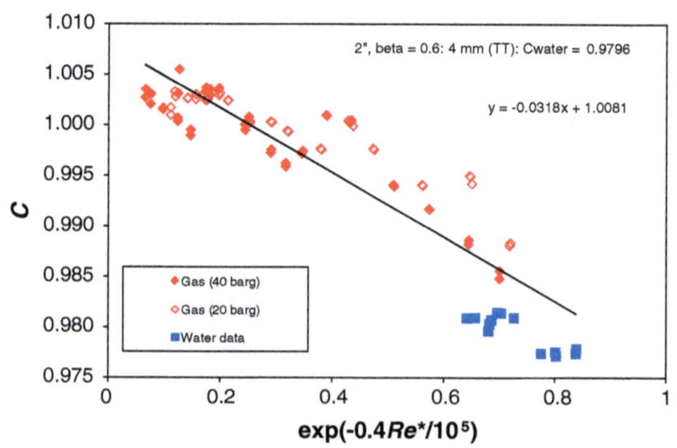

Fig. 7.A.17 Discharge coefficient for Venturi tube 29621 with convergent angle 10.5°: 2″, $\beta = 0.6$, 4 mm triple-T tappings

Table 7.A.4 Equations fitted to water and gas data for 6″ and 2″ Venturi tubes with convergent angle 10.5° and 4 mm triple-T and 2 mm tappings

N.B. (inch)	β	Venturi shape	Serial no.	Type of tappings	C	$C - C_{water}$	Standard deviation
6	0.6	10.5° Convergent	29624	4 mm (triple-T)	$1.0001 - 0.0125x$	$0.0109 - 0.0125x$	0.00267
6	0.6	10.5° Convergent	29624	2 mm	$1.0360 - 0.0314x$	$0.0323 - 0.0314x$	0.00440
6	0.4	10.5° Convergent	29623	4 mm (triple-T)	$1.0005 - 0.0146x$	$0.0117 - 0.0146x$	0.00160
6	0.4	10.5° Convergent	29623	2 mm	$1.0111 - 0.0184x$	$0.0174 - 0.0184x$	0.00240
2	0.75	10.5° Convergent	29622	4 mm (triple-T)	$1.0113 - 0.0307x$	$0.0229 - 0.0307x$	0.00266
2	0.75	10.5° Convergent	29622	2 mm	$1.0189 - 0.0303x$	$0.0268 - 0.0303x$	0.00277
2	0.6	10.5° Convergent	29621	4 mm (triple-T)	$1.0081 - 0.0318x$	$0.0285 - 0.0318x$	0.00354
2	0.6	10.5° Convergent	29621	2 mm	$1.0326 - 0.0469x$	$0.0472 - 0.0469x$	0.00429

x represents $\exp(-0.4Re^*/10^5)$

triple-T data were very close to the mean of the single-tapping data, differing by only 0.07 % at most in terms of line fits.

- In the data of Shaw (1960) the static-hole errors in a straight-pipe are consistent for d_{tap}/d up to and including 0.0625 but differ for d_{tap}/d equal to 0.075 and 0.0875, where d here represents the diameter of the pipe in which the tapping is located.
- In Chap. 9 on nozzles the static-hole errors are consistent for d_{tap}/d up to and including 0.06 but differ for d_{tap}/d equal to 0.08.

The data from Figs. 7.A.12 and 7.A.13 (except those with single 4 mm tappings) together with all the data from Tables 7.A.2 and 7.A.4 (i.e. those for Venturi tubes with convergent angle 10.5°) have therefore been replotted in Figs. 7.A.18 and 7.A.19 as a function of d_{tap}/d.

If only data with tapping diameters of 4 mm or greater were considered, the standard deviation of the data would increase with d_{tap}/d and the slope would increase in magnitude for d_{tap}/d greater than about 0.08. However, some of the 2 mm points are inconsistent with the other points.

For all the 2 mm data there is a sudden expansion in tapping diameter from 2 mm to a higher value at some depth h. For the two 2 mm points with the smallest standard deviation, which are also the two points with the smallest magnitude of slope, h/d_{tap} = 3.5 in one case and 8.5 in the other, both at the throat tappings; in both cases c/d_{tap} = 2. For the other five 2 mm points h/d_{tap} is either 18.5 or 19.5 at the throat tappings and the tapping does not just expand, but enters a ¼ BSP fitting;

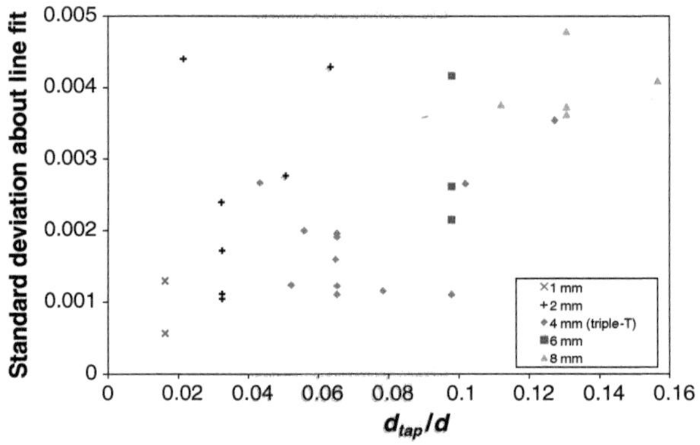

Fig. 7.A.18 The standard deviation of the water and gas discharge-coefficient data (shown by tapping diameter) about the line fit versus $\exp(-0.4Re^*/10^5)$, as a function of d_{tap}/d

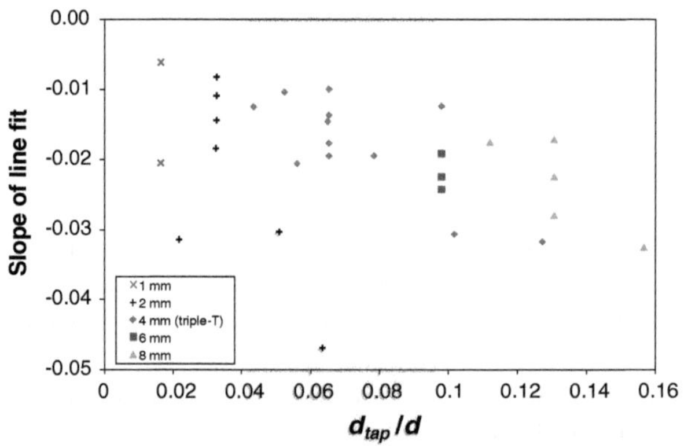

Fig. 7.A.19 The slope of the water and gas discharge-coefficient data (shown by tapping diameter) about the line fit versus $\exp(-0.4Re^*/10^5)$, as a function of d_{tap}/d

the four 2 mm points with the largest standard deviation and the largest magnitude of slope have h/d_{tap} = 18.5: it appears that this tapping arrangement can give surprising results.

Until an explanation is available it seems best to have $d_{tap} \geq 4$ mm, to avoid change in tapping diameter over as long a distance as possible and, where possible, to have $d_{tap}/d \leq 0.07$.

The analysis for Venturi tubes with convergent angle 10.5° which led to the text in ISO/TR 15377:2007 is presented in Sect. 7.4.2. All the 4 mm triple-T tapping data were included, with no restrictions due to d_{tap}/d.

An alternative analysis whose aim is to determine a curve for static-hole error is given in Appendix 7.C.

Appendix 7.B: Depth of Tappings: Tests at NEL

To investigate the effect of changing the depth of the tappings two 4″ Venturi tubes were used, one of which had a diameter ratio of 0.4, the other a diameter ratio of 0.65. Calibrations were carried out in gas at 20 barg. On each Venturi tube the tapping diameter was 4 mm, and the throat pressure tappings were of constant diameter for a depth of 94 mm and the upstream tappings for a depth of 53 mm. First the two Venturi tubes were calibrated with each pair of tappings considered separately. Then one of the throat tappings (A) was left unmodified while three of the throat tappings were drilled with a 9 mm drill so that for the first one (B) the remaining depth of 4 mm diameter tapping became 40 mm (expanding at that point to 9 mm), for the second one (C) the depth of 4 mm tapping became 20 mm and for the third one (D) the depth of 4 mm tapping became 10 mm. The Venturi tubes as modified were then recalibrated. The data from the unmodified tappings (A) are shown in Fig. 7.B.1: the first sets were taken when all the tappings were unmodified, the second sets when all the other tappings had been modified. The fact that the results with the unmodified tappings changed little suggests that the modifications to the other tappings did not have a significant effect on the flow as a whole. As the 4 mm depth was reduced to 40 mm so the discharge coefficient reduced by approximately 0.4 % (see Fig. 7.B.2, which shows the B tappings). As the 4 mm depth was reduced to 20 mm so the plotted data became less linear and the discharge coefficient increased from the initial value by approximately 0.15 % (see Fig. 7.B.3, which shows the C tappings). As the 4 mm depth was reduced to 10 mm so the plotted data became less linear and the discharge coefficient reduced from the initial value by approximately 0.55 % (see Fig. 7.B.4, which shows the D tappings). It is clear that tapping depth has an effect on

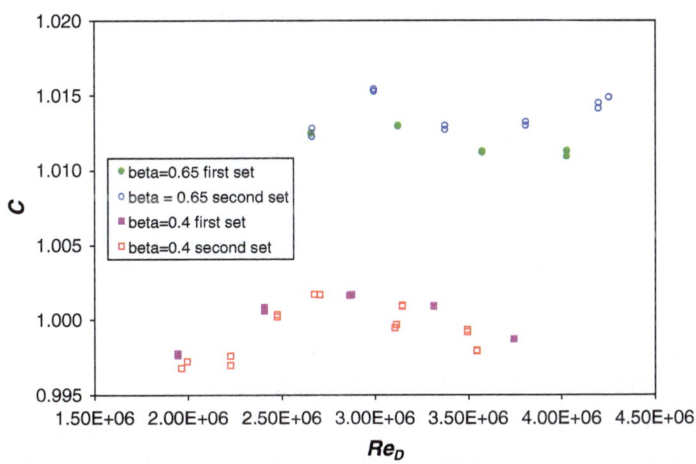

Fig. 7.B.1 Discharge coefficient: 4″ Venturi tubes: unmodified throat tappings (d_{tap} = 4 mm) of depth 94 mm

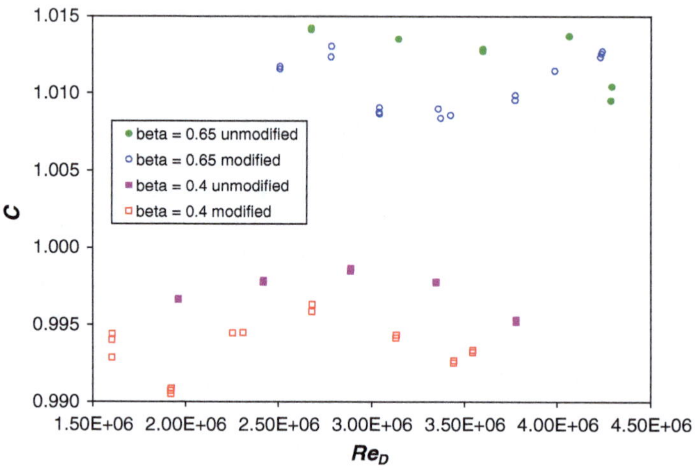

Fig. 7.B.2 Discharge coefficient: 4″ Venturi tubes: throat tappings (d_{tap} = 4 mm) modified from depth 94 mm to depth 40 mm

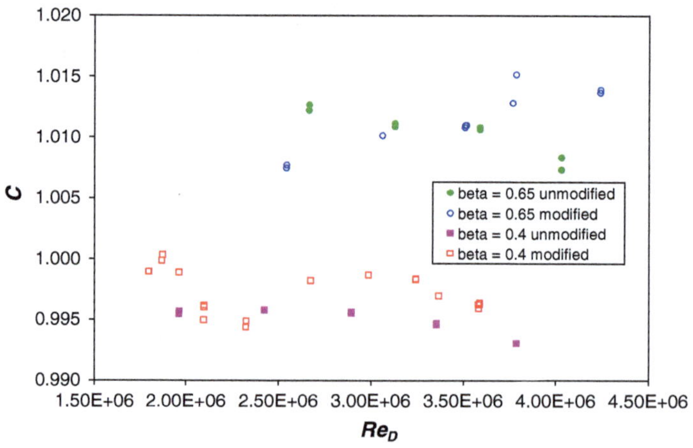

Fig. 7.B.3 Discharge coefficient: 4″ Venturi tubes: throat tappings (d_{tap} = 4 mm) modified from depth 94 mm to depth 20 mm

the discharge coefficient, but it is not easy to determine its exact effect. See the final paragraph of Appendix 7.C for further comments.

NOTE As stated in Sect. 7.3.4 Livesey et al. (1962) considered tappings of depth between 0.3 and 7.4d_{tap}, which beyond the parallel section of stated depth entered a chamber of diameter 14d_{tap}: they found that holes with a depth/diameter ratio less than 5 were sensitive to variations of tapping depth: if the depth/diameter ratio were 3 or less the static hole error was very different from that in a deep hole.

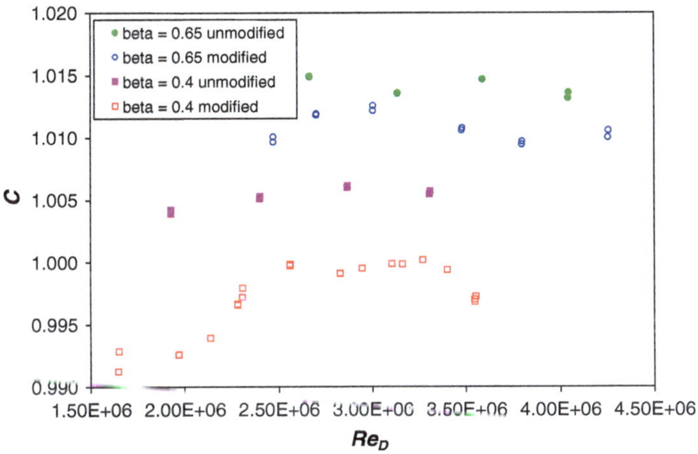

Fig. 7.B.4 Discharge coefficient: 4″ Venturi tubes: throat tappings (d_{tap} = 4 mm) modified from depth 94 mm to depth 10 mm

Shaw (1960) had a tapping on the wall entering a tapping of width $2d_{tap}$: provided the depth was greater than or equal to $1.5d_{tap}$ his results were generally unaffected by depth. In both cases the tapping-hole Reynolds number was much lower than in the Venturi tube data taken in gas and described here.

Appendix 7.C: Refitting the Data With Convergent Angle 10.5°

The aim of the work described in this appendix was to determine the static-hole error for Venturi tubes with convergent angle 10.5°. Accordingly all the data from the Venturi tubes considered in Sect. 7.4.2 were refitted, that is the Venturi tubes with convergent angle 10.5° and sharp corners together with the one which has a 10.5° convergent but with rounded corners of the radius required by an 'as cast' Venturi tube.

The data were fitted on the assumption that the data taken without tappings would have the discharge coefficient given by Eq. 7.14:

$$C = 1.0039 - 1.66\lambda_{th}.$$

Then it was assumed from Eq. 7.5 that the tapping terms were given by

$$\frac{\lambda_{th} f(Re_{tap,th}) - \lambda_{up}\beta^4 f(Re_{tap,up})}{8(1-\beta^4)}$$

where

$$f = \begin{cases} a - be^{-nRe_{tap}} & \text{for } Re_{tap} > 800 \\ \left(a - be^{-800n}\right)\frac{Re_{tap}}{800} & \text{for } Re_{tap} \leq 800. \end{cases} \tag{7.C.1}$$

and to fit Fig. 7.9 it was prescribed that $a - be^{-2000n} = 3.7$

From Eqs. 7.2 and 7.3

$$Re_{tap,up} = \sqrt{\frac{\lambda_{up}}{8}}\frac{d_{tap}}{D}Re_D \quad \text{and} \quad Re_{tap,th} = \sqrt{\frac{\lambda_{th}}{8}}\frac{d_{tap}}{d}Re_d$$

From the computational work

$$\lambda_{th} = (0.711 + 0.0624\ \lg(Re_d))\lambda_{sp} \tag{7.C.2}$$

where λ_{sp} was obtained from the Colebrook-White Equation (Schlichting 1960) (see also Sect. 1.5) in the following form:

$$\frac{1}{\sqrt{\lambda_{sp}}} = 1.74 - 2\lg\left(\frac{2\pi R_a}{d} + \frac{18.7}{Re_d\sqrt{\lambda_{sp}}}\right) \tag{7.C.3}$$

For the upstream tapping

$$\frac{1}{\sqrt{\lambda_{up}}} = 1.74 - 2\lg\left(\frac{2\pi R_{a,up}}{D} + \frac{18.7}{Re_D\sqrt{\lambda_{up}}}\right), \tag{7.C.4}$$

where the roughness $R_{a,up}$ was taken as 1.6 μm. The assumed upstream pipe roughness has little effect on the values of a and b.

It seems reasonable to suppose that in practice Venturi tube i will differ from the theoretical value by its own constant, ΔC_i, due to any defects in the tappings. So C was taken as

$$C = 1.0039 - 1.66\lambda_{th} + \frac{\lambda_{th}f(Re_{tap,th}) - \lambda_{up}\beta^4 f(Re_{tap,up})}{8(1-\beta^4)} + \Delta C_i \tag{7.C.5}$$

Considering only the data with $d_{tap}/d \leq 0.07$ and requiring that for each Venturi tube the mean deviation over the water data (where the tapping terms are most well known) is equal to 0, the best fit to f is given by

$$f = \begin{cases} 10.09 - 7.06e^{-0.00005\,Re_{tap}} & \text{for } Re_{tap} > 800 \\ 0.00413\,Re_{tap} & \text{for } Re_{tap} \leq 800. \end{cases} \tag{7.C.6}$$

ΔC_i varies from −0.0054 to 0.0006, and the deviations of the data from Eq. 7.C.5 are shown in Fig. 7.C.1.

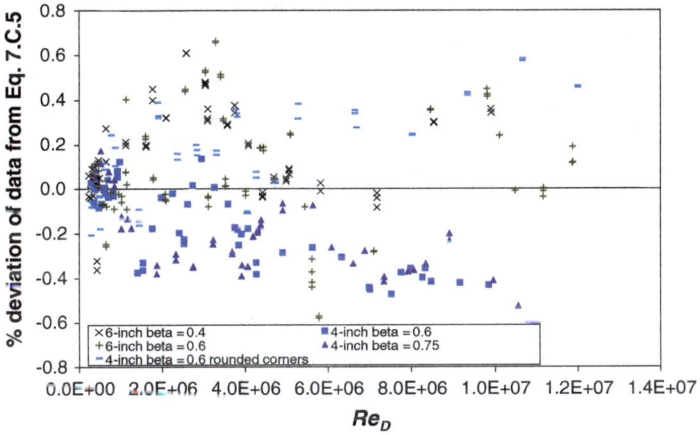

Fig. 7.C.1 Percentage deviation of the data with $d_{tap}/d \leq 0.07$ for Venturi tubes with convergent angle 10.5° from Eq. 7.C.5 with f as in Eq. 7.C.6 and ΔC_i chosen so that the mean deviation over the water data is 0

The effect of increasing the coefficient of λ_{th} by 30 % and recalculating f to obtain a revision to Eq. 7.C.6 is a change in f by 5 % in magnitude at most.

Considering the data with d_{tap} = 2 mm and requiring that for each Venturi tube the mean deviation over the water data (where the tapping terms are most well known) is equal to 0, ΔC_i varies from −0.0015 to 0.0130, and the deviations of the data from Eq. 7.C.5 are shown in Fig. 7.C.2. The data are plotted against throat

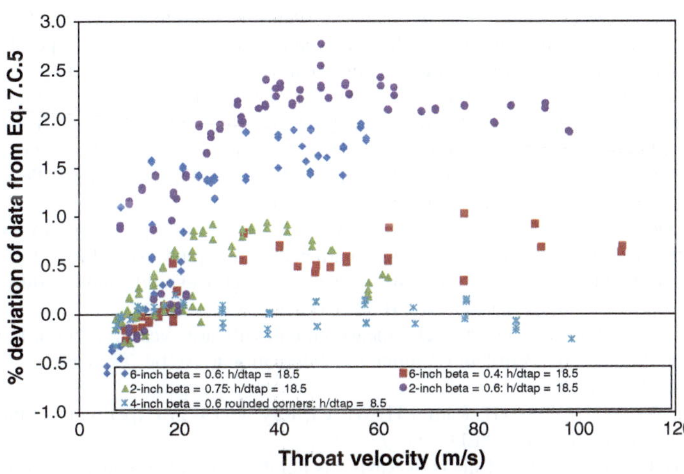

Fig. 7.C.2 Percentage deviation of the data with d_{tap} = 2 mm for Venturi tubes with convergent angle 10.5° from Eq. 7.C.5 with f as in Eq. 7.C.6 and ΔC_i chosen so that the mean deviation over the water data is 0 (values of h/d_{tap} are for the throat tappings)

velocity. It should be noted that for the 6″ $\beta = 0.6$ Venturi tube the edge radius of the 2 mm throat tapping was about 0.4 mm.

The data with $h/d_{tap} = 18.5$ at the throat tappings are very different from the data in Fig. 7.C.1: they give a surprisingly large positive deviation. However, the data with $h/d_{tap} = 8.5$ at the throat tappings are in good agreement with the data in Fig. 7.C.1: moreover, they are consistent with what might have been expected from Figs. 7.B.2 and 7.B.3.

References

Allen CM, Hooper, LJ (1932) Piezometer investigation. ASME Hydraulics Div: HYD-54-1, 54: 1–16

Flack RD Jr (1978) An experimental investigation of static pressure hole errors in transonic flow with pressure gradients. In: Proceedings of Southeast Seminar on Thermal Science: pp 364–378

Franklin RE, Wallace JM (1970) Absolute measurements of static-hole error using flush mounted transducers. J Fluid Mech 42(1):33–48

Geach DS, Jamieson AW (2005) Wet gas measurement in the southern North Sea. In: Proceedings of 23rd North Sea Flow Meas Workshop, Norway

Gibson JJ, Reader-Harris MJ, Gilchrist A (1999) CFD analysis of the static hole error caused by tappings in venturi meters operating in high-pressure gas. In: Proceedings of 3rd ASME/JSME joint fluids engineering conference, San Francisco, FEDSM99-7149. American Society of Mechanical Engineers, New York

ISO (1991) Measurement of fluid flow by means of pressure differential devices—Part 1: orifice plates, nozzles and Venturi tubes inserted in circular cross-section conduits running full. International Organization for Standardization, Geneva. ISO 5167-1

ISO (2003) Measurement of fluid flow by means of pressure differential devices inserted in circular cross-section conduits running full—Part 4: Venturi tubes. International Organization for Standardization, Geneva. ISO 5167-4:2003

ISO (2007) Measurement of fluid flow by means of pressure differential devices—Guidelines for the specification of orifice plates, nozzles and venturi tubes beyond the scope of ISO 5167. International Organization for Standardization, Geneva. ISO/TR 15377:2007

Jamieson AW, Johnson PA, Spearman EP, Sattary JA (1996) Unpredicted behaviour of Venturi flowmeter in gas at high Reynolds numbers. In: Proceedings of 14th North Sea Flow Meas Workshop, Peebles: paper 5. National Engineering Laboratory, East Kilbride, Glasgow

Lindley D (1966) Venturimeters and boundary layer effects. PhD thesis, Cardiff: Dept. of Mech. Eng, Univ. Coll. of South Wales and Monmouthshire

Livesey JL, Jackson JD, Southern CJ (1962) The static hole error problem. An experimental investigation of errors for holes of varying diameters and depths. Aircr Eng 34:43–47

McKeon BJ, Smits AJ (2002) Static pressure correction in high Reynolds number fully developed turbulent pipe flow. Meas Sci Tech 13:1608–1614

Metz WP, Bruchie JD, Fritz RH (1991) Gas measurement with large venturi tubes at Prudhoe Bay, Alaska. In: Society of Petroleum Engineers: International Arctic Technology Conference, Anchorage, Alaska

Moulden TH, Wu FG, Collins HJ, Ramm H, Wu CI, Ray R (1977) Experimental study of static pressure orifice interference. AEDC TR-77–57

Plentovich EB, Gloss BB (1982) Effects of Reynolds number on orifice induced pressure error. In: Proceedings of 12th aerodynamic testing conference. American Institute of Aeronautics and Astronautics, Reston, Virginia

Plentovich EB, Gloss BB (1986) Orifice-induced pressure error studies in Langley 7 by 10 foot high-speed tunnel. NASA Tech Paper 2545

Rainbird WJ (1967) Errors in measurement of mean static pressure of a moving fluid due to pressure holes, Quarterly Bulletin. Division of Mechanical Engineering. National Research Council, Canada, Rep DME/NAE No 3

Ray AK (1956) On the effect of orifice size on static pressure reading at different Reynolds numbers. Ingenieur-Archiv 24(3): 171–181 (translated by Skan S W as Aeronautical Research Council, F.M. 2479)

Rayle RE (1959) Influence of orifice geometry on static pressure measurements. ASME paper 59-A-234

Reader-Harris MJ, Brunton WC, Dickson IR, Hodges D, Nicholson IG (1999) An improved discharge coefficient equation for Venturi tubes. Report No 345/99 on project no WSDC51 for NMSPU, DTI. National Engineering Laboratory, East Kilbride, Glasgow

Reader-Harris MJ, Brunton WC, Gibson JJ, Hodges D, Nicholson IG (2001) Discharge coefficients of Venturi tubes with standard and non-standard convergent angles. Flow Meas Instrum 12(2):135–145

Reader-Harris MJ, Brunton WC, Hodges D, Nicholson IG (2002) Venturi tubes: improved shape. In: Proceedings of 20th North Sea Flow Meas Workshop, St Andrews. National Engineering Laboratory, East Kilbride, Glasgow

Reader-Harris MJ, Gibson J, Hodges D, Nicholson IG, Rushworth R (2005) Venturi tubes with a 10.5° convergent angle: development of a discharge coefficient equation. In: Proceedings of 13th international flow measurement conference Flomeko 2005, Peebles. National Engineering Laboratory, East Kilbride, Glasgow

Reader-Harris M J, Hodges D (2001) Venturi tubes—design optimization and investigation of high discharge coefficient phenomenon. Report no. 279/2001 on project no FDDP01. National Engineering Laboratory, East Kilbride, Glasgow

Reader-Harris MJ, Hodges D (2006) Venturi tubes in different gases. In: Proceedings of 5th International South East Asia Hydrocarbon Flow Meas Workshop, Kuala Lumpur

Reader-Harris MJ, Rushworth R, Gibson J (2004) Installation effects on Venturi tubes of convergent angle 10.5°. In: Proceedings of 22nd North Sea Flow Meas Workshop, St Andrews. National Engineering Laboratory, East Kilbride, Glasgow

Repik EU, Kuzenkov VK (1990) Measuring static pressure with minimum error by means of drainage orifices. Inzhenerno-Fizicheskii Zhurnal 59(2): 203–207

Roshko A (1955) Some measurements of flow in a rectangular cutout. National Advisory Committee for Aeronautics, Technical Note 3488, Washington

Schlichting H (1960) Boundary layer theory. McGraw-Hill, New York

Shaw R (1960) The influence of hole dimensions on static pressure measurements. J Fluid Mech 7: 550–564, previously issued as The measurement of static pressure. Aeronautical Research Council, F.M. 2792, 2nd March 1959

Spencer EA (1956) The M.E.R.L.-design Venturimeters. Fluids note no 40. Mechanical Engineering Research Laboratory, East Kilbride, Glasgow

Stratford BS, Ascough JC (1962) A recent attempt at accurate airflow measurement by Pitot-static traverse. In: Proceedings of symposium at National Engineering Laboratory, Flow Measurement in Closed Conduits, I (paper A-6), pp 67–81. HMSO, Edinburgh, September 1960)

Van Weers T, Van Der Beek MP, Landheer IJ (1998) Cd—factor of classical Venturi's: gaming technology? In: Proceedings of 9th international conference flow measurement, FLOMEKO, Lund, Sweden: 203-207

Zogg H, Thomann H (1971) Errors in static pressure measurements due to protruding pressure taps. J Fluid Mech 54(3):489–494

Chapter 8
Installation Effects

Abstract Most flowmeters are affected by how and where they are installed. Generally meters are calibrated (or designed in accordance with ISO 5167) to give a predictable performance when installed where the flow profile approximates to a fully developed flow profile at the Reynolds number of the flow. A flow profile that is disturbed by an upstream fitting may cause a change in the performance. All differential-pressure meters change their performance if flow disturbance is introduced, with orifice plates being particularly affected. Orifice plates do, however, have the advantage over many other meters in that the effect of installation is often known. There are huge amounts of data on installation effects on orifice plates, but very limited data on installation effects on Venturi tubes. The effect of upstream fittings and pipework is considered in terms of peakiness of profile, asymmetry and swirl. The basis on which the straight lengths in ISO 5167:2003 were derived is described. In practice it is necessary to follow ISO 5167-2:2003/ISO 5167-4:2003 to provide minimum straight lengths. Where it is not possible to conform to Table 3 of ISO 5167-2:2003/Table 1 of ISO 5167-4:2003, alternative options, some including the use of flow conditioners, are described. The basis and use of the flow conditioner test in ISO 5167-1:2003 are described. Limits on pulsation are described.

8.1 Introduction

Most flowmeters are affected by how and where they are installed. Generally meters are calibrated (or designed in accordance with ISO 5167) to give a predictable performance when installed where the flow profile approximates to a fully developed flow profile at the Reynolds number of the flow. A flow profile that is disturbed by an upstream fitting may cause a change in the performance. All differential-pressure meters change their performance if flow disturbance is introduced, with orifice plates being particularly affected.

M. Reader-Harris, *Orifice Plates and Venturi Tubes*,
Experimental Fluid Mechanics, DOI 10.1007/978-3-319-16880-7_8

Orifice plates do, however, have the advantage over many other meters in that the effect of installation is often known. The effect of upstream fittings and pipework can be considered in terms of peakiness of profile, asymmetry and swirl. For both orifice plates and Venturi tubes the effect of a peaked profile (due, for instance, to a roughened pipe) is to reduce the pressure drop for a given flowrate and thus increase the discharge coefficient (the more flow there is in the centre of the pipe the less the pressure drop). By the same argument, the effect of a flattened profile is to reduce the discharge coefficient. An asymmetric flow reduces the discharge coefficient because it requires more energy (hence differential pressure) to move an asymmetric flow through a constriction than the same flow arranged symmetrically. The effect of swirl is more complex. Swirl is nearly always accompanied by a change in axial velocity profile; on an orifice plate the effect of swirl itself is to increase the discharge coefficient [e.g. (McHugh et al. 1981)], presumably by increasing the diameter of the *vena contracta*; on a Venturi tube there is negligible effect of swirl itself [see Gibson and Reader-Harris (2007)], because the fluid continues to fill the whole throat. Swirl is accompanied by a flattened and generally asymmetric velocity profile, which as stated earlier reduces the discharge coefficient.

Most of this section covers upstream straight lengths. There is a little information on downstream straight lengths and on the effect of pulsating flow. Some items that might be termed installation effects but that are specific to orifice plates are in Chap. 2, e.g. the effect of a reversed orifice plate.

8.2 Upstream Straight Lengths

8.2.1 General

The most important type of installation effect is that of upstream flow profile (i.e. velocity (and turbulence) profile). This is controlled by having sufficient upstream straight length that the profile at the meter is sufficiently close to the fully developed profile in a long straight pipe (see Sect. 1.4).

Where a disturbance to the flow profile creates an asymmetric flow profile this may be expected to become close to a fully developed profile after about 50D. Where a disturbance to the flow profile creates a swirling flow this can, however, take around 200D to approach fully developed flow (see Appendix 8.A). This is of course a very approximate estimate as the fluid viscosity, the pipe roughness and the severity and nature of the swirl can make a huge difference to these estimates. A cautious baseline for research work is to employ a correctly installed flow straightener/conditioner (i.e. one with its required upstream length) followed by 70D of straight pipe.

Sometimes if there is a coating on an orifice plate the presence of a swirling flow can be detected from a visible pattern on the orifice plate (the effect of the coating itself on the discharge coefficient is described in Sect. 2.2.3).

According to 7.3 of ISO 5167-1:2003 (ISO 2003a) it is sufficient (but not necessary) that when the flow is measured using ISO 5167 the swirl angle be less than 2° at each point in the pipe and the axial profile be within 5 % of that which would be achieved in swirl-free flow after 100D of straight pipe. This specification is not very helpful, since to carry out the necessary measurements requires the use of a Pitot tube, laser Doppler anemometry or equivalent methods: such measurements are very rarely made. Sets of data in which both velocity profiles and shifts in discharge coefficient are measured are rare: e.g. Ferron (1963) and Reader-Harris et al. (1995a). The latter paper was based on work done for the Headers Consortium (Reader-Harris et al. 1995b). In the report for the Consortium H^* (there without the asterisk) is defined as the maximum percentage difference in u/u_{cl} from that in a fully developed flow over the range $r'/R \leq 0.96$, where u/u_{cl} is the ratio of the velocity at a point to that on the centre-line, r' is the distance from the centre line to a point and R is the pipe radius; it was then stated that 'from both experiment and computation it can be seen that $H^* \leq 5$ does not guarantee shifts less than 0.25 % for $\beta = 0.6$. It may be the case that limits on H^* and on turbulence levels are necessary to guarantee sufficiently small shift in discharge coefficient; nevertheless a rule that H^* should not exceed 5 is quite close to the required criterion for an acceptable velocity profile for orifice metering in installations where specific straight lengths have not been determined.' The criterion in 7.3 of ISO 5167-1:2003 is probably sufficient (although not always necessary) for a Venturi tube and for a small β orifice plate, but it is not always sufficient for a large β orifice plate.

In practice required straight lengths are determined by undertaking laboratory tests: the errors due to particular upstream fittings are measured, and then the necessary lengths are chosen so that the errors due to installation are not more than a specified figure.

The effect of velocity profile on discharge coefficient has been analysed by Morrison et al. (1992, 1993, 1995) who considered the relationship between shift in discharge coefficient and axial momentum flux, radial moment of axial momentum flux and two higher order moments. Although such studies are useful they have never enabled the time-consuming task of measuring the effect of individual fittings to be avoided.

The effect of Reynolds number on required straight lengths is generally negligible (some very strongly swirling flows may be an exception to this rule).

There is an immense quantity of data on installation effects on orifice plates. Some information and analysis are presented here.

There are very limited data on Venturi-tube installation effects; most of the available data are presented here.

Section 8.2 covers the work leading to Table 3 of ISO 5167-2:2003 (ISO 2003b) and Table 1 of ISO 5167-4:2003 (ISO 2003c) and what users should do if their installation is not covered in those tables.

8.2.2 Definitions

The measurement of length between a fitting and an orifice plate is the distance between the downstream end of the fitting and the upstream face of the orifice plate. The definition of the downstream end of the fitting varies in different papers, but for ISO 5167-2:2003 it is clear: it is the end of the curved portion of a bend or of the tapered portion of a contraction or an expansion (see Note 1 of Table 3 of ISO 5167-2:2003); the weld neck should be considered to be part of the straight length.

The measurement of length between a fitting and a Venturi tube is the distance between the downstream end of the fitting and the upstream tapping plane of the Venturi tube. The former should be taken to be the end of the curved portion of a bend or of the tapered portion of a contraction or an expansion (see Table 1 of ISO 5167-4:2003); the weld neck should be considered to be part of the straight length.

The shift in discharge coefficient for a particular installation condition is determined by subtracting the mean discharge coefficient obtained with that installation from the baseline discharge coefficient obtained in good flow conditions.

8.2.3 Orifice Plates

8.2.3.1 History

This is a well-worked field. Early data were collected in 1926–27 in the US at Columbus and Buffalo. Approximate models were produced [e.g. Oosthuizen (1964)]. Data up to 1995 are summarized in Spearman et al. (1995). The BHRA data in Martin (1986) based on the work of Irving (1976, 1977, 1978a, b) and Moore (1980a, b) were taken in 8″ pipe and are the most comprehensive up to 1995. Zimmermann (1996) collected a large set of data.

The data collected in the EEC Orifice Plate project at the same time as the basic data described in Chap. 5 are summarized in Sattary (1991).

Most of the best modern data have been collected in Studzinski et al. (2001a). Further modern work is in Morrow (2005), Studzinski et al. (2000, 2001b) and Weiss et al. (2002).

8.2.3.2 The Pattern of the Data

Although some later sets of data have smaller uncertainty, the BHRA sets are very useful to show the pattern of the data. The data using flange tappings downstream of a single bend of radius 1.5*D* are shown in Fig. 8.1.

Bends used in flow measurement are not normally specially manufactured; so the results are to some extent dependent on the details of the bend. The results vary

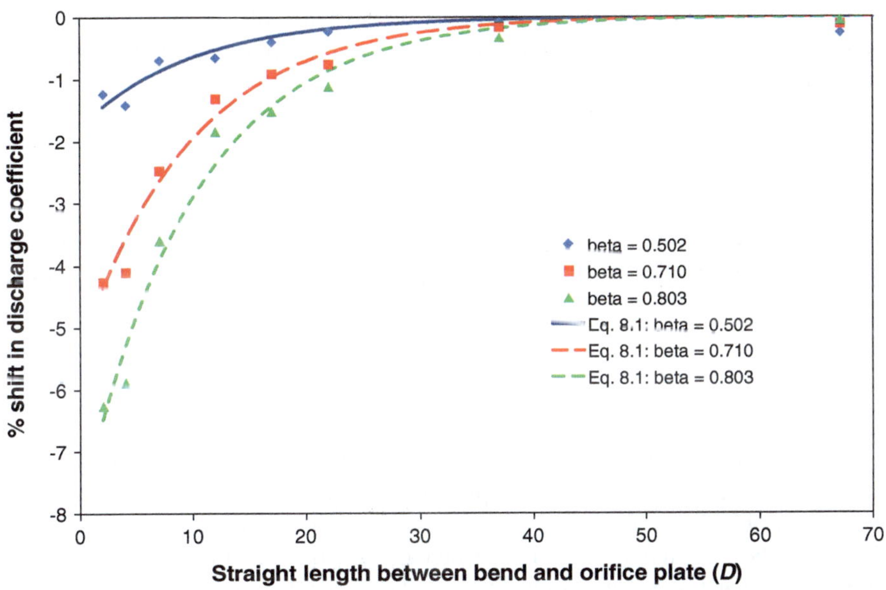

Fig. 8.1 Effect of a single bend (radius 1.5*D*) on the orifice discharge coefficient in an 8″ pipe using flange tappings (Martin 1986)

particularly widely downstream of two bends in perpendicular planes, since the swirl depends on the separation of the curved portions of the bends and the curved portions of the bends may be separated by weldnecks. However, the BHRA bends were made out of glass-reinforced plastic laid up on hardwood moulds so that their geometry could be completely described by the radius ratio (Irving 1977). It appears that the bends only consisted of curved portions; so where two bends in perpendicular planes were separated by 0*D*, there was no straight section between the curved portions (Irving 1978a).

A fit to the data in Fig. 8.1 (on minimizing errors while varying the coefficient and exponents) is

$$S = -16\beta^{3.2}\mathrm{e}^{-0.10x/D} \tag{8.1}$$

where S is the percentage shift in discharge coefficient and x is the distance between the nearest fitting and the orifice plate. Because the flow is asymmetric and non-swirling S is negative. Equation 2.5 on the effect of pipe roughness shows a dependence on $\beta^{3.5}$. A dependence on β raised to a power of about 3–4 is typical for velocity profile effects on orifice plates. The velocity profile term in the Reader-Harris/Gallagher (1998) Equation in ISO 5167-2:2003 has a dependence on $\beta^{3.5}$ (see Eq. 5.22a).

The BHRA data using flange tappings downstream of two bends in perpendicular planes with no separation between the curved portions of the bends are shown in Fig. 8.2.

A fit to the data of the form

$$S = a\mathrm{e}^{-bx/D} - c\beta^{3.2}\mathrm{e}^{-fx/D}$$

was tried to represent a swirling flow which also has an accompanying disturbed velocity profile. b and f were approximately equal. This suggests that the velocity profile is decaying as slowly as the swirl, and the best fit for the data for $x/D \geq 7$ was

$$S = (5 - 14.7\beta^{3.2})\mathrm{e}^{-0.039x/D} \tag{8.2}$$

Swirl has its largest effect around $\beta = 0.3$–0.4 (Shen 1991) when a swirler with blades at an angle of 20°–30° to the pipe axis is used to create the swirl. See also Brennan et al. (1989). The effect of swirl for $\beta = 0.2$ was found by Reader-Harris (2002b) to be very much smaller than that for $\beta = 0.4$: see Fig. 8.3. It should be emphasized that Eq. 8.2 was only fitted to data over the range of β from 0.50 to 0.80.

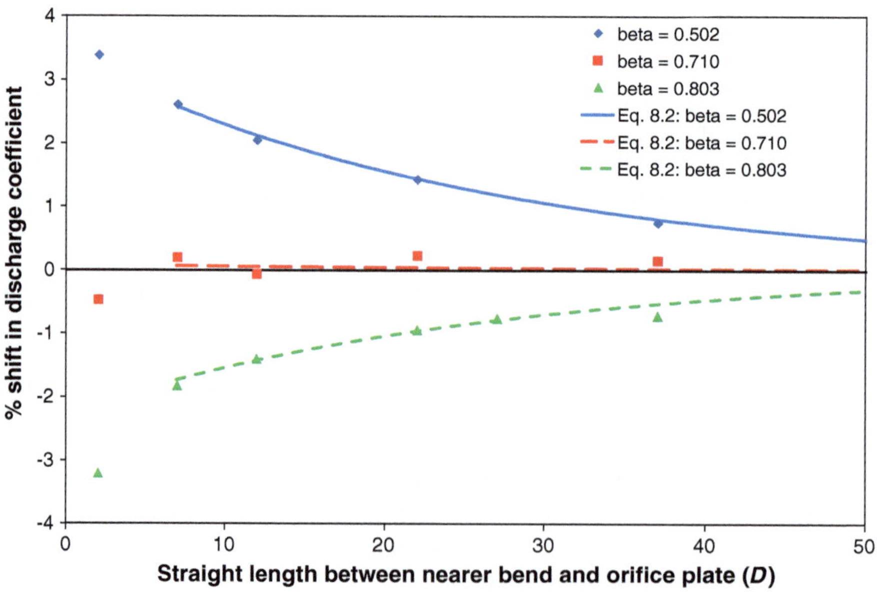

Fig. 8.2 Effect of two bends (radius 1.5D) in perpendicular planes (no separation) on the orifice discharge coefficient in an 8″ pipe using flange tappings (Martin 1986)

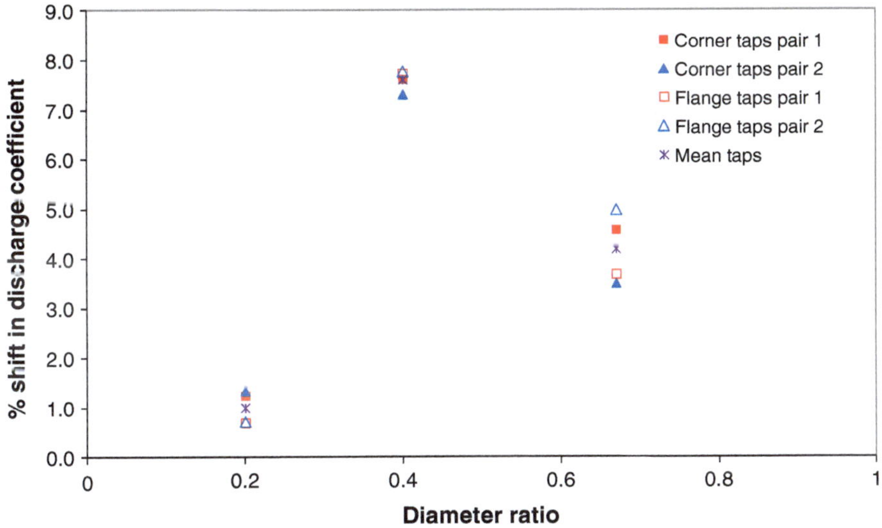

Fig. 8.3 Shift in discharge coefficient 18*D* downstream of a swirl generator (18*D* downstream of the swirl generator the swirl angle was 28° maximum) (Reader-Harris 2002b)

Detailed measurements of the velocity and turbulence profiles downstream of single bends and of two bends in perpendicular planes were made by Mattingly and Yeh (1992). The paper contains an interesting graph showing how much the swirl is reduced by including a short section of straight pipe between two bends in perpendicular planes: including 2.4 diameters of straight pipe between bends reduced the swirl to about 20 % of that obtained with no straight pipe between the bends.

The BHRA data using flange tappings downstream of two bends in perpendicular planes with 5*D* and 11*D* separation between the bends are shown in Figs. 8.4 and 8.5, respectively.

A fit to the data of the form

$$S = a\mathrm{e}^{-0.039x/D} - c\beta^{3.2}\mathrm{e}^{-fx/D}$$

was tried to represent a swirling flow which also has a disturbed velocity profile. The best fits gave values of f quite close to 0.1; so f was fixed at 0.1, and the fits were as follows:

$$S = 0.7\mathrm{e}^{-0.039x/D} - 10.7\beta^{3.2}\mathrm{e}^{-0.10x/D} \quad \text{for a separation of } 5D \tag{8.3}$$

$$S = 0.3\mathrm{e}^{-0.039x/D} - 11.4\beta^{3.2}\mathrm{e}^{-0.10x/D} \quad \text{for a separation of } 11D \tag{8.4}$$

The enormous reduction in swirl due to separating the bends is clear.

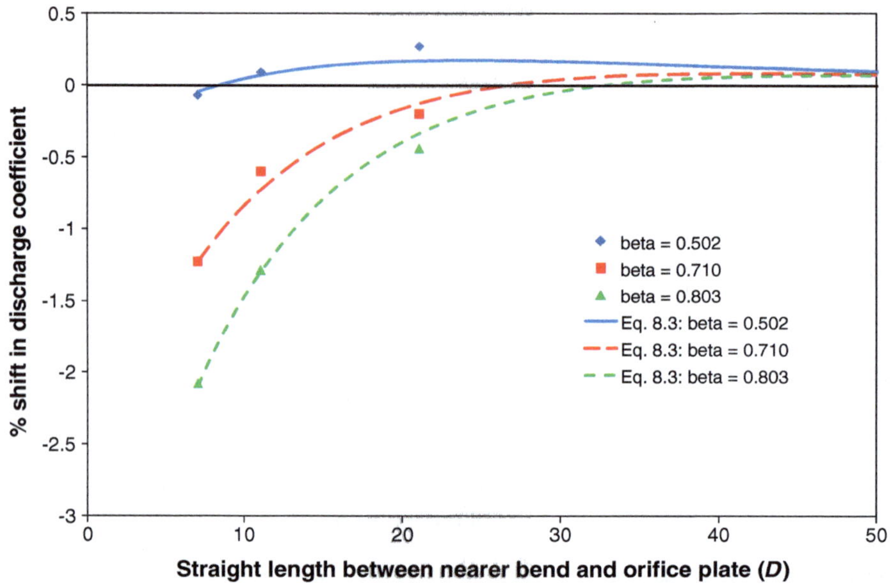

Fig. 8.4 Effect of two bends (radius 1.5*D*) in perpendicular planes (5*D* separation) on the orifice discharge coefficient in an 8″ pipe using flange tappings (Martin 1986)

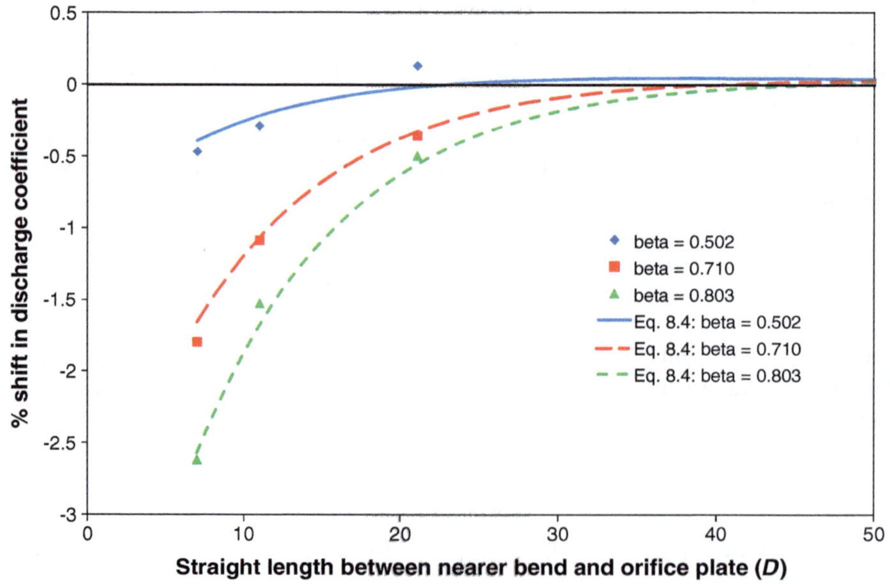

Fig. 8.5 Effect of two bends (radius 1.5*D*) in perpendicular planes (11*D* separation) on the orifice discharge coefficient in an 8″ pipe using flange tappings (Martin 1986)

8.2.3.3 The Straight Lengths in ISO 5167-2:2003

Even given a large population of data, deciding on acceptable lengths is not straightforward. It might be good to choose the straight lengths so that the effect of upstream installation is less than 0.1 %, but in practice the lengths would be very long and the uncertainty of differences in measurements of discharge coefficient for a very good water facility might be 0.05 %, with the result that it would be difficult to obtain the lengths accurately. Installation effects of more than 0.25 % would make obtaining an overall discharge-coefficient uncertainty of 0.5 % very difficult and lead to unacceptable bias in many cases. Moreover, there is discussion as to whether all the installation-effect data should lie within the acceptable limit or only a fit to the data needs to lie within it. Real bends are used for most experiments, and they are not exactly geometrically similar to one another.

The solution used by ISO 5167-2:2003 was put forward by API: all the data of sufficient quality are brought together; at least two high-quality sets of data should be used to determine each upstream length; the permissible upstream length chosen for the standard is a location at which data were taken; moreover, the data taken at that location and all the data with longer upstream lengths must lie within the acceptable band; the acceptable band is half the (expanded) uncertainty of the Reader-Harris/Gallagher (RG) Equation at infinite Reynolds number as calculated by API. The value of this (expanded) uncertainty is shown in Fig. 1–4 of API MPMS 14.3.1:1990 (API 1990) [Fig. 4 of API MPMS 14.3.1:2012 (API 2012)]: for example, the value is 0.51 % at $\beta = 0.2$, reducing to 0.44 % at $\beta = 0.5$ and 0.6 and then increasing to 0.47 % at $\beta = 0.67$ and 0.56 % at $\beta = 0.75$. By using this rule the maximum shift due to upstream installation should in practice often be significantly less than the maximum permitted value (i.e. half of the (expanded) uncertainty shown in Fig. 1–4 of API MPMS 14.3.1:1990). There are some exceptions where the rule is not precisely followed where the data are sparse. Longer lengths than those obtained by a strict application of the rules were sometimes considered wise. Whereas single-bend data are generally in good agreement with one another, data with two bends in perpendicular planes with a spacer less than $2D$ are not: the effect of small differences in spacing between the curved portions of the two bends is significant. Where modern data were not available to calculate straight lengths in accordance with the above rule the straight lengths used in earlier editions of ISO 5167 have been retained: this applies to 'Full bore ball valve or gate valve fully open', 'Abrupt symmetrical reduction' and 'Thermometer pocket or well of diameter $\leq 0.03D$'.

By way of example it is valuable to consider some straight lengths required for $\beta = 0.5$. They are given in Table 8.1. Both the current ISO and API standards [ISO 5167-2:2003 (ISO 2003b) and API MPMS 14.3.2:2000 (API 2000)] give shifts less than 0.2 % in absolute value: the API standard is the more conservative of the two. The earlier standards shown [ISO 5167-1:1991 (ISO 1991) and ANSI/API 2530:1985 (API 1985)] gave lengths that were too short, sometimes much too short.

It is worth pointing out that using the minimum length even in the current standards gives a bias. Moreover, the projects to obtain more accurate straight

Table 8.1 The straight lengths required by different standards upstream of an orifice plate of $\beta = 0.5$ and the corresponding shifts in discharge coefficient from the work in Sect. 8.2.3.2

	Single bend		2 bends in perpendicular planes (0*D* spacer)		2 bends in perpendicular planes (5*D* spacer)		2 bends in perpendicular planes (11*D* spacer)	
	Minimum length	% shift at min. length (Eq. 8.1) (%)	Minimum length	% shift at min. length (Eq. 8.2) (%)	Minimum length	% shift at min. length (Eq. 8.3) (%)	Minimum length	% shift at min. length (Eq. 8.4) (%)
ANSI/API 2530:1985	7*D**	−0.86	21*D*	1.53	21*D*	0.17	21*D***	−0.02***
ISO 5167-1:1991	14*D*	−0.43	40*D*	0.73	40*D*	0.13	40*D*	0.04
API MPMS 14.3.2:2000	30*D*	−0.09	95*D*	0.09	44*D*	0.11	44*D*	0.04
ISO 5167-2:2003	22*D*	−0.19	75*D*	0.19	44*D*	0.11	44*D*	0.04

*There was no value for a single bend: 7*D* was permitted for two in-plane bends with more than 10*D* separation

**There was no value for two bends in perpendicular planes with more than 10*D* separation: 21*D* was permitted for two bends in perpendicular planes with less than 10*D* separation

***Maximum error over all lengths ≥21*D* is 0.04 % at 39*D*

lengths and to include them in standards have been very worthwhile: using the straight lengths from older standards can lead to significant bias.

Table 3 of ISO/TR 12767:2007 (ISO 2007) gives predicted changes in discharge coefficient when the straight lengths are smaller than those permitted in ISO 5167-2:2003.

It may be possible to correct for the effect of installation by measuring the pressure rise from the upstream *D* tapping into the corner (Martin 1986; Humphreys and Maxwell 1989) or by measuring the pressure loss across the orifice plate (Steven 2008; Skelton et al. 2010): even if no correction is applied measuring an additional differential pressure (or better two differential pressures) checks that the data are consistent: see Sect. 4.10.

8.2.4 Venturi Tubes

8.2.4.1 Standard Venturi Tubes

General

There are very limited data on the effect of upstream fittings on Venturi tubes. Most of the data collected since 1970 are displayed here. The majority of them were taken in one project at NEL. All the experimental data were taken in water.

Three classical Venturi tubes with a machined convergent section had been made by ISA Controls Ltd for Shell UK Exploration and Production to meet the requirements of ISO 5167-1:1991. The pipe diameter was nominally 6″. They were designed not only to meet the Standard but to follow recommendations. In particular the Standard recommends that a divergent angle between 7° and 8° be used: 7½° was specified. The pressure tappings were 4 mm in diameter and were connected in triple-T arrangements. Over the complete database the throat Reynolds number, Re_d, was always in the range 1.9×10^5–2.0×10^6. More details can be found in Reader-Harris et al. (1997).

So that the effect of installing a bend at various distances upstream of the Venturi tubes could be measured without the results being corrupted by the introduction of steps at joins in the pipework, upstream lengths of 2*D*, 3.7*D* and 5*D* and a downstream length of 4*D* were manufactured with bores machined to ensure that the steps in diameter would not exceed 0.3 mm at the upstream flange of the Venturi tube and 0.5 mm at the flanges upstream of that and at the downstream flange of the Venturi tube; these tolerances are approximately what is required for orifice plates and significantly smaller than those required for Venturi tubes. The lengths of pipework were manufactured by boring out Schedule 80 pipe to the bore of a Schedule 40 pipe. The lengths of pipe and the Venturi tubes were dowelled to ensure concentricity; O rings were used to ensure that there would not be recesses or protruding gaskets. All these requirements ensured that measurements were not corrupted by secondary effects.

In addition to the shorter lengths of pipework already described an additional 20*D* length was manufactured by welding 18*D* of Schedule 40 pipe and 2*D* of pipe machined from Schedule 80 pipe to the bore of Schedule 40; any step at the weld was smoothed off; this length of pipe was used with the machined length closest to the Venturi tube and when used with the 3.7*D* and 5*D* lengths between it and the Venturi tube there was 11.2*D* of machined pipework, whose bore closely matched that of the Venturi tube, immediately upstream of the Venturi tube; the machined pipework was preceded by 18*D* of pipe of the same schedule with no steps or recesses or protruding gaskets throughout the whole 29.2*D* of pipe.

Three Venturi tubes of diameter ratio 0.4, 0.6 and 0.75 were initially calibrated for Shell with the 29.2*D* of pipe described above installed immediately upstream of the Venturi-tube upstream pressure tapping, and with a further 12*D* of pipe with the same nominal diameter upstream of the 29.2*D* of pipe. Upstream of the 12*D* of pipe was a tube bundle flow straightener. The Venturi tubes were then calibrated immediately before the installation effects tests described here. At the end of all the calibrations (except for the calibration 17.8*D* downstream of a single bend which took place one week after the final baseline) all three Venturi tubes were recalibrated again.

For each Venturi tube the slope of the calibration was positive for at least one calibration and negative for at least one calibration; so it seemed sensible simply to use the mean value for each calibration. The mean value of discharge coefficient used as the baseline for each Venturi tube was the average of the mean values from the second and third calibrations, which preceded and followed the installation-effects tests.

Calibrations Downstream of a Contraction and an Expansion

A contraction from 8″ to 6″ was installed upstream of each Venturi tube: the contraction was conical with a length of conical section of 350 mm, giving an included half-angle of about 4°. Upstream of the contraction there was 23*D* of 8″ pipe preceded by a tube bundle. The mean shifts are presented in Fig. 8.6.

An expansion from 4″ to 6″ was installed upstream of each Venturi tube: the expansion was conical with a length of conical section of 375 mm, giving an included half-angle of about 4°. Upstream of the expansion there was 48*D* of 4″ pipe preceded by a tube bundle. The mean shifts are presented in Fig. 8.7. The shifts are positive because the velocity profile is peaked.

Calibrations Downstream of Bends

For these NEL calibrations downstream of bends the bends were of radius 1.5*D* and each had short weldnecks of approximately 50 mm so that the bends could be as closely coupled as possible.

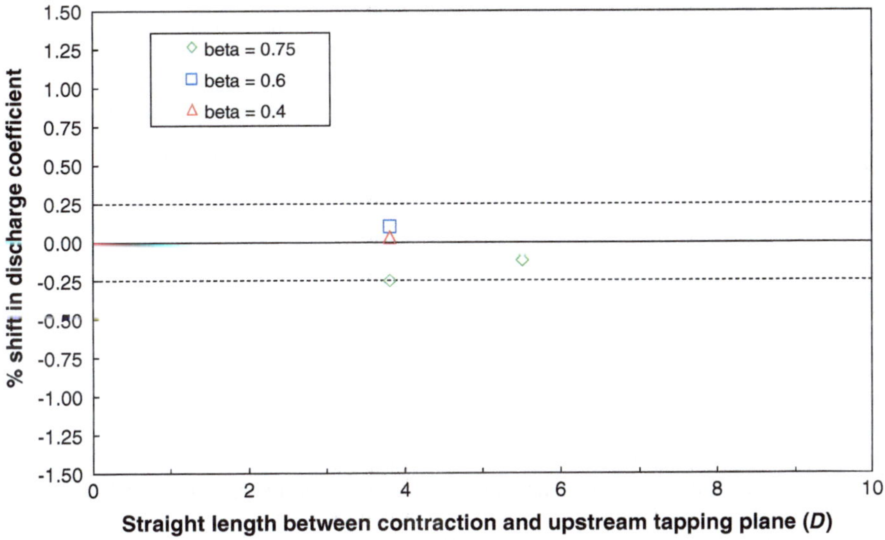

Fig. 8.6 Effect of an 8″–6″ contraction on the discharge coefficient of a standard Venturi tube

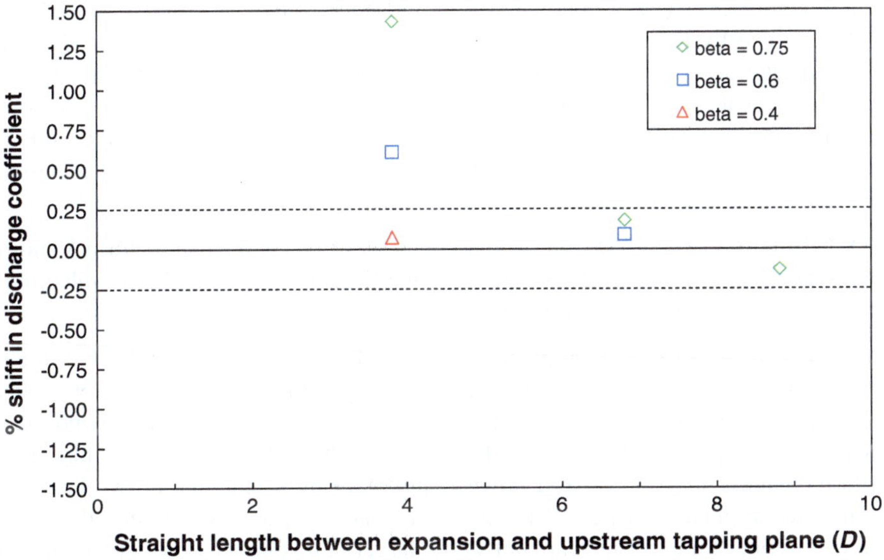

Fig. 8.7 Effect of a 4″–6″ expansion on the discharge coefficient of a standard Venturi tube

A single bend was installed upstream of each Venturi tube. Upstream of the bend there was 26*D* of straight pipe preceded by a Zanker Flow Conditioner. The mean shifts are presented in Fig. 8.8. These calibrations are in good agreement with the

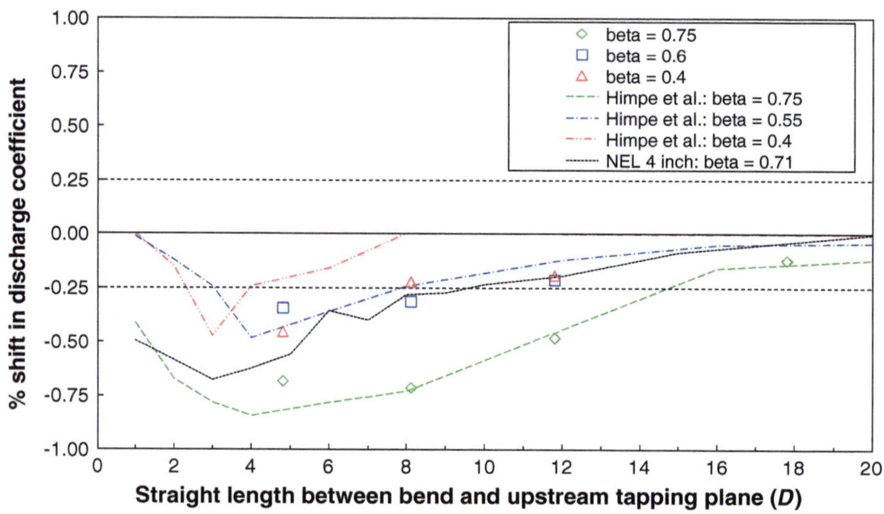

Fig. 8.8 Effect of a single bend (radius 1.5*D*) on the discharge coefficient of a standard Venturi tube

data of Himpe et al. (1994), particularly for the larger values of β, and those of NEL (1985) quoted by Kochen et al. (1989). Both Himpe et al. and NEL (1985) had a pipe diameter of 4″ and a bend of radius 1.5*D*. The shape of the curves, with shift in some cases near zero 1*D* downstream of the bend, is distinctive. Shifts are functions of axial velocity profile, turbulence and velocity components in non-axial directions: shifts do not always tend to zero monotonically with increasing separation between fitting and flowmeter.

NOTE Bradford et al. (2006) measured the effect of bends on a short-form Venturi tube (Halmi 1973), but in almost all cases the baseline was a predicted value, not one measured in good flow conditions.

Two bends in perpendicular planes were installed upstream of each Venturi tube. Upstream of the upstream bend there was the same 26*D* of straight pipe preceded by a Zanker Flow Conditioner as there was for the tests with a single bend. The mean shifts are presented in Fig. 8.9. In addition to those data shown Bluschke et al. (1966) obtained data for $\beta = 0.71$ downstream of a single bend and of two bends in perpendicular planes: they are very similar in shape to those shown here: the shifts are very similar for a single bend, but a little larger in magnitude downstream of two bends in perpendicular planes than those shown here.

Two bends in the same plane (S-configuration) were installed upstream of each Venturi tube. Upstream of the upstream bend there was 21*D* of straight pipe preceded by a tube bundle. The mean shifts are presented in Fig. 8.10.

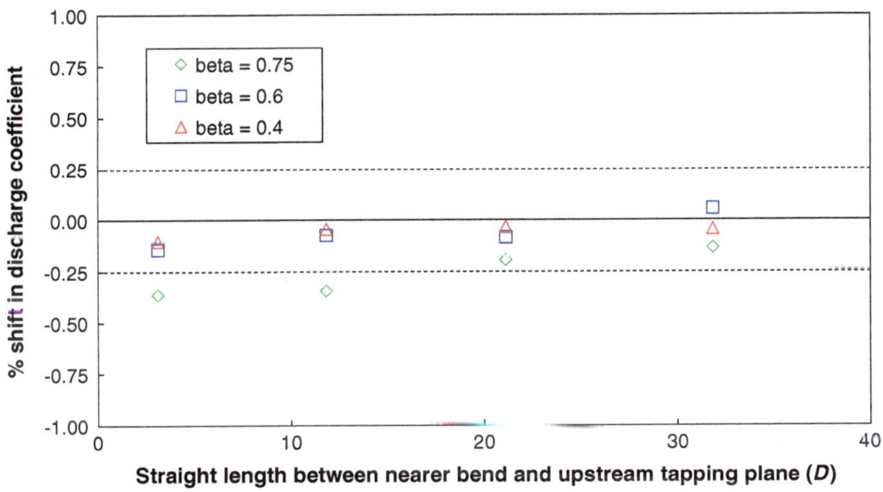

Fig. 8.9 Effect of two bends (radius 1.5*D*) in perpendicular planes on the discharge coefficient of a standard Venturi tube

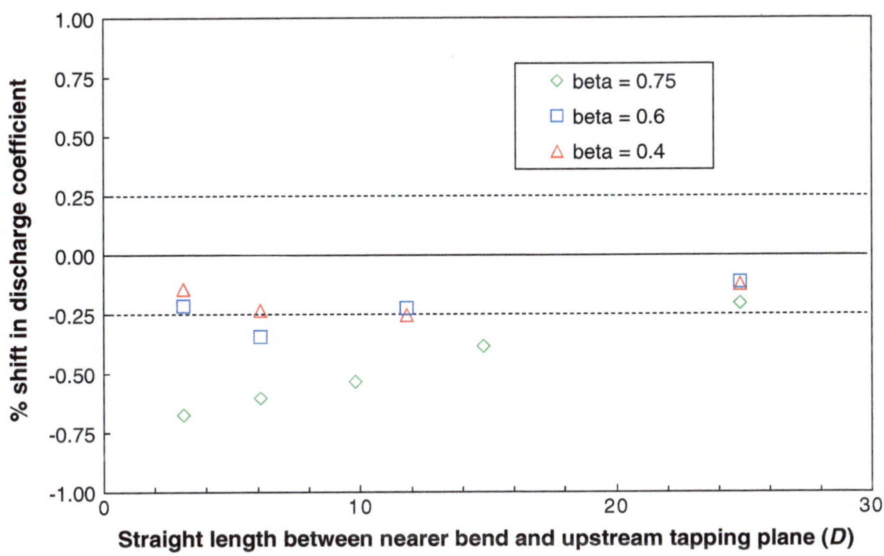

Fig. 8.10 Effect of two bends (radius 1.5*D*) in the same plane (S-configuration) on the discharge coefficient of a standard Venturi tube

Analysis

On the basis that straight lengths should be calculated so that either shifts in discharge coefficient should be smaller in magnitude than 0.25 % or alternatively, for 0.5 % additional uncertainty, 0.75 %, Table 1 in ISO 5167-4:2003 was obtained. As with the straight lengths for orifice plates this means that if the minimum length is selected a permanent bias due to installation may be built in. There are insufficient data for Venturi tubes to make it a requirement that there should be two sets of data for each straight length.

Although the required lengths of straight pipe downstream of two bends in perpendicular planes were in some cases smaller than the required straight lengths downstream of a single bend, for the lengths downstream of bend combinations it was decided that a length be used for standards purposes which was at least the length downstream of a single bend. This copes satisfactorily with the possibility that two bends with separation are used.

The other lengths in Table 1 of ISO 5167-4:2003 are based on data of Pardoe (1943): he obtained positive shifts in C downstream of a single bend and stated that $6D$ was sufficient downstream of a single bend even for $\beta = 0.8$, whereas downstream of two bends in perpendicular planes even $30D$ was not sufficient (at least for $\beta > 0.55$). An explanation for the difference between Pardoe's data and subsequent data is still required.

8.2.4.2 Venturi Tubes with Convergent Angle 10.5°

There is very limited information on required straight lengths for Venturi tubes with a convergent angle of 10.5°.

The only data that there are, on the effect of a single bend and of two bends in perpendicular planes, are shown in Figs. 8.11 and 8.12, respectively (Reader-Harris et al. 2004). The bends were of radius $1.5D$ and had a parallel section of length $0.8D$; so between the curved portions of the two bends in perpendicular planes was a straight length of $1.6D$. Upstream of the single bend there was a length of $20D$ of Schedule 40 pipe preceded by $25D$ of pipe of the same nominal diameter preceded by a flow conditioner.

The data in Fig. 8.11 for $x/D \geq 4.7$, where x is the distance between a fitting and the upstream pressure tappings of a Venturi tube, are well fitted by

$$S = -2.321 \frac{\beta^4}{1 - \beta^4} \mathrm{e}^{-0.064x/D} \tag{8.5}$$

This equation is shown on Fig. 8.11. This is similar to Eq. 8.1 but with a dependence on a higher power of β and with a slower decay with x/D.

A similar fit has been obtained to the data shown in Fig. 8.12: for $x/D \geq 7.7$

$$S = -1.719 \frac{\beta^4}{1 - \beta^4} \mathrm{e}^{-0.042x/D} \tag{8.6}$$

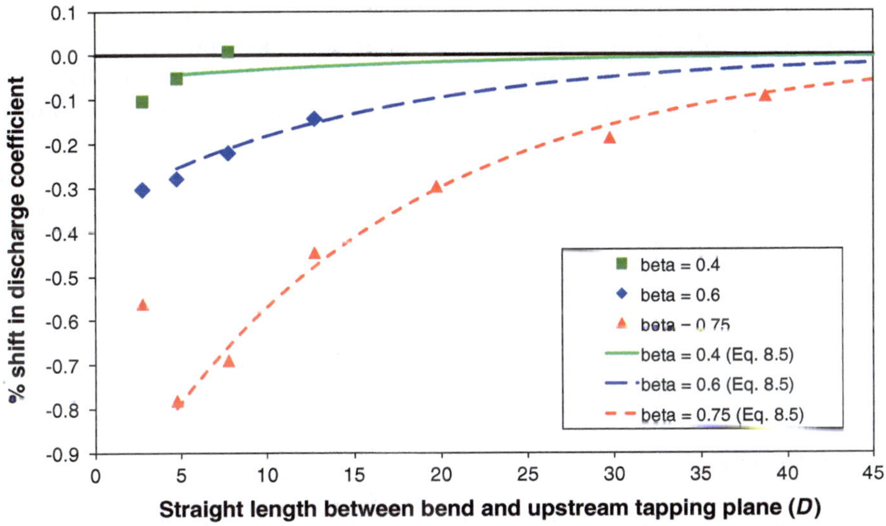

Fig. 8.11 Effect of a single bend (radius 1.5D) on the discharge coefficient of a Venturi tube with convergent angle 10.5°

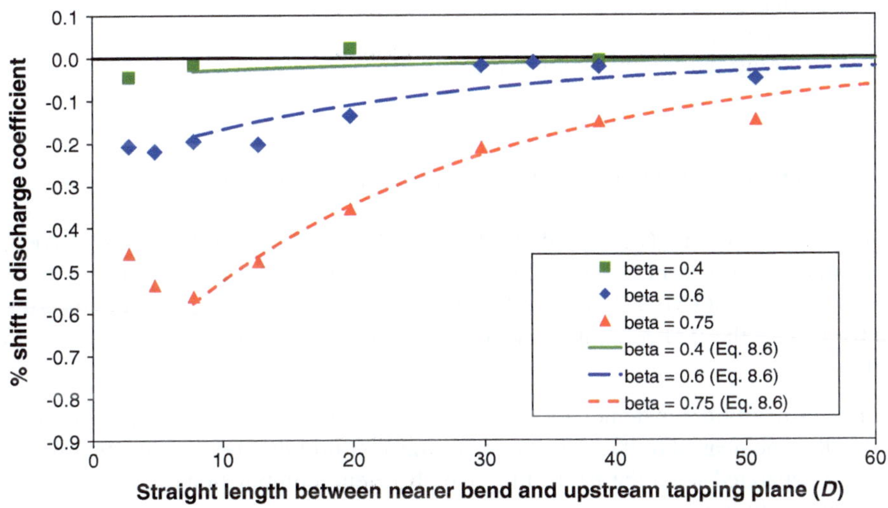

Fig. 8.12 Effect of two bends (radius 1.5D) in perpendicular planes on the discharge coefficient of a Venturi tube with convergent angle 10.5°

This equation is shown on Fig. 8.12. This has a similar decay with x/D to that in Eq. 8.2. The dependence on β is the same as in Eq. 8.5, since both equations describe the effect of velocity profile. For a Venturi tube swirl does not affect

C directly (because there is no vena contracta that can change in area with swirl) (Gibson and Reader-Harris 2006), but the associated velocity profile with reduced centre-line velocity and increased velocity near the wall does affect C; therefore Eq. 8.6 only has a term with strong dependence on β. The shift decays more slowly in swirling flow than in non-swirling flow.

The striking feature of the data in both Figs. 8.11 and 8.12 is that there is a systematic dependence on β, whereas for the standard Venturi tube such a pattern is lacking (especially in Fig. 8.8).

8.2.5 *What to Do if a Case is not Covered in Table 3 of ISO 5167-2:2003/Table 1 of ISO 5167-4:2003*

8.2.5.1 General

If a case is not covered directly in Table 3 of ISO 5167-2:2003/Table 1 of ISO 5167-4:2003 there are five ways of solving this problem:

- the upstream installation is a combination of fittings;
- a flow conditioner is used;
- a specific test is done;
- CFD is carried out;
- engineering judgement is employed.

8.2.5.2 The Upstream Installation is a Combination of Fittings

This is described in 6.2.8 of ISO 5167-2:2003 and 6.2.8 of ISO 5167-4:2003. Examples are given in 6.2.9 of ISO 5167-2:2003 and 6.2.9 of ISO 5167-4:2003. These parts of ISO 5167-2:2003 and ISO 5167-4:2003 are based on the engineering judgement of the ISO committee members, not on experimental data. They work as follows:

(a) Upstream fittings do not alter the length required downstream of the first fitting (the one nearest to the orifice plate or the Venturi tube) because the first fitting is separated from the next fitting (see b). Bend combinations are treated as a single fitting.
(b) The separation between the first fitting (fitting 1) and the next fitting (fitting 2) is based on the straight length in Table 3 of ISO 5167-2:2003/Table 1 of ISO 5167-4:2003 required for fitting 2 and is not a function of the actual β but is fixed: half the straight length required for fitting 2 with an orifice plate of $\beta = 0.67$ or with a Venturi tube of $\beta = 0.7$ is specified (the specified upstream straight length for a large β orifice plate or Venturi tube gives an indication of the distance required for flow profiles to decay towards fully developed).

(c) If there is a full-bore valve as fitting 1 it does not need to be placed near the middle of the length between the orifice plate or Venturi tube and fitting 2. The valve can be placed adjacent to fitting 2, since it will not interact strongly with fitting 2. This does not lead to a reduction in overall upstream length but may be convenient. A match-bored valve with no steps is treated as part of the straight pipework, not as a full-bore valve as in Table 3 of ISO 5167-2:2003/Table 1 of ISO 5167-4:2003.
(d) Under ISO 5167-1:1991 it was sometimes possible to reduce the allowable overall upstream length by inserting a valve between an orifice plate and fitting 1 (which then became fitting 2). This was unwise, and it was corrected in ISO 5167:2003; so it is not possible by inserting an additional fitting (as fitting 1) to reduce the distance required between the original fitting (now fitting 2) and the orifice plate or Venturi tube.
(e) The apparently fiddly rules about which pipe diameter to use (where an expansion or contraction is one of the upstream fittings) are based on the fact that expansions have much worse effects on swirl angle and velocity profile than contractions do.

8.2.5.3 A Flow Conditioner Is Used

General

A flow conditioner may be used. Even where a straight length is given in Table 3 of ISO 5167-2:2003/Table 1 of ISO 5167-4:2003 it may be possible to reduce the length upstream by using a flow conditioner.

A generic test is specified in 7.4.1 of ISO 5167-1:2003. In outline, to be acceptable in any installation a flow conditioner must give a shift from the baseline of less than 0.23 % in

(a) good flow conditions,
(b) strong asymmetry (i.e. downstream of a 50 % closed gate valve or equivalent) and
(c) strong swirl (i.e. downstream of a device that produces a maximum swirl angle of at least 24° 18D downstream of it).

The expectation is that if a flow conditioner gives good performance in these conditions it will give good performance in any conditions that might occur in the field (for the prescribed level of swirl for the test see below). 0.23 % is the value of maximum permissible shift for $\beta = 0.67$ as in Sect. 8.2.3.3 .

Most of the test work is carried out at one Reynolds number in one pipe size, but the flow conditioner must also give a shift less than 0.23 % for (a) or (b) or (c) at a significantly different Reynolds number and in a significantly different pipe size from the Reynolds number and pipe size in which the main test work was carried out.

Since the effect of velocity profile (other than swirl) increases with β, a test with $\beta = 0.67$ is sufficient for all smaller β for good flow conditions and for asymmetric non-swirling flow. For the strongly swirling case, in addition to the test with $\beta = 0.67$, a test with $\beta = 0.4$ is required as that is close to being the worst choice of β for an orifice plate (for a Venturi tube, swirl itself has a much smaller effect than velocity profile).

In API MPMS 14.3.2:2000 (API 2000) the generic test is similar, but the flow conditioner must also give a shift less than 0.23 % downstream of two bends in perpendicular planes; this was not required in ISO 5167-1:2003 because if a flow conditioner passes the test in good flow conditions, strong asymmetry and strong swirl, it almost always passes the test downstream of two bends in perpendicular planes. In API MPMS 14.3.2:2000 the test in a second pipe size must not be carried out in good flow conditions.

With an Orifice Plate

Early work on flow conditioners was carried out by Sprenkle and Courtright (1957), who had three perforated plates in series about one diameter apart, and by Zanker (1962), who designed a thin plate with holes of carefully chosen size to which a honeycomb with square passages was attached. Information can be found in Spearman et al. (1995). There was an enormous amount of work carried out. A distinction can be made between flow conditioners, which aim to produce a good profile, and flow straighteners, which simply remove swirl. A flow straightener, therefore, is only effective with a small β orifice plate (McHugh et al. 1984). Typical flow straighteners are tube bundles, AMCA Flow Straighteners, which consist of a honeycomb with square meshes, and étoiles, which consist of (eight) radial vanes at equal angular spacing: they are described in C.2 of ISO 5167-1:2003.

To obtain better results it was necessary to change from flow straighteners to flow conditioners. Mitsubishi (Akashi et al. 1979) showed that perforated plates did not have to be thick to remove swirl. The thickness just had to be roughly as large as the hole diameter. Humphreys and Wood (1984) looked at a $\beta = 0.8$ orifice with short (4*D*–12*D*) overall upstream length, but more work has been undertaken with longer upstream lengths in order to obtain the minimum lengths at which orifice discharge-coefficient equations can be used with no additional uncertainty.

A flow straightener tends to have a pressure loss coefficient of about 0.25–0.75, where the pressure loss coefficient K is given by the following equation:

$$K = \frac{\Delta p_c}{\frac{1}{2}\rho \bar{u}^2},$$

where Δp_c is the pressure loss across the flow straightener/conditioner and $\bar{u}$ is the mean axial velocity of the fluid in the pipe. A perforated-plate flow conditioner tends to have a pressure loss coefficient of about 2–3, because it is likely that a pressure loss coefficient of about 2–3 is required to produce a good downstream velocity profile.

There were tests within the EEC Orifice Project on tube bundles, Zanker Flow Conditioners and Sprenkle Conditioners [see, for instance, Smith (1986) and Sattary (1991)]. There were tests in the US soon after the work on determining the discharge-coefficient equation [e.g. Mattingly and Yeh (1991), Morrow et al. (1991) and Scott et al. (1991, 1993)]; these tests were done particularly on tube bundles (with 19 or 7 tubes) and on Zanker Flow Conditioners (including the honeycomb). Laws (1990) developed an improved perforated plate conditioner. Tests were carried out at NEL for Amoco, the aim being to examine the performance of different flow conditioners downstream of a header [see Lake and Reid (1992)]. A header is a common upstream installation and has a different effect from a bend combination with the same separation but no changes in diameter: in particular the swirl downstream may be significantly larger than that downstream of two bends in perpendicular planes. Williamson et al. (1993) measured swirl angles exceeding 22° downstream of a header. Nova improved the Laws design of flow conditioners (Karnik 1995). See also Wilcox et al. (1990), Laws and Ouazzane (1994, 1995) and Langsholt and Erdal (1998).

The tests conducted at NEL on the effectiveness of flow conditioners for the Headers Consortium built upon the work that had been done for Amoco: the results were presented in Reader-Harris et al. (1995a). One configuration of the header involved a vertical 12″ inlet, a horizontal header of nominal diameter 18″ (D_1), and a horizontal meter tube of nominal diameter 10″ (D). The distance between the centreline of the meter tube and the centreline of the inlet was 2.9 D_1. The measured velocity profile and swirl angle 4D downstream of the meter tube inlet are shown in Figs. 8.13 and 8.14.

A Gallagher Conditioner (Gallagher et al. 1994a, b; Gallagher and LaNasa 1995), which consisted of a perforated plate preceded by a tube bundle (see B.2 of ISO 5167-2:2003), was mounted with the upstream flange of its perforated plate 4D from the header so that the tube bundle lay within the 4D of pipe: downstream of the perforated plate there was 9D of pipe to the point at which the velocity profile and swirl angle were measured. The results are shown in Figs. 8.13 and 8.14.

The effectiveness of the Gallagher Conditioner is clear. The Headers Consortium investigated both velocity profiles downstream of flow conditioners and shifts in discharge coefficient, but a conclusive statement on their relationship was not achieved.

The results from the Headers Consortium tests and other tests gave guidance as to the sort of test of flow conditioners that should be included in ISO 5167. For example, the swirl level required for the test resulted from the measured swirl downstream of the header in the Headers Consortium test, as this is the highest swirl that has been measured in a real flow-measurement installation (not downstream of a swirl generator, for example).

Before the test (as described above with its two slightly different forms) was included in API MPMS 14.3.2 and then ISO 5167 the test was examined to check

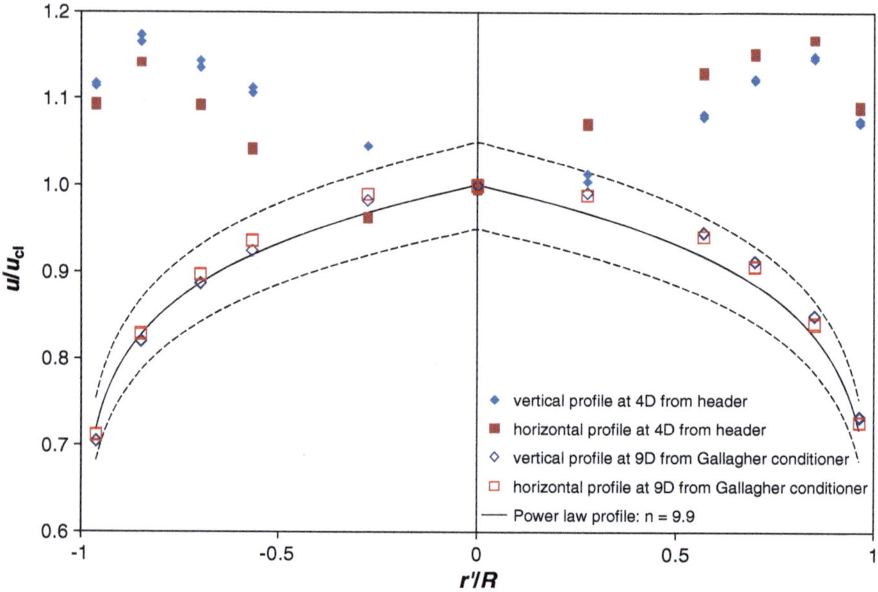

Fig. 8.13 Velocity profiles in a meter tube: (i) $4D$ downstream of the header and (ii) $9D$ downstream of the perforated plate of a Gallagher Flow Conditioner (the perforated plate was $4D$ downstream of the header)

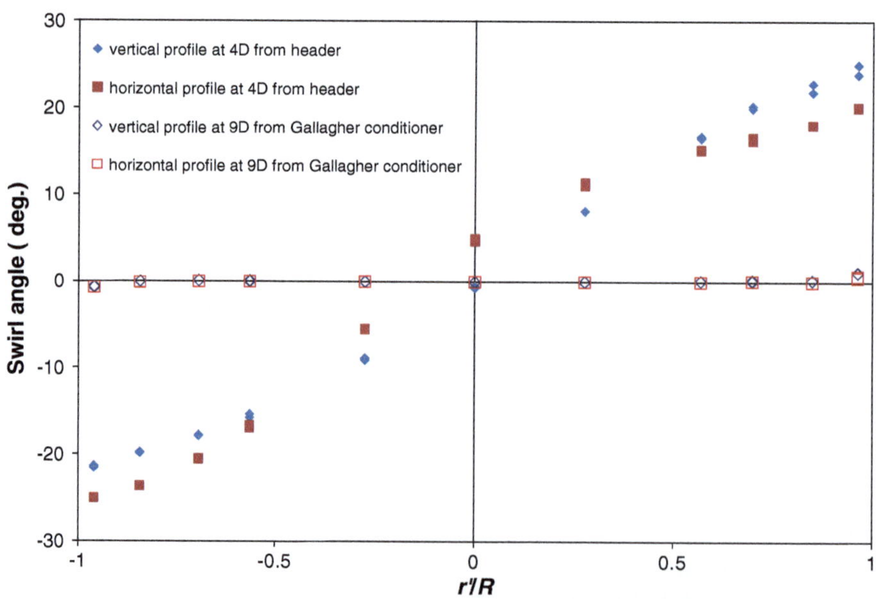

Fig. 8.14 Measurements of swirl angle in a meter tube (i) $4D$ downstream of the header and (ii) $9D$ downstream of the perforated plate of a Gallagher Flow Conditioner (the perforated plate was $4D$ downstream of the header)

that there were conditioners that would pass the test. The test results are given in Morrow (1997). Successful results were obtained with a Gallagher Conditioner and with NOVA's design of K-Lab Perforated Plate (now manufactured by Canada Pipeline Accessories), which enabled them to be included in ISO 5167 with an overall upstream length of 17*D*. The compliance test results are given in ISO 5167-2:2003. Successful results, but with an overall length of 30*D*, were obtained with a tube bundle. Even to achieve that length there are restrictions on the design of the tube bundle. The very tight restrictions on the location of the tube bundle are due to the fact that it passes the test at a 'sweet spot', where the discharge-coefficient error crosses zero.

In Europe at any rate the Zanker Flow Conditioner Plate (one with the same hole pattern as the original Zanker Flow Conditioner as in ISO 5167-1:1991, but with greater thickness and no honeycomb: see C.3.2.6 of ISO 5167-1:2003, and Fig. 8.18 for a picture) has proved preferable to the tube bundle. Zanker and Goodson (1999) showed that the Zanker Flow Conditioner Plate passed the test in ISO 5167-1:2003 provided that the overall upstream length is at least 17*D* and the downstream face of the Zanker Flow Conditioner Plate is at least 7.5*D* from the orifice plate and at least 8.5*D* from the upstream fitting (ISO 5167-2:2003, 6.3.3.3). If the overall length is 17*D* the results of the tests required for ISO 5167-1:2003 carried out by Daniel Industries and CEESI are shown in Fig. 8.15. It is striking that

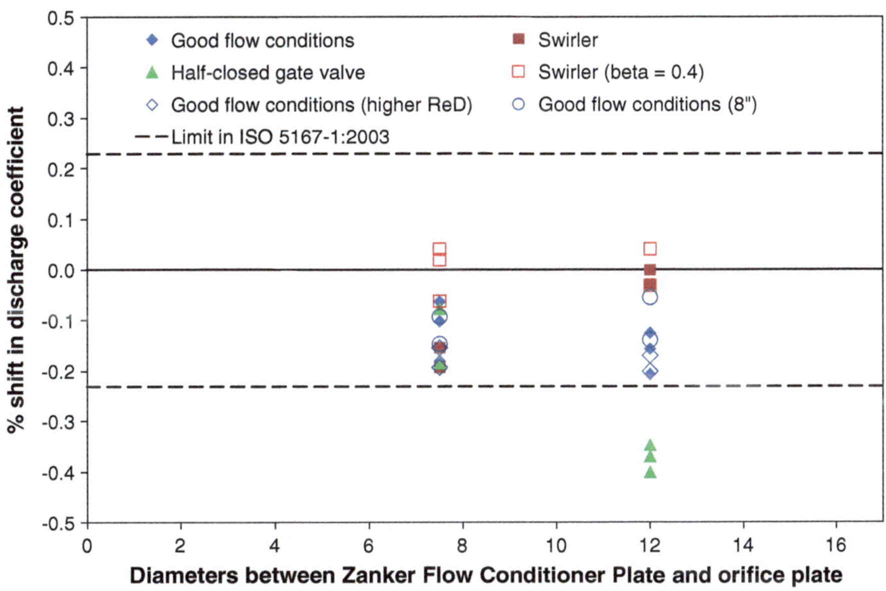

Fig. 8.15 Shift in discharge coefficient due to upstream fittings in a 17*D* meter tube: all data were taken in 4″ pipe with $\beta = 0.67$ at lower Re_D (i.e. in water) unless otherwise stated; multiple identical symbols correspond to flange tappings in different circumferential locations: water data by Daniel Industries, gas (higher Re_D) data by CEESI

in a 17D upstream length the Zanker Flow Conditioner Plate cannot be installed as close as 5D to a half-closed gate valve (i.e. with the Zanker Flow Conditioner Plate 12D from the orifice plate).

A Zanker Flow Conditioner Plate is much more expensive than a tube bundle. It is very likely that the Zanker Flow Conditioner Plate is over-specified. However, the experiments required to relax the specification limits would be very laborious and have not been performed: CFD would be helpful. The limits must not be relaxed too much, because, for example, chamfering the Zanker Flow Conditioner Plate gave poorer results than those obtained with a standard Zanker Flow Conditioner Plate (Reader-Harris 2002a): see below. In the absence of experiments to relax the specifications it is necessary to manufacture perforated plates that are as similar as possible to those used for the tests which led to the permitted locations in ISO 5167-2:2003.

With a Venturi Tube

Tests were carried out at NEL using a Zanker Flow Conditioner Plate as in C.3.2.6 of ISO 5167-1:2003 (Reader-Harris et al. 2003b). The calculated shifts in discharge coefficient when the Zanker Flow Conditioner Plate was installed 3D upstream of the upstream tappings of the Venturi tube are given in Fig. 8.16. The data taken in good flow conditions are shown with the distance to the disturbance described as infinity. The data shown were taken in water in 4″ pipe. Given particular lengths upstream and downstream of the Zanker Flow Conditioner Plate no data were taken

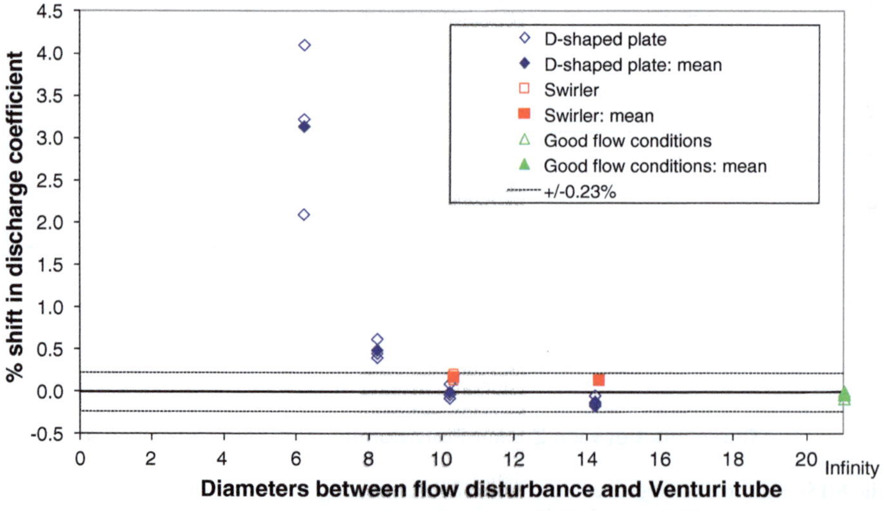

Fig. 8.16 Effect of a Zanker Flow Conditioner Plate 3D upstream of a 4″ Venturi tube ($\beta = 0.65$): multiple identical symbols correspond to tappings in different circumferential locations

with other upstream fittings if for one upstream fitting the test in ISO 5167-1:2003 had been failed.

It is striking how large the shift in discharge coefficient is when the D-shaped plate (causing asymmetric flow) is too close to the Zanker Flow Conditioner Plate. It is necessary to have at least 7*D* between the D-shaped plate and the Zanker Flow Conditioner Plate.

Data were then taken when the Zanker Flow Conditioner Plate was 7*D* upstream of the upstream tappings of the Venturi tube and are given in Fig. 8.17.

In essence (on the basis of these and other tests), the Zanker Flow Conditioner Plate met the compliance test upstream of a Venturi tube provided that there was at least 3*D* between the plate and the upstream pressure tapping of the Venturi tube and 7*D* between the plate and any upstream fitting. This gives a minimum overall upstream length of 10*D*. However, when applying the compliance test to flow conditioners with Venturi tubes at high Reynolds number, it is worth noting that the shift in discharge coefficient due to a flow conditioner does not have a single value for all $Re_D > 3 \times 10^6$. Moreover, the requirements of the compliance test in terms of range of friction factor (see 7.4.1.5 of ISO 5167-1:2003) may be too restrictive.

Note that the Zanker Flow Conditioner Plate is relatively close to the Venturi tube: this is a requirement, not a mistake. If a Zanker Flow Conditioner Plate is placed too close to a fitting it can prevent the asymmetry from decaying: this problem was also found in the orifice-plate tests (Fig. 8.15).

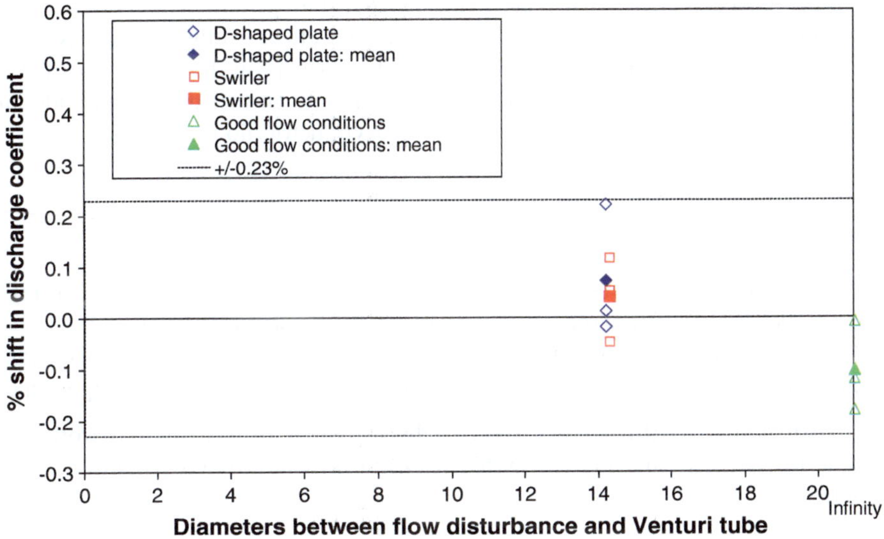

Fig. 8.17 Effect of a Zanker Flow Conditioner Plate 7*D* upstream of a 4″ Venturi tube ($\beta = 0.65$): multiple identical symbols correspond to tappings in different circumferential locations

Damage to Flow Conditioners

Flow conditioners can be damaged or blocked.

A 4″ Schedule 40 $\beta = 0.67$ orifice plate was calibrated downstream of a Zanker Flow Conditioner Plate which was in accordance with ISO 5167-2:2003 and, subsequently, downstream of two modified Zanker Flow Conditioner Plates in turn [see Reader-Harris et al. (2003a)]. The Zanker Flow Conditioner Plates were preceded by 41*D* of pipe machined (and honed) to be very close in diameter to the orifice meter, preceded by additional 4″ pipe of Schedule 10, preceded by a tube bundle. The flow conditioners were installed 8*D* upstream of the orifice plate since, following ISO 5167-2:2003, this is approximately the closest permissible location to the orifice plate and so modifications to the conditioner are likely to have particularly large effects. The first modification tested was a 1 mm 45° chamfer on the upstream side of each hole, to simulate erosion. The average hole size was a little less than 12 mm. The second modification tested was to use a plate in which an outside row of 4 holes in the standard drawing of the Plate (e.g. in Fig. 10 of ISO 5167-2:2003) was not drilled, to simulate a row of holes blocked by debris. The undrilled row was adjacent to Flange Tappings C. The flow conditioners are shown in Fig. 8.18, which also shows the orientation of the flow conditioners relative to the pressure tappings.

The shifts in discharge coefficient with the Zanker Flow Conditioner Plate and the two modified versions of the Zanker Flow Conditioner Plate were evaluated as described above and are given in Fig. 8.19. It is clear that partially blocking a plate has a much larger effect than chamfering it, even though the chamfer was quite large. In fact, because the chamfer had a beneficial effect, further work was done [and was described later by Reader-Harris et al. (2003a)] on the use of chamfered Zanker Flow Conditioner Plates: however, chamfering in fact did not provide an improvement, as it gave poor performance in asymmetric flow.

8.2.5.4 A Specific Test Is Done

A specific test of an installation can be carried out in accordance with 7.4.2 of ISO 5167-1:2003: if the discharge coefficient given in the standard is to be used and $\beta \leq 0.67$ it must be shown that the installation does not cause a shift in discharge coefficient of more than 0.23 % from the baseline obtained in a long straight pipe.

NOTE Obviously the flowmeter could be calibrated with its upstream pipework, but such a use would be outside ISO 5167, which at present only covers uncalibrated meters.

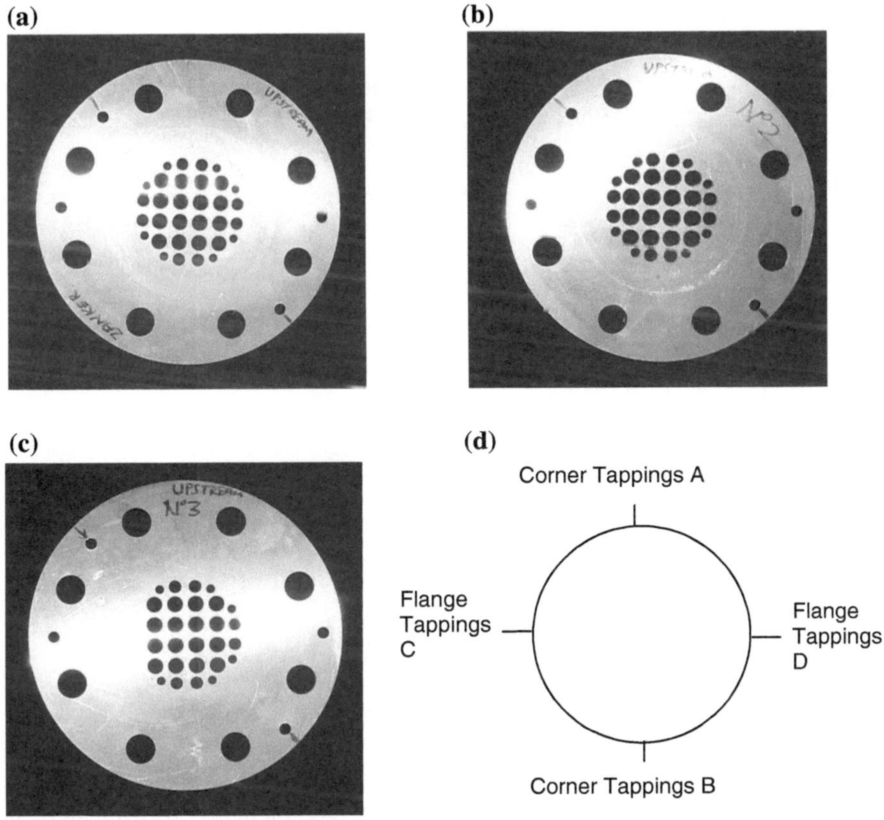

Fig. 8.18 The flow conditioners and pressure tapping locations (viewed from upstream): single pressure tappings (not a triple-T) were used. **a** Zanker Flow Conditioner Plate. **b** Chamfered Zanker Flow Conditioner Plate. **c** Partially blocked Zanker Flow Conditioner Plate. **d** Pressure tapping locations

8.2.5.5 CFD is Carried Out

CFD can be used to calculate the effect of an upstream installation, e.g. Barton et al. (2004); it is important to validate the CFD code by carrying out computations of known flows.

8.2.5.6 Engineering Judgement is Employed

By comparing an actual installation with ones that are not too different it may be possible to make an estimate of installation effects and thus of acceptability. An understanding of the physics of flow through the flowmeter is required so that wise judgement may be employed.

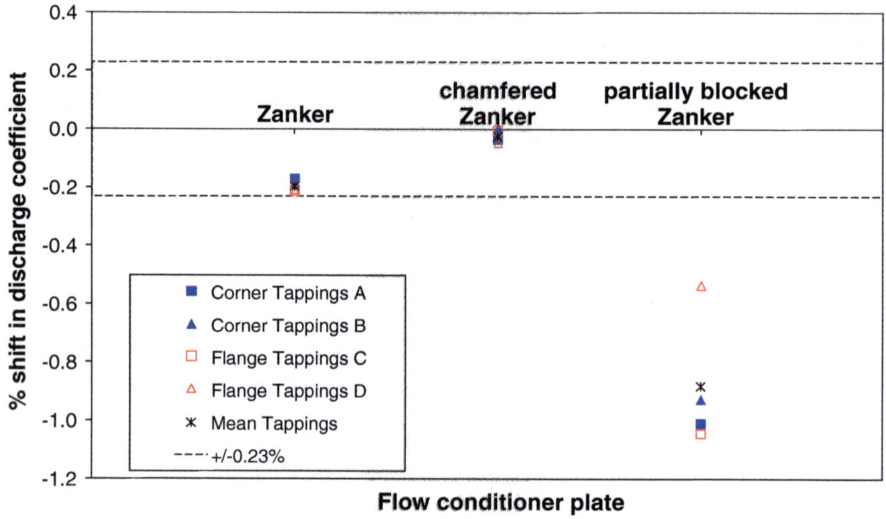

Fig. 8.19 Effect of modified Zanker Flow Conditioner Plates 8*D* upstream of an orifice plate ($\beta = 0.67$)

8.3 Downstream Straight Length

8.3.1 Orifice Plates

There appears to be a lack of data on the required straight lengths downstream of an orifice plate. Lengths in API 14.3.2:2000 are about 60 % of those stated in ISO 5167-2:2003. For gas flows the difference between the API and ISO lengths is of little importance, because the straight length has to be sufficient to include the thermowell 5*D*–15*D* downstream of the plate. Installing the thermowell too close to the downstream face of the plate leads to errors in the measurement of temperature (see Sect. 4.6.2).

8.3.2 Venturi Tubes

The requirement in Table 1 of ISO 5167-4:2003 is that fittings shall not be installed less than 4*d* downstream of the throat tapping plane. In many cases no straight length is required downstream of the Venturi tube itself in a liquid flow. In a gas flow there will be a need for a straight length to the thermowell (see 5.4.4.1 of ISO 5167-1:2003).

8.4 Pulsations

8.4.1 General

ISO 5167 does not cover pulsating flow. From 6.3.1 of ISO 5167-1:2003 the flow is considered as not being pulsating when

$$\frac{\Delta p'_{\mathrm{rms}}}{\overline{\Delta p}} \leq 0.10, \tag{8.7}$$

where $\overline{\Delta p}$ is the time-mean value of the differential pressure and $\Delta p'_{\mathrm{rms}}$ is the r.m.s. value of $\Delta p'$, the fluctuating component of the differential pressure.

$\Delta p'_{\mathrm{rms}}$ can only be measured accurately using a fast-response differential-pressure sensor, but it will not normally be necessary to check that this condition is satisfied. Pulsations can be caused by compressors, valves and obstructions that shed vortices, besides other causes.

More information on pulsating flow can be found in ISO/TR 3313:1998 (ISO 1998) (see Sect. 12.3.2.6). This is based on work of Mottram, in particular in Mottram (1974), Downing and Mottram (1975) and Mohammad and Mottram (1981). Generally in a pulsating flow if the true time-averaged differential pressure is measured the largest source of error is the square-root effect: i.e. the mean of the square root of the differential pressure is not equal to the square root of the mean differential pressure: square-root errors increase with the amplitude of the flow pulsation. At low Strouhal number (i.e. $fd/\bar{u}_{th}$, where f is the pulsation frequency and $\bar{u}_{th}$ the mean throat velocity), low $\frac{\Delta p'_{\mathrm{rms}}}{\overline{\Delta p}}$, and low static pressure it may be assumed that the standard differential-pressure-meter equation (Eq. 1.15) holds but with instantaneous values for the variables. Then if it is assumed that the flowrate is equal to its mean value plus a sinusoidal term, $\overline{\Delta p}$ and $\Delta p'_{\mathrm{rms}}$ can be calculated and provided that $\frac{\Delta p'_{\mathrm{rms}}}{\overline{\Delta p}}$ is small it can be shown that if $\overline{\Delta p}$ is used to calculate the flowrate the measurement error is approximately equal to

$$12.5\left(\frac{\Delta p'_{\mathrm{rms}}}{\overline{\Delta p}}\right)^2\%. \tag{8.8}$$

Where possible, taking the average of the square root of the differential-pressure readings rather than the square root of the average differential pressure is wise. According to Jungowski et al. (1990) for $\frac{\Delta p'_{\mathrm{rms}}}{\overline{\Delta p}}$ up to 0.35 'square root correction is effective and the residual pulsation error should not exceed 0.2 %'.

Impulse lines in pulsating flow are covered in Appendix 4.A.

8.4.2 Orifice Plates

In high-frequency pulsating flow assuming that the standard orifice equation applies but with instantaneous values for the variables no longer holds true: temporal inertia effects increase as the Strouhal number increases and for sine-wave pulsations become significant when $fd/\bar{u}_{th}$ exceeds 0.02, where f is the pulsation frequency and $\bar{u}_{th}$ the mean throat velocity. The effect of very severe pulsations has been measured by Earles and Zarek (1963). Pulsation effects on orifice plates at high pressure were measured by McBrien (1997): his results are different in significant respects from the results at lower pressures, but he states that if Eq. 8.7 holds the errors due to pulsation are less than 0.2 %. Studzinski et al. (1998) state that the measurement error 'does not always follow the square root relationship and an application of square root correction is not advisable because it may introduce an error instead of correcting it.'

The effect of acoustic noise was investigated by Norman et al. (1995). They used an upstream ball valve with a pressure drop across it to give pipe noise levels up to 170 dBA: a pipe noise level of 170 dBA gave a decrease in discharge coefficient of 0.4 % for flange tappings for a $\beta = 0.4$ orifice plate in a 24″ line; provided that the pipe noise level did not exceed 145 dBA the discharge coefficient did not change by more than about ±0.15 %; noise levels of 170 dBA could be reduced by 20 dBA by using an appropriately designed silencer; the conventional impulse lines did not create any errors in the measured differential pressures.

8.4.3 Venturi Tubes

There is a lack of information on pulsating flow through Venturi tubes.

8.5 Conclusions

There are huge amounts of data on installation effects on orifice plates, but very limited data on installation effects on Venturi tubes. The pattern of the data has been described and discussed. The basis on which the straight lengths in ISO 5167:2003 were derived has been described. In practice it is necessary to use ISO 5167-2:2003/ ISO 5167-4:2003 to provide minimum straight lengths.

Where it is not possible to conform to Table 3 of ISO 5167-2:2003/Table 1 of ISO 5167-4:2003, alternative options, some including the use of flow conditioners, have been described. The basis and use of the flow conditioner test in ISO 5167-1:2003 have been described.

Limits on pulsation have been described.

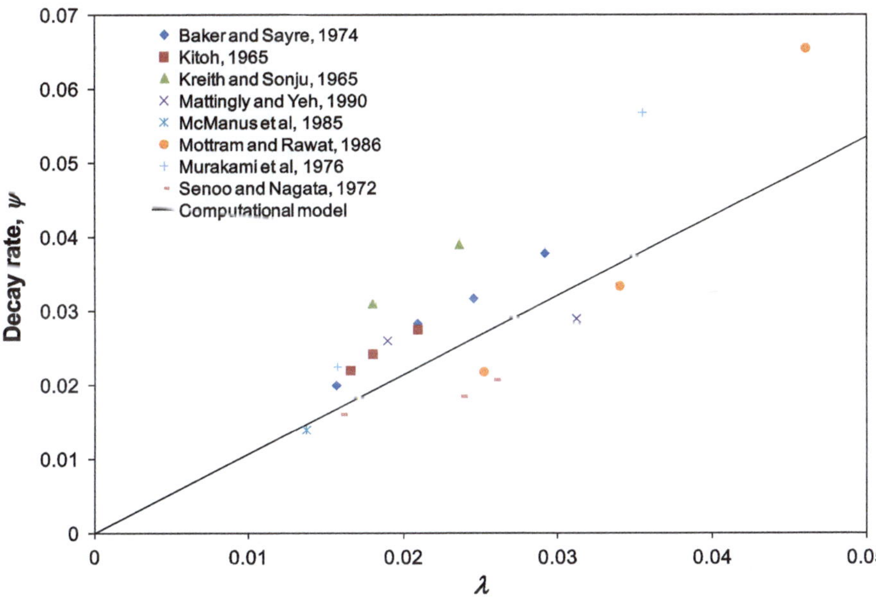

Fig. 8.A.1 The decay rate of swirl in a pipe

Appendix 8.A: Swirl Decay

Swirl decay was calculated by Reader-Harris (1994). Computational work showed that beyond a short distance downstream of a swirl-inducing installation the swirl is proportional to $\exp(-\psi x/D)$, where x is the distance down the duct, and that the decay rate ψ is equal to 1.07λ, where λ is the friction factor in the pipe (see Sect. 1.5). In Fig. 8.A.1 the decay rate in this computational model is compared with experimental data (Baker and Sayre 1974; Kitoh 1965; Kreith and Sonju 1965; Mattingly and Yeh 1990; McManus et al. 1985; Mottram and Rawat 1986; Murakami et al. 1976; Senoo and Nagata 1972).

On the basis of the computational model, if the pipe Reynolds number is 6×10^6 and R_a/D is 10^{-5}, after 200 diameters an 18° swirl (a typical value immediately downstream of two bends in perpendicular planes) will decay to 2°.

References

Akashi K, Watanabe H, Kenichi K (1979) Development of a new flow rectifier for shortening upstream straight pipe lengths for flowmeters. In: IMEKO Tokyo flow symposium: Society of Instrumentation and Control, paper 12b-5. Tokyo, Japan, pp 279–284

API (1985) Chapter 14—natural gas fluids measurement, Section 3—orifice metering of natural gas and other related hydrocarbon fluids. Manual of petroleum measurement standards, 2nd edn. ANSI/API 2530

API (1990) Manual of petroleum measurement standards Chapter 14—natural gas fluids measurement Section 3—concentric, square-edged orifice meters, Part 1—general equations and uncertainty guidelines. MPMS Chapter 14, Section 3, Part 1, 3rd edn. American Petroleum Institute, Washington DC

API (2000) Natural gas fluids measurement. Part 2—specification and installation requirements—concentric, square-edged orifice meters. API MPMS (Manual of Petroleum Measurement Standards) chapter 14.3.2:2000. American Petroleum Institute, Washington DC

Baker DW, Sayre CL Jr (1974) Decay of swirling turbulent flow of incompressible fluids in long pipes. In: Dowdell R (ed) Flow—its measurement and control in science and industry, vol 1. Instrument Society of America, Pittsburgh, pp 301–312 (Conference was in May 1971)

Barton N, Gibson J, Reader-Harris MJ (2004) Overview of CFD modelling of orifice plate flowmeters. In: Proceedings of 22nd North Sea Flow Meas Workshop, St Andrews, National Engineering Laboratory, East Kilbride, Glasgow

Bluschke H, Hucho W-H, Weber G (1966) Versuche über den Einfluβ von Rohrkrümmern auf die Durchfluβzahlen von Normventuridüsen und Klassischen Venturirohren. Brennst-Wärme-Kraft 18(2): 68–73

Bradford JE, Johnson MC, Gilbert JG (2006) Performance of Venturi meters installed downstream of bends. J AWWA 98(4):156–164 (April. American Water Works Association, Denver)

Brennan JA, McFaddin SE, Sindt CF, Kothari KM (1989) The influence of swirling flow on orifice and turbine flowmeter performance. Flow Meas Instrum 1:5–8

Downing PM, Mottram RC (1975) The effect of flow pulsations on orifice. In: Proceedings of Fluid Flow Measurement in the mid 1970s, National Engineering Laboratory, Glasgow

Earles SWE, Zarek JM (1963) Use of sharp-edged orifices for metering pulsating flow. Proc I Mech E 177(37):997–1024

Ferron AG (1963) Velocity profile effects on the discharge coefficient of pressure-differential meters. Trans ASME J Basic Eng 85:338–346

Gallagher JE, LaNasa PJ (1995) Field performance of the Gallagher flow conditioner. In: Proceedings of 3rd International Symposium Fluid Flow Meas, paper II.2, San Antonio

Gallagher JE, LaNasa PJ, Beaty RE (1994a) Development of Gallagher flow conditioner. In: Proceedings of FLOMEKO '94, 7th international conference on flow measurement. National Engineering Laboratory Executive Agency, Glasgow

Gallagher JE, LaNasa PJ, Beaty RE (1994b) The Gallagher flow conditioner. In: Proceedings of 12th North Sea Flow Meas Workshop, paper 2.4, National Engineering Laboratory Executive Agency, Glasgow, Peebles

Gibson J, Reader-Harris MJ (2006) Swirling flow through Venturi tubes of convergent angle 10.5° and 21°. In: Proceedings of ASME 2nd joint US-European Fluids Engineering Summer Meeting, Miami

Gibson J, Reader-Harris MJ (2007) Swirling flow through Venturi tubes. In: Proceedings of 6th International South East Asia Hydrocarbon Flow Meas Workshop, Kuala Lumpur

Halmi D (1973) Metering performance investigation and substantiation of the "Universal Venturi Tube" (U.V.T.). Part 1- hydraulic shape and discharge coefficient. ASME paper 73-WA/FM-3

Himpe U, Götte B, Schatz M (1994) Influence of upstream bends on the discharge coefficient of classical Venturi tubes and orifice plates. Flow Meas Instrum 5(3):209–216

Humphreys JS, Maxwell AD (1989) An installation-tolerant orifice-plate package. NEL flow measurement memo DA/382

Humphreys JS, Wood IM (1984) Progress report on several flow straighteners operating in identical hydraulic conditions. National Engineering Laboratory: flow measurement memo DA/344

Irving SJ (1976) Effect of system layout on the discharge coefficient of orifice plates. Part I Literature review. BHRA report TN 1352. BHRA, Cranfield, Bedfordshire

Irving SJ (1977) Effect of system layout on the discharge coefficient of orifice plates. Part II Experimental investigation into the effect of isolated fittings. BHRA RR 1424. BHRA, Cranfield, Bedfordshire

Irving SJ (1978a) Effect of system layout on the discharge coefficient of orifice plates. Part III Experimental investigation into the effects of bend combinations. BHRA RR 1462. BHRA, Cranfield, Bedfordshire

Irving SJ (1978b) Effect of system layout on the discharge coefficient of orifice plates. Part IV Methods for predicting meter errors. BHRA RR 1463. BHRA, Cranfield, Bedfordshire

ISO (1991) Measurement of fluid flow by means of pressure differential devices—part 1: orifice plates, nozzles and Venturi tubes inserted in circular cross-section conduits running full. ISO 5167-1:1991. International Organization for Standardization, Geneva

ISO (1998) Measurement of fluid flow in closed conduits—guidelines on the effects of flow pulsations on flow measurement instruments. ISO/TR 3313:1998. International Organization for Standardization, Geneva

ISO (2003a) Measurement of fluid flow by means of pressure differential devices inserted in circular cross-section conduits running full—part 1: general principles and requirements. ISO 5167-1:2003. International Organization for Standardization, Geneva

ISO (2003b) Measurement of fluid flow by means of pressure differential devices inserted in circular cross-section conduits running full—part 2: orifice plates. ISO 5167-2:2003. International Organization for Standardization, Geneva

ISO (2003c) Measurement of fluid flow by means of pressure differential devices inserted in circular cross-section conduits running full—part 4: Venturi tubes. ISO 5167-4:2003, International Organization for Standardization, Geneva

ISO (2007) Measurement of fluid flow by means of pressure differential devices—guidelines on the effect of departure from the specifications and operating conditions given in ISO 5167. ISO/TR 12767:2007. International Organization for Standardization, Geneva

Jungowski W, Studzinski W, Szabo JL (1990) Orifice meter performance under pulsating flow conditions. In: Proceedings of 2nd International Symposium Fluid Flow Meas, Calgary, pp 255–269

Karnik U (1995) A compact orifice meter/flow conditioner package. In: Proceedings of 3rd International Symposium Fluid Flow Meas, San Antonio

Kitoh D (1965) Experimental study of turbulent swirling flow in a straight pipe. J Fluid Mech 225:445–479

Kochen G, Smith DJM, Umbach H (1989) Installation effects on Venturi tube flowmeters. Intech Engineer's Notebook, pp 41–43

Kreith F, Sonju OK (1965) The decay of a turbulent swirl in a pipe. J Fluid Mech 22:257–271

Lake W T, Reid J (1992) Optimal flow conditioner. In: Proceedings of 10th North Sea Flow Meas Workshop, paper 1.3, National Engineering Laboratory, Glasgow

Langsholt M, Erdal A (1998) Design parameters for perforated plate flow conditioners. In: Proceedings of FLOMEKO '98

Laws EM, Ouazzane AK (1994) Compact installations for orifice plate flow meters. In: Proceedings of FLOMEKO '94, 7th international conference on flow measurement, National Engineering Laboratory Executive Agency, Glasgow

Laws EM, Ouazzane AK (1995) Flow conditioning for orifice plate meters. In: Proceedings of 3rd International Symposium Fluid Flow Meas, paper I.3, San Antonio

Laws EM (1990) Flow conditioning—a new development. Flow Meas Instrum 1(3):165–170

Martin CNB (1986) Effects of upstream bends and valves on orifice plate pressure distributions and discharge coefficients. NEL Report no 702. National Engineering Laboratory, Glasgow

Mattingly GE, Yeh TT (1990) Effects of pipe elbows and tube bundles on 50 mm orifice meters. In: Proceedings of Seminar on installation effects on flow metering, paper 3.1, Glasgow

Mattingly GE, Yeh TT (1991) Effects of pipe elbows and tube bundles on selected types of flowmeters. Flow Meas Instrum 2(1):4–13

Mattingly GE, Yeh TT (1992) Flowmeter installation effects due to several elbow configurations. In: Proceedings of 2nd International Symposium Fluid Flow Meas, Calgary, pp 271–283

McBrien RK (1997) High pressure pulsation effects on orifice meters. 1997 ASME Fluids Engineering Division Summer Meeting, FEDSM97-3700

McHugh A, Dyet D, Kinghorn FC (1981) Prediction of the effect of swirling flow on orifice plate measurements. NEL Report no 676. National Engineering Laboratory, East Kilbride, Glasgow

McHugh A, Kinghorn FC, Dyet WD (1984) Efficiency of an étoile flow straightener in non-symmetric swirling flow upstream of orifice plates. NEL Report no 692. National Engineering Laboratory

McManus SE, Bateman BR, Brennan JA, Vazquez Pantoja I, Mann D (1985) The decay of swirling gas flow in long pipes. In: Proceedings of American Gas Association operating section distribution/transmission conference, Boston, Massachusetts

Mohammad WA, Mottram RC (1981) High frequency pulsation effects on orifice meter accuracy. In: International Conference on Advances in Flow Meas Techniques, Warwick, paper C3. BHRA, Cranfield, Bedfordshire

Moore AE (1980a) Effect of system layout on the discharge coefficient of orifice plates. Part V BHRA report RR 1597. BHRA, Cranfield, Bedfordshire

Moore AE (1980b) Effect of system layout on the discharge coefficient of orifice plates. Part VI Experimental investigation into the effects of valve/bend combinations. BHRA report RR 1626. BHRA, Cranfield, Bedfordshire

Morrison GL, Hall KR, Macek ML, Ihfe LM, DeOtte RE, Hauglie JE (1993) Upstream velocity profile effect upon orifice flow meters. Flow Measurement for the Utilities, Amsterdam

Morrison GL, Hauglie JE, DeOtte RE (1995) Beta ratio axisymmetric flow distortion effects upon orifice flow meters. In: Proceedings of 3rd International Symposium Fluid Flow Meas, San Antonio

Morrison G L, Panak D L, DeOtte RE (1992) Numerical study of the effects of upstream flow condition upon orifice flow meter performance. In: ASME 11th International Conference Offshore Mechanics and Arctic Engineering, American Society of Mechanical Engineers, Calgary, Alberta, Washington DC

Morrow TB (1997) Orifice meter installation effects: development of a flow conditioner performance test. Report prepared by SwRI as GRI report No GRI-97/0207, Gas Research Institute, Chicago

Morrow TB (2005) Metering research facility program: additional studies of orifice meter installation effects and expansion factor. GRI report GRI-04/0246 on SwRI project No 18.10315. Gas Research Institute, Chicago

Morrow TB, Park JT, McKee RJ (1991) Determination of installation effects for a 100 mm orifice meter using a sliding vane technique. Flow Meas Instrum 2(1):14–20

Mottram RC (1974) The measurement of pulsating flow using orifice plate meters. In: Dowdell R (ed) Flow its measurement and control in science and industry, vol 1. Instrument Society of America, Pittsburgh, pp 197–208 (Conference was in May 1971)

Mottram RC, Rawat MS (1986) The swirl damping properties of pipe roughness and the implications for orifice meter installation. In: Proceedings of Flow Measurement in the mid-80s, vol 2. National Engineering Laboratory, East Kilbride, Glasgow, paper 6.1

Murakami M, Kito O, Katayama Y, Iida AY (1976) An experimental study of swirling flow in pipes. Bull JSME 19:118–126

NEL (1985) The influence of the proximity of a 90 degree bend upon the performance of a 100 mm Venturimeter of diameter ratio 0.7069. Report VEDE/01 for VDI/VDE. National Engineering Laboratory, East Kilbride, Glasgow

Norman R, Graham P, Drew WA (1995) Effects of acoustic noise on orifice meters. In: Proceedings of 3rd International Symposium Fluid Flow Meas, San Antonio, Texas

Oosthuizen PH (1964) An approximate method of predicting the effect of small changes in the initial velocity profile on the coefficient of discharge of an orifice plate. South African Mechanical Engineer, pp 238–243

Pardoe WS (1943) The effect of installation on the coefficients of Venturi meters. Trans ASME 6:337–349

Reader-Harris MJ (1994) The decay of swirl in a pipe. Int J Heat Fluid Flow 13(3):212–217

Reader-Harris MJ (2002a) Compliance testing of flow conditioners with differential pressure meters. Report No 2002/77 on project No FDDP02. NEL, East Kilbride, Glasgow

Reader-Harris MJ (2002b) The effect on orifice plates of diameter steps in upstream pipework and of swirl. Report No 2002/75 on project No FDDP02. NEL, East Kilbride, Glasgow

Reader-Harris MJ, Brunton WC, Nicholson IG, Rushworth R (2003a) Ageing effects on orifice metering. In: Proceedings of 21st North Sea Flow Meas Workshop, Norway

Reader-Harris MJ, Brunton WC, Nicholson IG, Rushworth R, Hodges D (2003b) The compliance test for flow conditioners as applied to a Zanker flow conditioner plate with a Venturi tube. In: Proceedings of FLOMEKO 2003, Groningen. Gasunie Research, Groningen, The Netherlands

Reader-Harris MJ, Brunton WC, Sattary JA (1997) Installation effects on Venturi tubes. In: Proceedings of ASME Fluids Engineering Division Summer Meeting, FEDSM97-3016. American Society of Mechanical Engineers, Vancouver, New York

Reader-Harris MJ, Rushworth R, Gibson J (2004) Installation effects on Venturi tubes of convergent angle 10.5°. In: Proceedings of 22nd North Sea Flow Meas Workshop, St Andrews, National Engineering Laboratory, East Kilbride, Glasgow

Reader Harris MJ, Sattary JA, Woodhead E (1995a) The use of flow conditioners to improve flow measurement accuracy downstream of headers. In: Proceedings of 3rd International Symposium Fluid Flow Meas, San Antonio, Texas

Reader-Harris MJ, Woodhead E, Sattary JA, McEwen D (1995b) Flow conditions downstream of headers. Report HCP001 (223/94) for the Headers Consortium, National Engineering Laboratory, East Kilbride, Glasgow

Sattary JA (1991) EEC orifice plate programme—installation effects. Flow Meas Instrum 2(1): 21–33

Scott JL, Sindt CF, Lewis MA (1993) Flow conditioner location effects in orifice flowmeters. NIST technical note 1356. National Institute of Standards and Technology, Boulder, Colorado

Scott JL, Sindt CF, Lewis MA, Brennan JA (1991) The effects of flow conditioners and tap location on orifice flowmeter performance. NIST Technical Note 1352. National Institute of Standards and Technology, Boulder, Colorado

Senoo Y, Nagata T (1972) Swirl flow in long pipes with different roughness. Bull JSME 15:1514–1521

Shen JJS (1991) Characterization of swirling flow and its effects on orifice metering. SPE 22865. In: 66th annual technical conference and exhibition, Society of Petroleum Engineers, Dallas, Texas

Skelton M, Barrons S, Ayre J, Steven R (2010) Developments in the self-diagnostic capabilities of orifice plate meters. In: Proceedings of 28th International North Sea Flow Meas Workshop, St Andrews. NEL, East Kilbride, Glasgow

Smith DJM (1986) The effects of flow straighteners on orifice plates in good flow conditions. In: Proceedings of Flow Measurement in the mid 80s, National Engineering Laboratory, East Kilbride, Glasgow

Spearman EP, Sattary JA, Reader-Harris MJ, Rhodes FS (1995) The effect of upstream installations on orifice meter discharge coefficients. Flow measurement memo FL/445. National Engineering Laboratory Executive Agency, East Kilbride, Glasgow

Sprenkle RE, Courtright NS (1957) Straightening vanes for flow measurement. ASME paper No 57-A-76. Presented at the ASME winter annual meeting, New York

Steven R (2008) Diagnostic methodologies for generic differential pressure flow meters. In: Proceedings of 26th International North Sea Flow Meas Workshop, St Andrews. NEL, East Kilbride, Glasgow

Studzinski W, Karnik U, Geerligs J, McBrien R (1998) Pulsation effects on orifice meter performance. In: Proceedings of 9th international conference flow meas, FLOMEKO, Lund, Sweden, pp 417–422

Studzinski W, Karnik U, LaNasa P, Morrow T, Goodson D, Husain Z, Gallagher J (2001a) White paper on orifice meter installation configurations with and without flow conditioners. White paper prepared for API 14.3 Part 2. GRI report GRI 99/0262, Gas Research Institute, Chicago

Studzinski W, Weiss M, Attia J (2000) Effect of reducers, expanders and a gate valve on orifice meter performance. NRTC report #01426, Nova Research and Technology Corporation, Calgary

Studzinski W, Weiss M, Geerligs J, Attia J (2001b) Effect of reducers, expanders, a gate valve and two bends in perpendicular planes on orifice meter performance. In: Proceedings of Flow Measurement 2001 international conference, paper 3.1. National Engineering Laboratory, Peebles, East Kilbride, Glasgow

Weiss M, Studzinski W, Attia J (2002) Performance evaluation of orifice meter standards for selected T-junction and elbow installations. In: Proceedings of 5th International Symposium Fluid Flow Meas, paper 5.1, Washington, DC

Wilcox PL, Weberg T, Erdal A (1990) Short gas metering systems using K-lab flow conditioners. In: Proceedings of 8th North Sea Flow Meas Workshop, National Engineering Laboratory, East Kilbride, Glasgow (also paper of identical title at 2nd International Symposium Fluid Flow Meas, Calgary, 1990)

Williamson ID, Botros KK, Price GR (1993) Flow characteristics and orifice meter error caused by upstream headers in multi-run meter stations. ASME Fluids Eng Div 181:1–11

Zanker K, Goodson D (1999) Qualification of a flow conditioning device according to the new AGA procedure. In: Proceedings of 4th International Symposium Fluid Flow Meas, Denver

Zanker KJ (1962) The development of a flow straightener for use with orifice-plate flowmeters in disturbed flows. Flow measurement in closed conduits, II (paper D-2), pp 395–415. HMSO, Edinburgh. (Proceedings of symposium at National Engineering Laboratory, September 1960)

Zimmermann H (1996) Einfluβ einer drallbehafteten Strömung auf die Durchfluβzahl α einer Normblende. Projekt 06/92. Fachhochschule Mannheim: Hochschule für Technik und Gestaltung

Chapter 9
Nozzle Discharge Coefficient

Abstract Four ASME (PTC 6) flow nozzles were manufactured together with upstream and downstream pipework in two pipe sizes. Throat and wall tappings were used. The flow nozzles were calibrated in water and in high pressure nitrogen at 20 and 60 barg. The wall-tapping data are in excellent agreement with the discharge-coefficient equation in ISO 5167-3:2003: all the data for $Re_d > 10^6$ fit the equation with twice the r.m.s. deviation equal to 0.39 %. The data extrapolate well from water to higher Reynolds number. The equation in ASME PTC 6:1996 fits all the NEL throat-tapping data for $Re_d > 10^6$ with twice the r.m.s. deviation equal to 0.61 %. This figure is significantly larger than that for the wall-tapping data: in some cases there are humps in the data as there are when Venturi tubes are used. The conclusion is that more use of nozzles with wall tappings instead of throat-tapped nozzles is very desirable. However, the data show that it is possible to extrapolate from a throat Reynolds number just above 10^6 to high Reynolds number with throat-tapped nozzles using consistent equations for static-hole error. Another conclusion, which is very relevant to this book, is that the issues with Venturi tubes at high Reynolds number are due to the use of pressure tappings in high-velocity regions.

9.1 Introduction

Throat-tapped flow nozzles are extremely important in themselves as they are commonly used in power-station performance tests. As the efficiency of power stations has major financial implications, the accuracy of measurements is vital. Nozzles are included in this book because they can have tappings on either the nozzle itself or the wall, and so the effect of the different tapping locations can be seen and provide wisdom on the interpretation of data from Venturi tubes.

Nozzles with wall tappings are standardized in ISO 5167-3:2003 (ISO 2003), whereas nozzles with throat tappings are used in accordance with ASME PTC 6:1996 (ASME 1996). However, although Sect. 4.8.15 of ASME PTC 6:1996 states

M. Reader-Harris, *Orifice Plates and Venturi Tubes*,
Experimental Fluid Mechanics, DOI 10.1007/978-3-319-16880-7_9

that a satisfactory nozzle calibration will, among other things, have a mean difference from a particular calibration curve of not more than 0.25 %, in practice many nozzles fail to achieve the requirements of ASME PTC 6. There are, therefore, problems between manufacturers, users and calibration laboratories.

The performance of Venturi tubes in high-pressure gas proved to be different from what had been expected on the basis of calibrations in water at lower Reynolds number (see Chap. 7). Throat-tapped flow nozzles are generally calibrated in water and extrapolated to the higher Reynolds numbers found when they are used in high-pressure high-temperature water or gas. Problems with extrapolating Venturi-tube calibrations led to doubts about extrapolating throat-tapped-nozzle calibrations.

Therefore NEL had some throat-tapped nozzles and associated pipework manufactured (see an example of the nozzles as Fig. 9.1) and calibrated them in water and gas to compare the results with those given in ASME PTC 6 and thus to see how well extrapolation from a water calibration to higher Reynolds number works. Details of this project are given in Reader-Harris et al. (2005). This work was published as Reader-Harris et al. (2007).

Nozzles with throat tappings require an upstream tapping on the pipe wall; the pipes were also manufactured with downstream wall tappings, and the data taken with the upstream and downstream wall tappings were used to check the uncertainty claim for the discharge-coefficient equation for low-ratio long-radius nozzles in ISO 5167-3:2003.

The work described here is concerned with high Reynolds number flow through nozzles (throat Reynolds number, Re_d, mainly greater than 10^6): there is an extensive literature on lower Reynolds numbers: for instance, Hall (1959) considers laminar-turbulent boundary-layer transition and its consequences in the range of Re_d from 10^5 to 10^6.

Fig. 9.1 4″ throat-tapped nozzle for use in 8″ pipe

9.2 Manufacture

9.2.1 General

Four nozzles, 2″, 3″, 3.5″ and 4″ in diameter, and upstream and downstream pipework of both 6″ and 8″ diameter were manufactured in stainless steel. The design was in accordance with the requirements of ASME PTC 19.5:1971 (ASME 1971) (see Fig. 9.2). The nozzle is installed concentrically with the pipe. The design almost meets the requirements for a low-ratio long radius nozzle in ISO 5167-3:2003, but it is not possible to meet ISO 5167 and ASME PTC 19.5 simultaneously.

Since the nozzles can be installed in pipework of either diameter it is appropriate to divide the data by nozzle diameter (with some reference to pipe diameter) rather than to divide the data by diameter ratio (or pipe diameter) as has been done for the Venturi tubes in Chap. 7 and the orifice plates in Chap. 5.

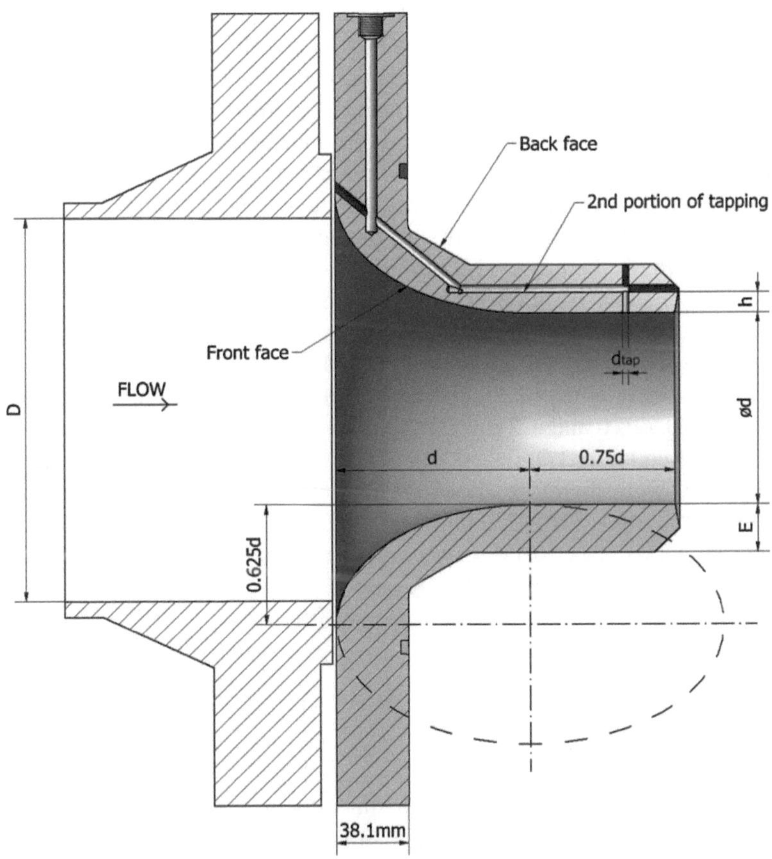

Fig. 9.2 ASME long-radius nozzle: the centre of the throat tapping is 0.25d from the downstream end of the parallel section of the throat; E = 0.25d (but for the 2″ nozzle E = 16 mm)

There need be no confusion in this chapter between nozzle diameters and pipe diameters: at NEL the nozzle diameters were 2″ to 4″ and pipe diameters 6″ and 8″; actual nozzle diameters were within 1.02 % of the nominal values.

NOTE Where comparison is made with hot-water data from NMIJ in Sects. 9.6 and 9.7 the nozzle diameters at NMIJ were 4″ and 6.5″ and pipe diameters 8″ and 14″.

9.2.2 Pipework and Nozzles

The 6″ and 8″ pipework consisted of an upstream length of 4*D* and a downstream length of 3*D*. The pipework was of the internal diameter of Schedule 40 with ANSI Class 600 flanges for a design pressure of 70 barg. In the 6″ pipework the flanges adjacent to the nozzles were 8″ Class 600 so that the 2″ and 3″ nozzles could be used in both 6″ and 8″ pipes. The pipework was machined from Schedule 80 seamless pipe, was dowelled, and used 'O' rings so that there were negligible steps between the new pipework and the machined pipework upstream and downstream of the package, which had been manufactured previously. In water, at least 11*D* of machined pipe was used upstream of the nozzle; upstream of the nozzle there was at least 29*D* of Schedule 40 pipe (including the machined section), preceded by at least another 23*D* of pipe of the same nominal diameter. In gas in 8″, there was 6*D* of machined pipe upstream of the nozzle; there was 27*D* of Schedule 40 pipe upstream of the nozzle (including the machined section) preceded by a flow conditioner. In gas in 6″, there was 11*D* of machined pipe upstream of the nozzle; there was 37*D* of Schedule 40 pipe upstream of the nozzle (including the machined section). In each case the pipework of Schedule 40 was smoothly blended into the machined pipe. Diagrams are given in Reader-Harris et al. (2005).

The nozzles had values of R_a, the arithmetic mean deviation of the roughness profile, of 0.5, 0.7, 0.6 and 1.2 μm for the 2″, 3″, 3.5″ and 4″ nozzles respectively, giving $10^5 R_a/d$ equal to 1.0, 0.9, 0.7 and 1.2, respectively, where d is the nozzle diameter. They were therefore about 10 times smoother than the maximum permitted value in ISO 5167-3:2003, but just outside the maximum permitted roughness in ASME PTC 6:1996. According to Fig. 4.6 of ASME PTC 6:1996, $10^5 R_a/d$ equal to 1 gives a hydraulically smooth surface for a throat-tapped nozzle for $Re_d < 1.1 \times 10^7$.

9.2.3 Nozzle Tappings

Since tapping diameter and tapping depth affect the discharge coefficient of Venturi tubes, tappings of different diameter and depth were installed in the nozzle throat in accordance with Table 9.1. h is the tapping depth (to the first point at which the

Table 9.1 Tapping diameters and depths for the nozzles

Nozzle	Tapping diameter d_{tap} (mm)	Tapping depth h (mm)	h/d_{tap}	Second portion of the tapping within the nozzle
2″	3 mm IN	6.0	2.0	Half way between front and back
2″	3 mm OUT	8.0	2.7	
2″	4 mm	8.0	2.0	
3″	3 mm	7.6	2.5	Half way between front and back
3″	4 mm	10.0	2.5	
3.5″	3 mm IN	9.1	3.0	Half way between front and back
3.5″	3 mm OUT	13.5	4.5	
3.5″	4 mm IN	9.2	2.3	Half way between front and back
3.5″	4 mm OUT	13.5	3.4	
4″	3 mm	10.7	3.6	Half way between front and back
4″	6 mm	15.0	2.5	

tapping intersects the tapping perpendicular to it, called here the second portion of the tapping), and d_{tap} is the tapping diameter: see Fig. 9.2. The tappings were inserted using spark erosion (EDM) to ensure a sharp edge. There is a substantial literature on the effect of fabrication technique, e.g. Brunkalla (1985). The tappings with the smaller values of h are referred to as 'IN'; those with the larger values of h are referred to as 'OUT'. '(IN)' and '(OUT)' are given as labels to different nozzle tappings that are of the same depth.

The nozzle design is only in accordance with Figure II-III-14 of ASME PTC 19.5:1971 where the second portion of the tapping is half way between the front and back faces of the nozzle.

The 2″ nozzle had one 3 mm tapping installed so that the second portion of the tapping lay half way between the front and back faces of the nozzle; this gave $h/d_{tap} = 2.0$. Since this value of h/d_{tap} was quite small, the 2″ nozzle had one 3 mm tapping installed with the second portion of the tapping as close as was considered reasonable to the back face of the nozzle: this gave $h/d_{tap} = 2.7$. The largest practical tapping diameter for which h/d_{tap} was at least 2 was 4 mm; so two 4 mm tappings were inserted with $h/d_{tap} = 2$ and the second portion of the tapping close to the back face of the nozzle.

The 3″ nozzle had two pairs of tappings so that the effect of variability in tapping manufacture could be seen, since where the tappings were of the same diameter they were intended to be identical. For the 3 mm tappings the second portion of each tapping lay half way between the front and back faces of the nozzle. The 4 mm tappings were designed to give $h/d_{tap} = 2.5$.

The 3.5″ nozzle had two 3 mm tappings and two 4 mm tappings. In each tapping size, one tapping was of such a depth that the second portion of the tapping lay half way between the front and back faces of the nozzle, the other was such that the second portion of the tapping lay near the back face of the nozzle to ensure that an effect of tapping depth could be measured.

The 4″ nozzle had two 3 mm tappings and two 6 mm tappings. The aim here was to see the effect of different tapping diameters. For both the 3 mm tappings the second portion of the tapping lay half way between the front and back faces of the nozzle and still gave $h/d_{tap} = 3.6$. The second portion of each of the 6 mm tappings was near the back face of the nozzle.

9.2.4 Wall Tappings

In the upstream sections, wall tappings were provided 1*D* upstream of the nozzle (and used for both throat-tapping and wall-tapping data). In the downstream sections, wall tappings were provided 0.5*D* downstream of the upstream face of the nozzle and, in the case of the 8″ pipe, also at a distance 0.39*D* downstream of the upstream face of the nozzle so that downstream wall tappings were located in accordance with ISO 5167-3:2003 for $\beta = 0.25$.

In every case, the upstream tapping was of the same diameter as the throat tapping and the downstream wall tapping and was in line with both. As a result, ten tappings were required in both the upstream and the downstream 8″ pipes.

9.3 Data

Data were collected in both water and high-pressure gas. The data were collected in water in NEL's water flow facility using water/air and water/mercury manometers. The temperature of the water was approximately 40 °C in order to increase the Reynolds number from that which would have been obtained at 20 °C.

Equations 1.15 and 1.13 were used:

$$q_m = \frac{C\varepsilon}{\sqrt{1-\beta^4}} \frac{\pi d^2}{4} \sqrt{2\rho_1 \Delta p} \tag{1.15}$$

$$\varepsilon = \sqrt{\left(\frac{\kappa \tau^{\frac{2}{\kappa}}}{\kappa - 1}\right)\left(\frac{1-\beta^4}{1-\beta^4 \tau^{\frac{2}{\kappa}}}\right)\left(\frac{1-\tau^{\frac{\kappa-1}{\kappa}}}{1-\tau}\right)}. \tag{1.13}$$

Both throat-tapping data (as in ASME PTC 6:1996) and wall-tapping data (using upstream and downstream wall tappings) were obtained. Because the two smaller nozzles are included twice within the data, the data taken in 8″ pipe have, on some occasions, been analysed together and the data taken in 6″ pipe analysed in comparison with those from the same nozzle in 8″ pipe. The data for $Re_d < 10^6$ tend to be more scattered and are not to be used within an analysis under ASME PTC 6. The mean deviations of the discharge coefficients in 6″ pipe from those in 8″ pipe for the data for $Re_d > 10^6$ are given in Table 9.2.

Table 9.2 Mean deviations of water data in 6″ pipe from those in 8″ pipe for $Re_d > 10^6$

Nozzle diameter	Tappings	Mean deviation (%)
2″	Wall	−0.01
2″	Throat	−0.10
3″	Wall	0.08
3″	Throat	0.01

The calibrations in high-pressure gas (nitrogen) were undertaken at 20 and 60 barg in the NEL recirculating loop using a bank of four 4″ turbine meters as reference. The gas temperature at the nozzle throughout the tests was in the range 17.92 ± 0.22 °C. The results are, in general, a function of Reynolds number, but data taken with throat tappings tend also to have peaks at particular values of throat velocity.

9.4 Wall-Tapping Data: Analysis

All the wall-tapping data obtained in 8″ pipe in water are shown in Fig. 9.3. All the wall-tapping data obtained in 8″ pipe in water or gas are shown in Fig. 9.4. To linearize the data at least partially the data in Fig. 9.4 are replotted in Fig. 9.5 against $\exp(-0.2Re_d/10^6)$.

Since the data from the wall tappings were almost unaffected by tapping diameter, for each nozzle in 8″ pipe those from different tapping diameters are combined (from 3 and 4 mm tappings for the 2″, 3″ and 3.5″ nozzles, from 3 mm and 6 mm tappings for the 4″ nozzle); for the 2″ and 3″ nozzles in 6″ pipe only 3 mm tappings were used.

For each nozzle the range of $\Delta p/p_1$ for the two static pressures in gas was almost the same (the values of $\Delta p/(\kappa p_1)$ were about 7.5 % higher for the lower pressure

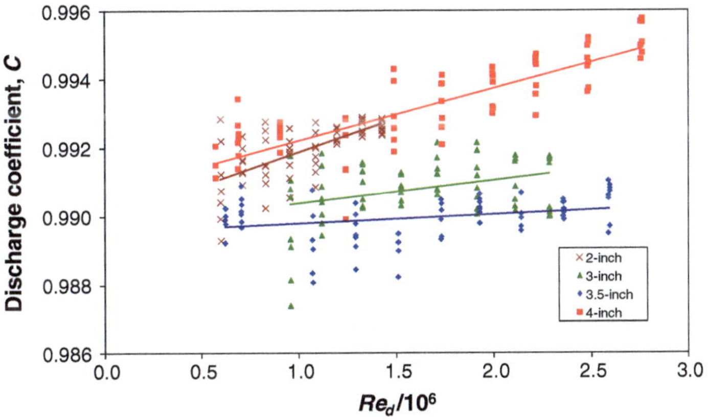

Fig. 9.3 Wall-tapping data taken in water in 8″ pipe as a function of Re_d

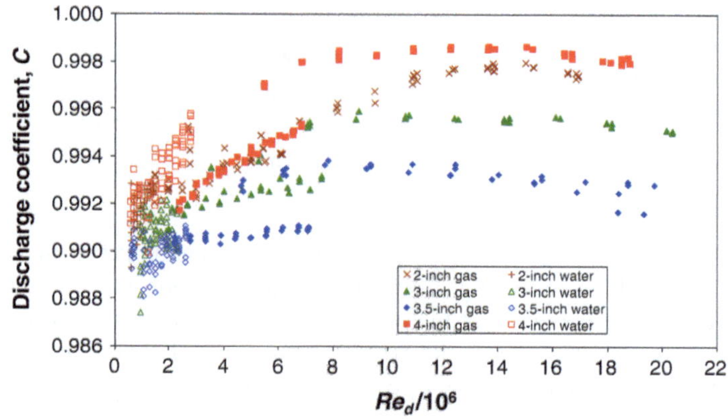

Fig. 9.4 Wall-tapping data taken in water and gas in 8″ pipe as a function of Re_d

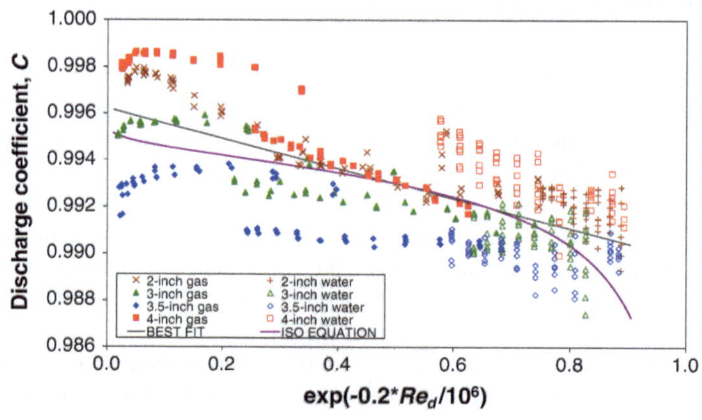

Fig. 9.5 Wall-tapping data taken in water and gas in 8″ pipe versus $\exp(-0.2Re_d/10^6)$

data; the maximum value of $\Delta p/p_1$ reduced with nozzle size, from 0.047 for the 2″ nozzle to 0.012 for the 4″ nozzle). So the decrease in discharge coefficient at the highest Reynolds numbers is probably not due to error in the expansibility-factor equation: it is more likely to be due to the reference meters.

NOTE At high Reynolds number as the Reynolds number increases, the friction loss on the nozzle wall reduces as a fraction of the differential pressure; so the discharge coefficient increases. Simultaneously the velocity profile becomes flatter, and this flattening causes the differential pressure to increase and the discharge coefficient to reduce. The former effect will be dominant on almost all nozzles. In contrast, for an orifice plate at moderate or high Reynolds number an increase in Reynolds number reduces the friction on the orifice plate itself as a fraction of the differential pressure; this reduces the size of the vena contracta and thus the discharge coefficient; moreover, the velocity profile becomes flatter, and this flattening causes the differential pressure to increase and the discharge coefficient to reduce. Therefore, for an orifice plate at moderate or high Reynolds number the discharge coefficient reduces with Reynolds number.

The best-fit equation to the data in Fig. 9.5 is

$$C = 0.9962 - 0.0064\exp(-0.2Re_d/10^6) \tag{9.1}$$

Twice the standard deviation of the data about this equation is 0.33 %.

The data are also compared in Fig. 9.5 with the discharge-coefficient equation in 5.2.6.2 of ISO 5167-3:2003, which is

$$C = 0.9965 - 0.00653\sqrt{\frac{10^6}{Re_d}} \tag{9.2}$$

Twice the r.m.s. deviation of the 8″ data from this equation is 0.40 % (0.38 % if only the data for $Re_d > 10^6$ are included). The wall-tapping data are in excellent agreement with this equation: of the 8″ data for $Re_d > 10^6$, 91 % lie within 0.25 % of the equation, and there is a mean deviation of 0.01 %.

The uncertainty of Eq. 9.2 is stated in 5.2.7.1 of ISO 5167-3:2003 to be 2.0 %; these data fit Eq. 9.2 much better than that. The main reason why the uncertainty in ISO 5167-3:2003 might be much larger than the 0.40 % obtained here (excluding the facility uncertainty) is the effect of nozzle roughness. These nozzles all had $R_a \approx 10^{-5}d$, whereas ISO 5167-3:2003 only requires $R_a \leq 10^{-4}d$. The effect on C of changing R_a from $10^{-5}d$ to $10^{-4}d$ is around 1 % at high Reynolds number [see Reader-Harris et al. (2005)]. Reducing the limit on nozzle roughness would, therefore, enable the uncertainty to be reduced.

The 6″ data are plotted in Fig. 9.6. For $Re_d > 10^6$ the ISO Equation fits the 6″ data with twice the r.m.s. deviation equal to 0.38 %.

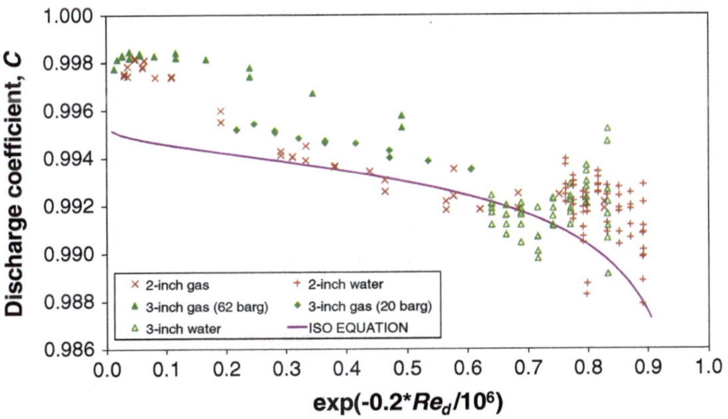

Fig. 9.6 Wall-tapping data taken in water and gas in 6″ pipe versus $\exp(-0.2Re_d/10^6)$

The data taken in gas could also be compared with those taken in water. There were two reasons for doing this: firstly, to check the performance of NEL's two facilities and, secondly, to assess the possibilities for extrapolation from water to gas. In terms of the former, the gas data in Figs. 9.4, 9.5 and 9.6 divide into those taken at around 20 barg and those at around 60 barg. This is best seen in Fig. 9.6 where the 3″ gas data have been separated into two sets; similar steps can be identified in Figs. 9.4 and 9.5. The data suggest that at high flowrates at 20 barg the reference flowrate in gas might be about 0.1 % low, whereas at medium flowrates at 60 barg the reference flowrate in gas might be about 0.1 % high. These are well within the claimed reference uncertainty of 0.35 %.

It is possible to analyse all the 8″ and 6″ data together. Equation 9.1, derived from the 8″ data, fits the 6″ data with twice the r.m.s. deviation equal to 0.29 % and all the data with twice the r.m.s. deviation equal to 0.33 %. Fitting all the wall-tapping data gives

$$C = 0.9965 - 0.0066\ \exp(-0.2Re_d/10^6) \tag{9.3}$$

Twice the standard deviation of the data about this equation is also 0.33 %.

Sometimes nozzles are calibrated in water and extrapolated to high Reynolds number. Given the narrow Reynolds number range in water it is best to use the slope from Eq. 9.3 and determine the constant from the water data. The results are shown in Table 9.3. The extrapolated equations fit the gas data well.

It is also possible to extrapolate from calibration in water to high Reynolds number parallel to Eq. 9.2, the equation in ISO 5167-3:2003. Only the data with $Re_d > 10^6$ were used. The results are shown in Table 9.4. The extrapolated equations fit the gas data well. The deviations between calculated values from extrapolation and measured values are shown in Fig. 9.7. 95 % of the extrapolated points have C within 0.24 % of the measured value.

Table 9.3 Extrapolated lines for wall-tapping data for each nozzle: $x = \exp(-0.2Re_d/10^6)$

Nozzle size (inch)	Pipe size (inch)	Extrapolated line based on water data	Twice the r.m.s. deviation (gas only) (%)
2	8	$0.9973 - 0.0066\ x$	0.17
3	8	$0.9956 - 0.0066\ x$	0.16
3.5	8	$0.9948 - 0.0066\ x$	0.29
4	8	$0.9980 - 0.0066\ x$	0.26
2	6	$0.9972 - 0.0066\ x$	0.19
3	6	$0.9966 - 0.0066\ x$	0.33

Table 9.4 Extrapolated lines for wall-tapping data for each nozzle: $x = \sqrt{10^6/Re_d}$

Nozzle size (inch)	Pipe size (inch)	Extrapolated line based on water data with ($Re_d > 10^6$)	Twice the r.m.s. deviation (gas only) (%)
2	8	0.9982 − 0.00653 x	0.24
3	8	0.9961 − 0.00653 x	0.21
3.5	8	0.9949 − 0.00653 x	0.20
4	8	0.9983 − 0.00653 x	0.34
2	6	0.9982 − 0.00653 x	0.29
3	6	0.9969 − 0.00653 x	0.45

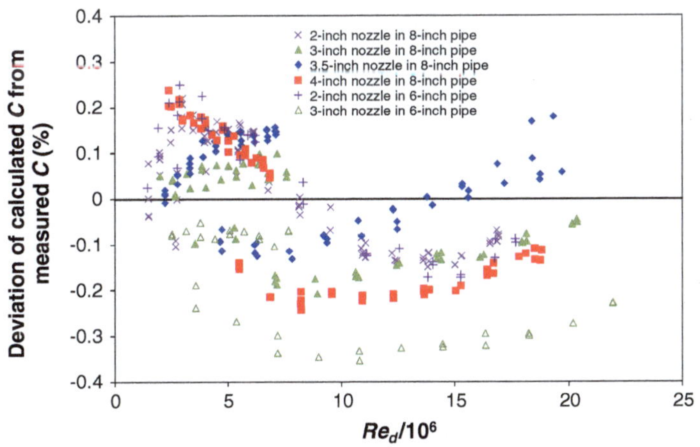

Fig. 9.7 Wall tappings: deviation of calculated values in gas (by extrapolation parallel to Eq. 9.2 from water data for $Re_d > 10^6$) from measured gas data

9.5 Throat-Tapping Data: Initial Analysis

All the throat-tapping data taken in water and gas are presented in Figs. 9.8, 9.9, 9.10, 9.11, 9.12 and 9.13.

These data show only a small effect of tapping depth (the variation in tapping depth is quite small), but a significant effect of tapping diameter. This can be seen in the best-fit lines for each nozzle in Table 9.5.

The equation in ASME PTC 6:1996 is plotted in each of Figs. 9.8, 9.9, 9.10, 9.11, 9.12 and 9.13:

$$C = 1.0054 - \frac{0.185}{Re_d^{0.2}}\left(1 - \frac{361{,}239}{Re_d}\right)^{0.8} \tag{9.4}$$

Twice the r.m.s. deviation from Eq. 9.4 for all the throat-tapping data for $Re_d > 10^6$ is 0.61 %. For the 8″ throat-tapping water data for $Re_d > 10^6$, 78 % lie

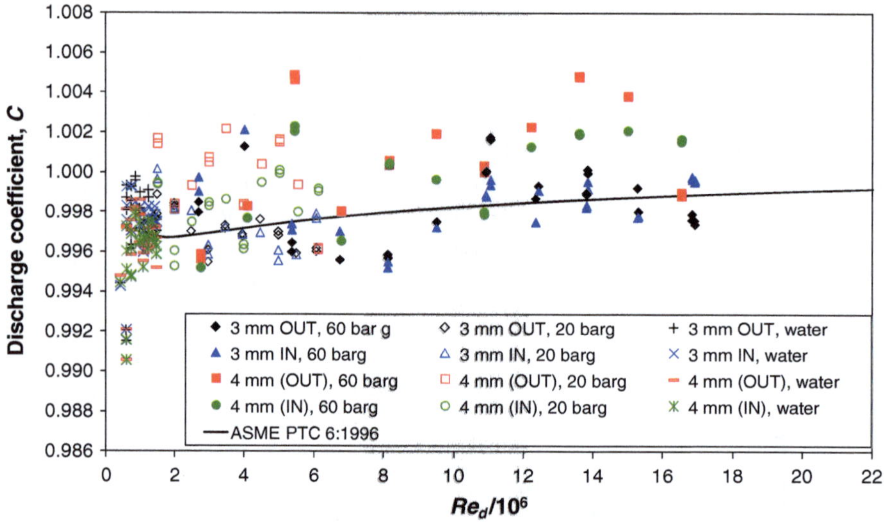

Fig. 9.8 Throat-tapping data taken in water and gas: 2″ nozzle in 8″ pipe

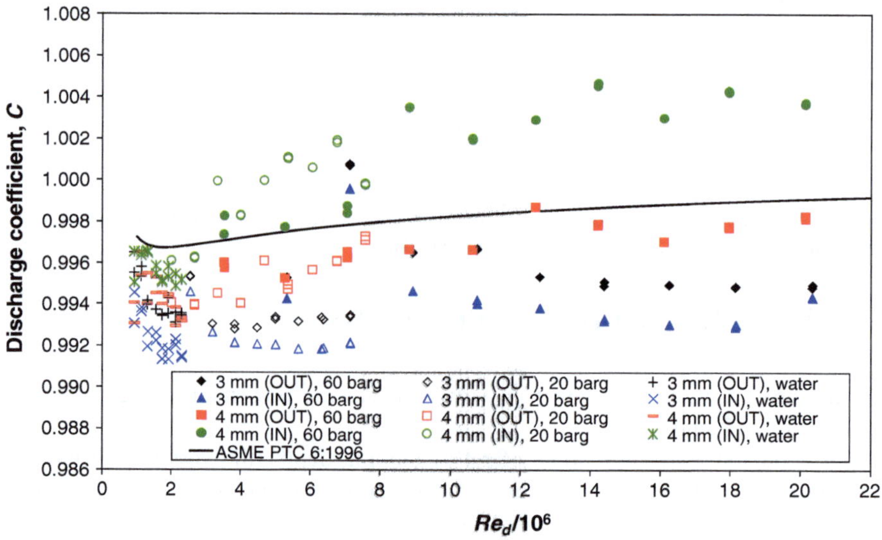

Fig. 9.9 Throat-tapping data taken in water and gas: 3″ nozzle in 8″ pipe

within 0.25 % of the ASME PTC 6 Equation, and there is a mean deviation of −0.12 %. In some cases extrapolation parallel to the PTC 6 Equation from the water data to the gas data (permitted by PTC 6) works well, but in other cases poorly: see Table 9.6.

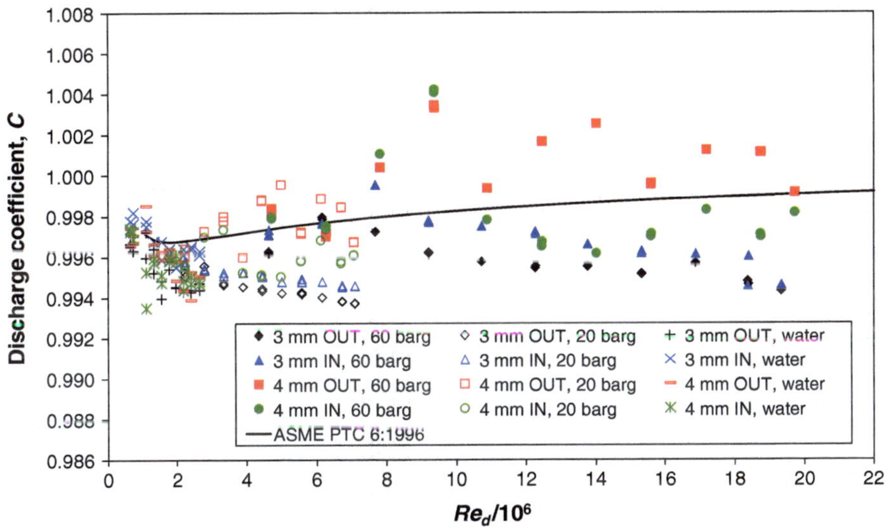

Fig. 9.10 Throat-tapping data taken in water and gas: 3.5″ nozzle in 8″ pipe

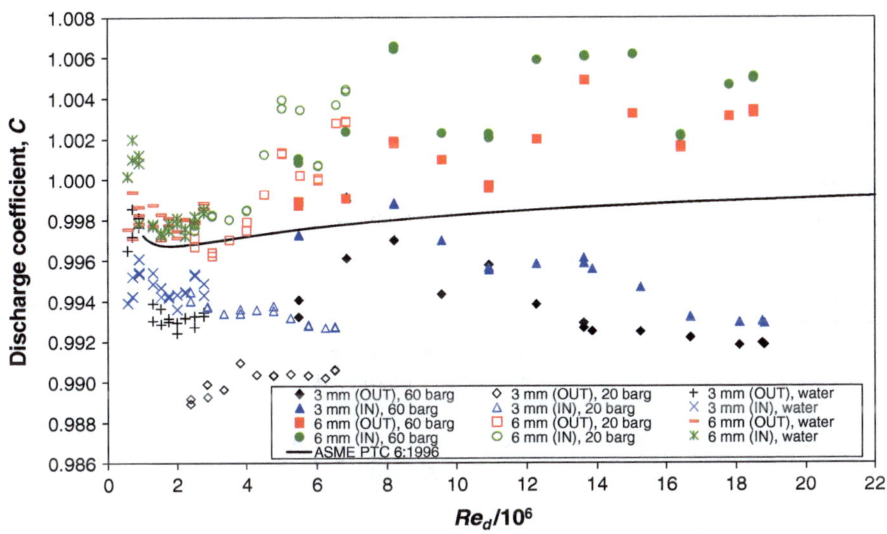

Fig. 9.11 Throat-tapping data taken in water and gas: 4″ nozzle in 8″ pipe

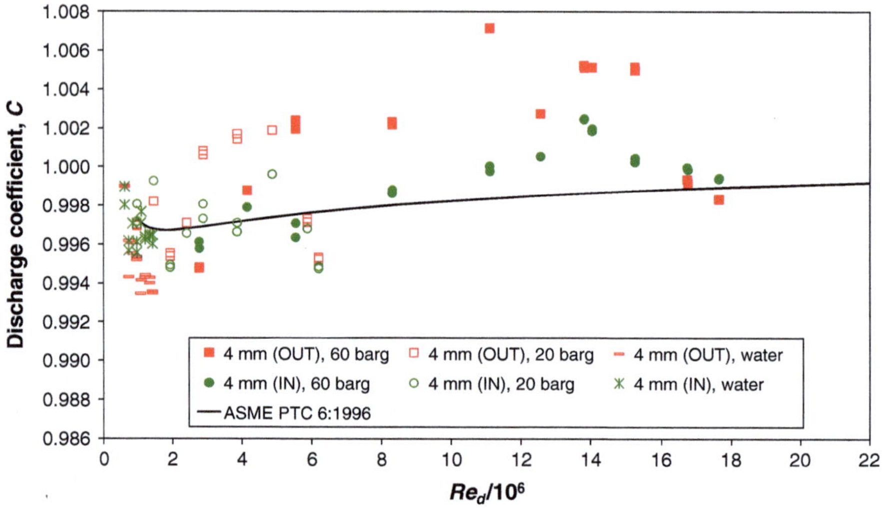

Fig. 9.12 Throat-tapping data taken in water and gas: 2″ nozzle in 6″ pipe

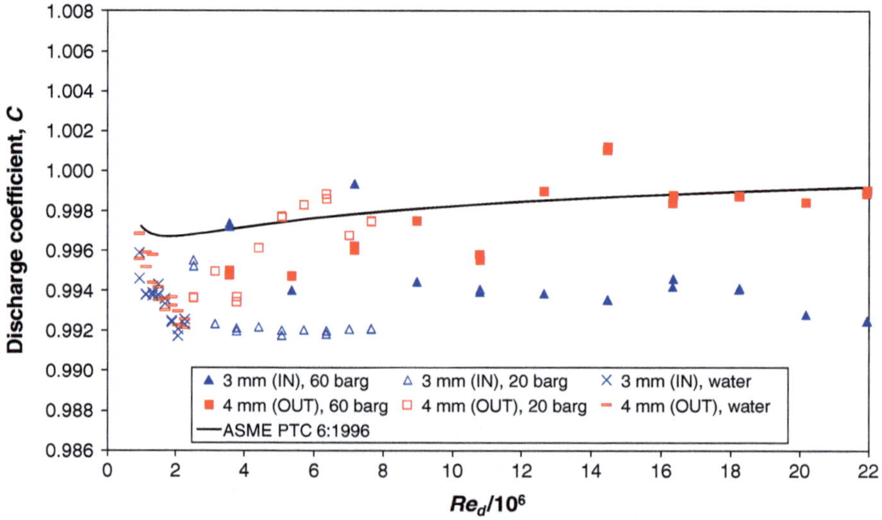

Fig. 9.13 Throat-tapping data taken in water and gas: 3″ nozzle in 6″ pipe

Table 9.5 Best-fit lines for throat-tapping data for each nozzle: $x = \exp(-0.2Re_d/10^6)$

Nozzle size (inch)	Tapping diameter	Pipe size (inch)	Best-fit line	Twice the r.m.s. deviation (gas only) (%)	Twice the r.m.s. deviation (all data) (%)
2	3 mm IN	8	$0.9983 - 0.0009\,x$	0.30	0.26
2	3 mm OUT	8	$0.9984 - 0.0014\,x$	0.32	0.28
3	3 mm (IN)	8	$0.9940 - 0.0023\,x$	0.40	0.34
3	3 mm (OUT)	8	$0.9955 - 0.0022\,x$	0.40	0.34
3.5	3 mm IN	8	$0.9960 + 0.0001\,x$	0.26	0.23
3.5	3 mm OUT	8	$0.9952 - 0.0001\,x$	0.22	0.21
4	3 mm (IN)	8	$0.9951 - 0.0011\,x$	0.39	0.33
4	3 mm (OUT)	8	$0.9929 - 0.0014\,x$	0.44	0.40
3	3 mm (IN)	6	$0.9938 - 0.0004\,x$	0.42	0.37
2	4 mm (IN)	8	$1.0011 - 0.0059\,x$	0.34	0.29
2	4 mm (OUT)	8	$1.0023 - 0.0066\,x$	0.51	0.44
3	4 mm (IN)	8	$1.0038 - 0.0115\,x$	0.24	0.24
3	4 mm (OUT)	8	$0.9976 - 0.0047\,x$	0.15	0.19
3.5	4 mm IN	8	$0.9983 - 0.0043\,x$	0.41	0.35
3.5	4 mm OUT	8	$1.0009 - 0.0073\,x$	0.30	0.31
2	4 mm (IN)	6	$1.0002 - 0.0054\,x$	0.33	0.31
2	4 mm (OUT)	6	$1.0038 - 0.0115\,x$	0.61	0.57
3	4 mm (OUT)	6	$0.9990 - 0.0073\,x$	0.27	0.28
4	6 mm (IN)	8	$1.0055 - 0.0115\,x$	0.35	0.30
4	6 mm (OUT)	8	$1.0029 - 0.0081\,x$	0.30	0.27

9.6 Hot-Water (NMIJ Throat-Tapping) Data

After the work described in this chapter was undertaken, work was undertaken at NMIJ in Japan (Furuichi et al. 2013a) with two 6.5″ ASME nozzles in a 14″ pipe (internal diameter 338.96 mm) in water at temperatures 20, 70 and 75 °C. Nozzle B had four 6 mm throat tappings, whereas Nozzle A had one 2 mm, one 3.5 mm, one 5 mm and one 6 mm throat tapping. The discharge coefficients using the four 6 mm tappings on Nozzle B differed by 0.13 % at most, and the discharge coefficient using the 6 mm tapping on Nozzle A differed by 0.08 % at most from that using one of the tappings on Nozzle B. The discharge coefficients using the tappings of different diameter differed from each other in a way similar to the NEL data. Further work was published by Furuichi et al. (2013b) showing data in an 8″ pipe on a 4″ ASME nozzle, which had four 3.5 mm tappings, all giving data in exceptionally good agreement with one another except for Tapping 3 which gave a discharge coefficient about 0.1 % below the other three sets.

Table 9.6 Extrapolated lines for throat-tapping data for each nozzle: $x = 0.185\ Re_d^{-0.2}\ (1 - 361{,}239/Re_d)^{0.8}$

Nozzle size (inch)	Tapping diameter	Pipe size (inch)	Best-fit line	Twice the r.m.s. deviation (gas only) (%)	Twice the r.m.s. deviation (%)
2	3 mm IN	8	$1.0060 - x$	0.31	0.27
2	3 mm OUT	8	$1.0061 - x$	0.33	0.29
3	3 mm (IN)	8	$1.0008 - x$	0.40	0.34
3	3 mm (OUT)	8	$1.0026 - x$	0.40	0.33
3.5	3 mm IN	8	$1.0050 - x$	0.40	0.34
3.5	3 mm OUT	8	$1.0036 - x$	0.30	0.27
4	3 mm (IN)	8	$1.0032 - x$	0.45	0.38
4	3 mm (OUT)	8	$1.0017 - x$	0.58	0.50
3	3 mm (IN)	6	$1.0018 - x$	0.46	0.40
2	4 mm (IN)	8	$1.0051 - x$	0.49	0.41
2	4 mm (OUT)	8	$1.0049 - x$	0.78	0.65
3	4 mm (IN)	8	$1.0042 - x$	0.89	0.70
3	4 mm (OUT)	8	$1.0030 - x$	0.22	0.21
3.5	4 mm IN	8	$1.0037 - x$	0.45	0.39
3.5	4 mm OUT	8	$1.0043 - x$	0.55	0.47
2	4 mm (IN)	6	$1.0051 - x$	0.39	0.36
2	4 mm (OUT)	6	$1.0024 - x$	1.32	1.20
3	4 mm (OUT)	6	$1.0024 - x$	0.53	0.46
4	6 mm (IN)	8	$1.0064 - x$	0.84	0.71
4	6 mm (OUT)	8	$1.0066 - x$	0.48	0.41

9.7 Throat-Tapping Data: Further Analysis

9.7.1 General

The discharge-coefficient data using throat tappings were fitted on the assumption that the wall shear stress, τ_w, was given by Eq. 9.5:

$$\frac{\tau_w}{\frac{1}{2}\rho\bar{u}_{th}^2} = \frac{\lambda_{th}}{4} = (2\lg Re_x - 0.65)^{-2.3} \tag{9.5}$$

where, following ISO 80000-2:2009, $\log_{10}$ is written lg, $\bar{u}_{th}$ is the mean velocity in the throat, λ_{th} is the friction factor in the throat,

$$Re_x = \frac{x\bar{u}_{th}}{\nu} = \frac{x}{d}Re_d \tag{9.6}$$

$$\text{and}\, x = d\left(1 - \frac{361{,}239}{Re_d}\right). \tag{9.7}$$

Equations 9.5 and 9.6 are based on a boundary layer on a flat plate at zero incidence a distance x downstream of the leading edge. Equation 9.5 is based on the logarithmic velocity distribution law and is given in the note to Eq. (21.16) of Schlichting (1960).

The distance from the beginning of the parallel portion of the throat to the throat tapping is $0.75d$; however, the boundary layer begins before the parallel section; moreover, it begins as a laminar boundary layer before transition to a turbulent boundary layer. So it is not obvious what value of x should be used. Equation 9.7 is consistent with ASME PTC 6, and has been found satisfactory in the work below.

One way of calculating the discharge coefficient that would be obtained without pressure tappings, C_{notap}, is to assume that the effective area of the nozzle throat is reduced by the presence of a boundary layer whose width is the displacement thickness: so

$$C_{notap} = 1 - 4\frac{\delta^*}{d},$$

where δ^* is the displacement thickness. If this is calculated based on the 1/7th-power velocity distribution law [see Eqs. (21.6) and (21.8) of Schlichting (1960)]

$$C_{notap} = 1 - 0.185\frac{\frac{x}{d}}{Re_x^{0.2}} = 1 - 0.185\frac{\left(\frac{x}{d}\right)^{0.8}}{Re_d^{0.2}} \tag{9.8}$$

It might be more consistent to use a formula based on the logarithmic velocity distribution law, but Eq. 9.8 would be consistent with ASME PTC 6 if the tapping terms were constant, and a revised version of Eq. 9.8 will be determined below.

Then it was assumed from Eq. 7.5 that the increase in discharge coefficient due to the tappings is given by

$$\frac{\lambda_{th} f(Re_{tap,th}) - \lambda_{up}\beta^4 f(Re_{tap,up})}{8(1-\beta^4)}$$

where

$$f = \begin{cases} a - be^{-nRe_{tap}} & \text{for } Re_{tap} > 800 \\ \left(a - be^{-800\,n}\right)\frac{Re_{tap}}{800} & \text{for } Re_{tap} \le 800 \end{cases} \tag{9.9}$$

and to fit Fig. 7.9 it was prescribed that $a - be^{-2000n} = 3.7$; from Eqs. 7.2 and 7.3 $Re_{tap,up} = \sqrt{\frac{\lambda_{up}}{8}}\frac{d_{tap}}{D}Re_D$ and $Re_{tap,th} = \sqrt{\frac{\lambda_{th}}{8}}\frac{d_{tap}}{d}Re_d$; and for the upstream tapping

$$\frac{1}{\sqrt{\lambda_{up}}} = 1.74 - 2\lg\left(\frac{2\pi R_{a,up}}{D} + \frac{18.7}{Re_D\sqrt{\lambda_{up}}}\right), \tag{9.10}$$

where $R_{a,up}$ was taken as 1.6 μm, λ_{up} is the friction factor in the upstream pipe and Re_D is the pipe Reynolds number.

The assumed upstream pipe roughness has little effect on the value of b and thus the value of a.

It seems reasonable to suppose that in practice nozzle i will differ from the theoretical value by its own constant ΔC_i, due to any defects in the tappings. So C was taken as

$$C = C_{notap} + \frac{\lambda_{th} f(Re_{tap,th}) - \lambda_{up}\beta^4 f(Re_{tap,up})}{8(1-\beta^4)} + \Delta C_i \tag{9.11}$$

9.7.2 *Analysis of NMIJ Data*

The NMIJ data are considered first, because they were taken in an incompressible fluid. Only data for $Re_d > 10^6$ were used, and no graphs of errors are presented as the data were digitized and the errors are so small as to be largely due to digitization. n was taken to be 0.00005, as this was the best fit for Eq. 9.14. The data taken with a 2 mm tapping in a 6.5″ nozzle were considered first, because they have small values of $Re_{tap,th}$ at high Re_d. If $Re_d < 7 \times 10^6$, $Re_{tap,th} < 3000$, and the choice of b (and thus a) is not very important. An excellent fit to the NMIJ data with 2 mm tappings for $Re_d < 7 \times 10^6$ is obtained with

$$C_{notap} = 0.9986 - 0.185\frac{\frac{x}{d}}{Re_x^{0.2}} = 0.9986 - 0.185\frac{\left(\frac{x}{d}\right)^{0.8}}{Re_d^{0.2}} \tag{9.12}$$

with x given by Eq. 9.7 and ΔC_i taken as 0 for this set of tappings: with $b = 13.5$ the standard deviation of the data about the equation is 0.021 %; it is at most 0.05 % for $6 < b < 22$. Then given Eq. 9.12 all the NMIJ data for $Re_d < 7 \times 10^6$ were fitted: for $Re_{tap,th} < 9000$ the optimum value of b was 12.41 with ΔC_i ranging from 0.0000 to 0.0020, where for each set of tappings the mean deviation over the data for $Re_{tap,th} < 3000$ (where the tapping terms are most well known) is equal to 0, i.e. f was given by

$$f = \begin{cases} 14.93 - 12.41\, e^{-0.00005\, Re_{tap}} & \text{for } 9000 > Re_{tap} > 800 \\ 0.00376\, Re_{tap} & \text{for } Re_{tap} \le 800. \end{cases} \tag{9.13}$$

This equation for f was then used for the data with 2 mm tappings for $Re_{tap,th} < 9000$, i.e. over the remaining range of the data: Eqs. 9.11, 9.12 and 9.13 with ΔC_i taken as 0 for this set of tappings fit all the data with 2 mm tappings with the standard deviation of the data about the equation being 0.019 %; hence Eq. 9.12 can be used over the range of the data, i.e. $Re_d < 1.43 \times 10^7$. Then given Eq. 9.12 all the NMIJ data were fitted, and for $Re_{tap,th} < 16{,}300$ the optimum value of b was

12.68 with ΔC_i ranging from 0.0000 to 0.0020, where for each set of tappings the mean deviation over the data for $Re_{tap,th} < 3000$ (where the tapping terms are most well known) is equal to 0, i.e. f was given by

$$f = \begin{cases} 15.17 - 12.68\,e^{-0.00005\,Re_{tap}} & \text{for } 16{,}300 > Re_{tap} > 800 \\ 0.00373\,Re_{tap} & \text{for } Re_{tap} \le 800. \end{cases} \quad (9.14)$$

Equation 9.14 is almost identical to Eq. 9.13. The standard deviation of the NMIJ data about Eq. 9.11 with Eqs. 9.12 and 9.14 is 0.025 %.

9.7.3 Application to NEL Data

The NEL data for $Re_d > 10^6$ were then examined in relation to the analysis in Sect. 9.7.2: only those data for which the nozzle was hydraulically smooth according to Fig. 4.6 of ASME PTC 6:1996 were included. Using Eq. 9.11 with Eqs. 9.12 and 9.14 over the range of $Re_{tap,th}$ for which Eq. 9.14 was derived (i.e. $Re_{tap,th} < 16{,}300$) gave the errors in Figs. 9.14, 9.15 and 9.16. ΔC_i was chosen so that for each set of tappings the mean deviation over the water data was equal to 0, since in comparison with the gas data the water data had smaller uncertainty and the value of f was better known. The data were divided into three groups:

- those with $d_{tap} \ge 4$ mm and $d_{tap}/d \le 0.07$ (see Fig. 9.14)
- those with $d_{tap} \ge 4$ mm and $d_{tap}/d > 0.07$ (see Fig. 9.15)
- those with $d_{tap} = 3$ mm (see Figs. 9.16 and 9.17).

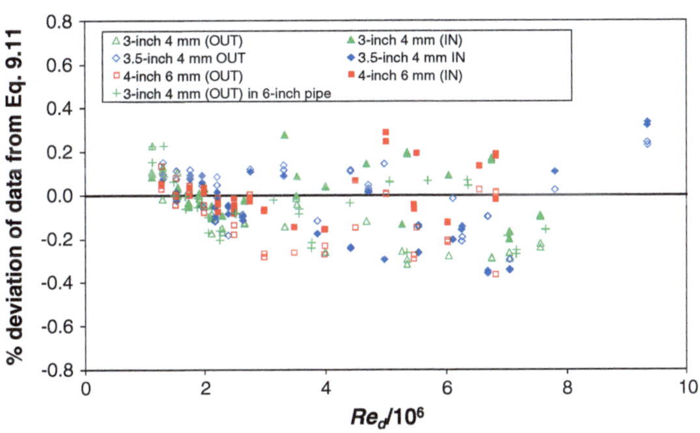

Fig. 9.14 % deviation of the throat-tapping data with $Re_{tap,th} < 16{,}300$ in smooth nozzles with $d_{tap} \ge 4$ mm and $d_{tap}/d \le 0.07$ from Eq. 9.11 with Eqs. 9.12 and 9.14 and ΔC_i chosen so that the mean deviation over the water data is 0

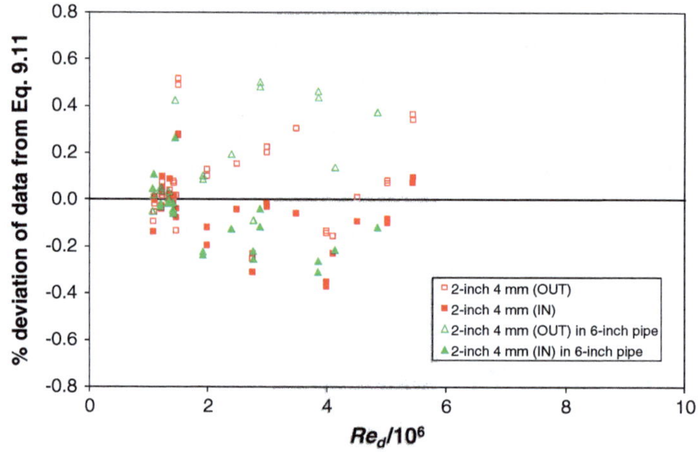

Fig. 9.15 % deviation of the throat-tapping data with $Re_{tap,th} < 16{,}300$ in smooth nozzles with $d_{tap} \geq 4$ mm and $d_{tap}/d > 0.07$ from Eq. 9.11 with Eqs. 9.12 and 9.14 and ΔC_i chosen so that the mean deviation over the water data is 0

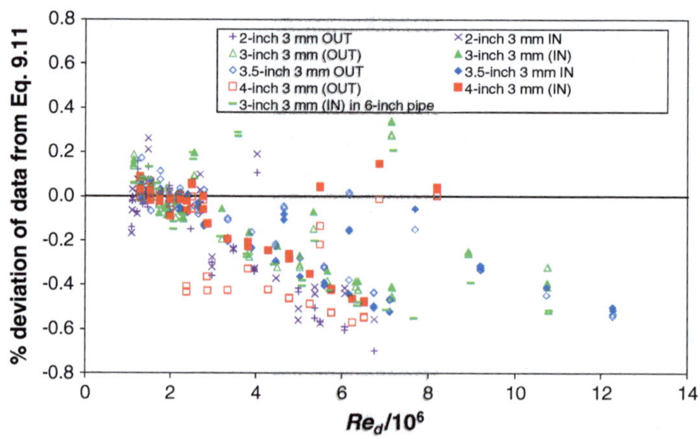

Fig. 9.16 % deviation versus Re_d of the throat-tapping data with $Re_{tap,th} < 16{,}300$ in smooth nozzles with $d_{tap} = 3$ mm from Eq. 9.11 with Eqs. 9.12 and 9.14 and ΔC_i chosen so that the mean deviation over the water data is 0

For the division by d_{tap}/d see Sect. 7.A.7. In Sect. 7.A.7 it was also found that some of the data with 2 mm tappings were different from the general pattern for data with $d_{tap}/d \leq 0.07$; here one feature of the 3 mm tappings is that the tappings expanded into 4 mm tappings at the point where the tapping turns through 90° (after between $2d_{tap}$ and $4.5d_{tap}$) (the 3 mm tappings became 4 mm after the first bend and then 6 mm after the third bend). The 4 mm tappings did not change diameter till they became 6 mm after the third bend from the nozzle (see Fig. 9.2); so for the 3.5″

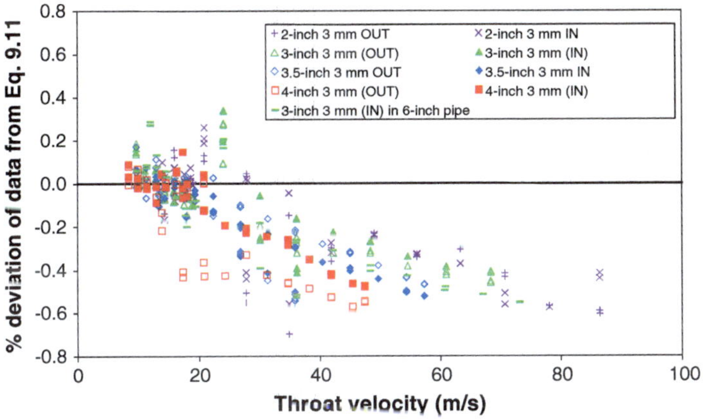

Fig. 9.17 % deviation versus throat velocity of the throat-tapping data with $Re_{tap,th} < 16{,}300$ in smooth nozzles with $d_{tap} = 3$ mm from Eq. 9.11 with Eqs. 9.12 and 9.14 and ΔC_i chosen so that the mean deviation over the water data is 0

nozzle the 4 mm OUT and IN tappings expanded to 6 mm $36d_{tap}$ and $35d_{tap}$ from the nozzle wall, respectively; for the 3″ and 2″ nozzles the 4 mm tappings expanded to 6 mm $29d_{tap}$ and $19d_{tap}$ from the nozzle wall, respectively. The 6 mm tappings did not change diameter till they reached the ¼ BSP fitting (beyond the pipe wall), $43d_{tap}$ from the nozzle wall.

For the data with $d_{tap} \geq 4$ mm and $d_{tap}/d \leq 0.07$ (in Fig. 9.14) there is excellent agreement between the static-hole error predicted from the NMIJ hot water data and the NEL water and high-pressure gas data.

For the data in Fig. 9.14 twice the r.m.s. deviation is 0.32 %; ΔC_i ranges from −0.0048 to −0.0014.

For the data in Fig. 9.15 ΔC_i ranges from −0.0057 to −0.0030. The two pairs of tappings have very different performance; performance is less consistent than with $d_{tap}/d \leq 0.07$. These 4 mm tappings expanded to 6 mm $19d_{tap}$ from the nozzle wall.

For the data in Fig. 9.16 ΔC_i ranges from −0.0053 to −0.0010. The deviations in Fig. 9.16 are in fact a function of throat velocity: see Fig. 9.17. In Fig. 7.B.4 data with $h/d_{tap} = 2.5$ also had a negative deviation.

9.7.4 Analysis of NEL Data

The NEL data for $Re_d > 10^6$ with $d_{tap} \geq 4$ mm and $d_{tap}/d \leq 0.07$ and smooth nozzles (according to ASME PTC 6) were analysed. Equation 9.12 was used as determined in Sect. 9.7.2, but b was determined from the NEL data: the optimum value of b was 9.20 with ΔC_i ranging from −0.0043 to −0.0006, where for each set of tappings the mean deviation over the water data is equal to 0, i.e. f was given by

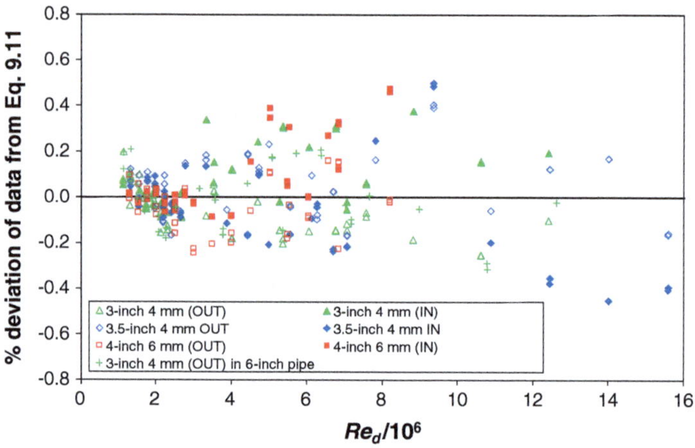

Fig. 9.18 % deviation of the throat-tapping data in smooth nozzles with $d_{tap} \geq 4$ mm and $d_{tap}/d \leq 0.07$ from Eq. 9.11 with Eqs. 9.12 and 9.15 and ΔC_i chosen so that the mean deviation over the water data is 0

$$f = \begin{cases} 12.02 - 9.20\,e^{-0.00005\,Re_{tap}} & \text{for } 24{,}500 > Re_{tap} > 800 \\ 0.00398\,Re_{tap} & \text{for } Re_{tap} \leq 800 \end{cases} \tag{9.15}$$

The deviations are shown in Fig. 9.18.

For the data in Fig. 9.18 twice the standard deviation is 0.32 %.

If Eqs. 9.12 and 9.15 were applied to the NEL data for $Re_d > 10^6$ with $d_{tap} \geq 4$ mm and $d_{tap}/d \leq 0.07$ without excluding any points on the basis of the roughness of the nozzle twice the r.m.s. deviation would be 0.34 %.

9.8 Conclusions

Four ASME flow nozzles were manufactured for NEL together with upstream and downstream pipework in two pipe sizes. Throat-tapping and wall-tapping data were collected. The flow nozzles were calibrated in water and in high-pressure nitrogen at 20 and 60 barg.

The wall-tapping data are in excellent agreement with the discharge-coefficient equation in ISO 5167-3:2003: it fits all the data for $Re_d > 10^6$ with twice the r.m.s. deviation equal to 0.39 %. Since this figure is so much smaller than the uncertainty in ISO 5167-3:2003, brief text was submitted for inclusion in ISO/TR 9464:2008. The data extrapolate well from water to higher Reynolds number and the agreement between the two NEL facilities (water and high-pressure gas recirculating loop) used was very good.

The equation in ASME PTC 6:1996 fits all the NEL throat-tapping data for $Re_d > 10^6$ with twice the r.m.s. deviation equal to 0.61 %. This figure is significantly larger than that for the wall-tapping data: in some cases there are humps in the data as there are with Venturi tubes. The conclusion is that more use of nozzles with wall tappings rather than throat-tapped nozzles is very desirable. The former are much cheaper and easier to manufacture, and give better performance, at least in gas (it is possible to fit the throat-tapping data taken in hot water with very small uncertainty: there are no humps in the data in that case, although the data are different from what would have been expected from ASME PTC 6:1996).

The problem is that the use of throat-tapped nozzles is very entrenched in certain industries. Even in 1975 Wyler and Benedict (1975) wrote 'The data do not support the claim that the throat tap nozzle is a more accurate flow metering device than the pipe wall tap nozzle'.

It is possible to extrapolate from a throat Reynolds number just above 10^6 to high Reynolds number with throat-tapped nozzles using consistent equations for static-hole error. The equation determined from the NMIJ hot water data applies for the NEL gas and water data with both $d_{tap} \geq 4$ mm (i.e. when there is no sudden expansion in tapping diameter close to the nozzle) and $d_{tap}/d \leq 0.07$ (i.e. where the throat-tapping diameter is not too large a fraction of the throat diameter) over the range of $Re_{tap,th}$ for which it was derived. Moreover, an equation for static-hole error derived from the NEL data can be used to extrapolate from a throat Reynolds number just above 10^6 to high Reynolds number for $d_{tap} \geq 4$ mm and $d_{tap}/d \leq 0.07$ over a wider range of $Re_{tap,th}$. Equations 9.14 and 9.15 for static-hole error in this chapter have been included in Fig. 7.7. Either $d_{tap} = 3$ mm (for which there is a sudden expansion in tapping diameter close to the nozzle wall) or larger values of d_{tap}/d give less good results.

The other conclusion, and the one more relevant to this book, is that the issues with Venturi tubes at high Reynolds number are due to the use of pressure tappings in high-velocity regions. On the basis of the nozzle results it seems reasonable to expect that the discharge coefficient for Venturi tubes has the same dependence on Reynolds number for both incompressible and compressible flow with deviations for compressible flow largely due to throat velocity effects. Data obtained in incompressible flow with throat-tapped nozzles can be fitted remarkably well.

References

ASME (1971) Interim supplement 19.5 on instruments and apparatus, Part II of fluid meters. ASME 19.5:1971. American Society of Mechanical Engineers, New York

ASME (1996) Performance test code 6 on steam turbines. ASME PTC 6:1996. American Society of Mechanical Engineers, New York

Brunkalla RL (1985) Effects of fabrication technique on the discharge coefficient of a throat tap flow nozzle. In: Jt ASME/IEEE power generation conference, Milwaukee, Wisconsin, 85-JPGC-PTC-3

Furuichi N, Cheong K-H, Terao Y, Nakao S, Fujita K, Shibuya K (2013a) Experimental and theoretical analysis of discharge coefficient of throat tap nozzle based on PTC 6 for wide range Reynolds number. In: Proceedings of the ASME 2013 Fluids Engineering Division Summer Meeting FEDSM2013, Incline Village, Nevada: FEDSM2013-16154

Furuichi N, Cheong KH, Terao Y, Nakao S, Fujita K, Shibuya K (2013b) Experimental investigation of discharge coefficient and tapping error of PTC 6 flow nozzle using high Reynolds number calibration rig. In: Proceedings of FLOMEKO 2013, Paris

Hall GW (1959) Application of boundary layer theory to explain some nozzle and Venturi flow peculiarities. Proc Inst Mech Eng 173(36):837–870

ISO (2003) Measurement of fluid flow by means of pressure differential devices inserted in circular cross-section conduits running full—Part 3: Nozzles and Venturi nozzles. International Organization for Standardization, Geneva, ISO 5167-3:2003

Reader-Harris MJ, Gibson J, Hodges D, Nicholson IG, Rushworth R (2005) The performance of flow nozzles at high Reynolds number. Report no 2005/260 on project no FEDP03, National Engineering Laboratory, East Kilbride, Glasgow

Reader-Harris MJ, Gibson J, Hodges D, Nicholson I G, Rushworth R (2007) The performance of flow nozzles at high Reynolds number. In: Proceedings of FLOMEKO 14, Sandton, South Africa

Schlichting H (1960) Boundary layer theory. McGraw-Hill, New York

Wyler JS, Benedict RP (1975) Comparisons between throat and pipe wall tap nozzles. J Eng Gas Turbines Power 97(4):569–573 (Trans ASME)

Chapter 10
Orifice Plates with Drain Holes

Abstract A drain hole through an orifice plate is appropriate where some liquid is introduced to a gas pipeline over a short period of time, but thereafter the flow is dry. In this case there is no need for wet-gas measurement; however, it is essential that there is no error during the dry-gas flow due to a pool of liquid that never passes through the orifice. Installing a drain hole can be regarded as equivalent to increasing the area of the orifice in a predictable way: an equation is given in ISO/TR 15377:2007. However, recent data in this chapter show that the existing drain-hole equation in ISO/TR 15377:2007 is unsatisfactory. The data on drain holes have a surprisingly strong dependence on the circumferential location of the pressure tappings, although very little dependence on Reynolds number. A new analysis is presented based on the assumption that, although the differential pressure is disturbed by the presence of the drain hole, the discharge coefficient of the orifice (distinct from the drain hole) is unaffected by the drain hole's presence provided that the mean differential pressure around the pipe is used. On this basis an equation has been produced for the corrected orifice diameter taking account of the drain hole. More data would be useful; nevertheless it is very desirable to amend ISO/TR 15377:2007, given that for the data collected here its equation leads to flowrate errors up to nearly 2 % in magnitude, whereas the new equation gives errors less than 0.25 % in magnitude.

10.1 Introduction

Where a gas flow is continuously wet the instructions of Chap. 11 should be followed. However, a drain hole is appropriate where some liquid is introduced to the pipeline over a short period of time, but thereafter the gas flow is dry. In this case there is no need for wet-gas measurement; the problem is that during periods in which there is some liquid flow a pool of liquid may build up against the upstream face of a standard orifice plate (without a drain hole) and thus the metering accuracy of a standard orifice plate is undermined even during the time when the flow is dry.

M. Reader-Harris, *Orifice Plates and Venturi Tubes*,
Experimental Fluid Mechanics, DOI 10.1007/978-3-319-16880-7_10

A drain hole ensures that liquid does not build up against the plate. Another installation where a drain hole was found appropriate was one where without a drain hole there was very significant accumulation of dirt; a new orifice plate with a drain hole allowed small particles of dirt to pass through the plate and hence solved the problem.

While drain-hole plates are a cost-effective way of measuring gas with a very low liquid content, they are not as accurate in single-phase flow as the standard design. As the extent of this inaccuracy is not well documented and as industry is sceptical of the existing formula, drain-hole plates are not as widely used as they might be: new data have therefore been needed to give confidence in their use.

ISO/TR 11583:2012 (ISO 2012) covers wet-gas flow using orifice plates without drain holes. There is no accepted correlation for over-reading for orifice plates with drain holes in wet gas, and one is probably not needed.

Figure 10.1 shows a picture of an orifice plate with a drain hole. The plate was fitted between the flanges using the bolt holes around the plate circumference. The dashed line visible in Fig. 10.1 marks out the area of the plate that is exposed to the fluid within the pipe.

The drain hole is a smooth circular hole through the plate located at the bottom of the pipe and with the bottom of the hole aligned with the bottom edge of the upstream pipe. The upstream edge of the drain hole should be sharp. The diameter of the downstream pipe at the orifice plate should be very similar to the upstream pipe: if the downstream pipe were much smaller than the upstream it would partially or completely block the drain hole. Alignment of the upstream pipe, the orifice plate and the downstream pipe is particularly important when there is a drain hole: the dowel holes in Fig. 10.1 achieve this alignment.

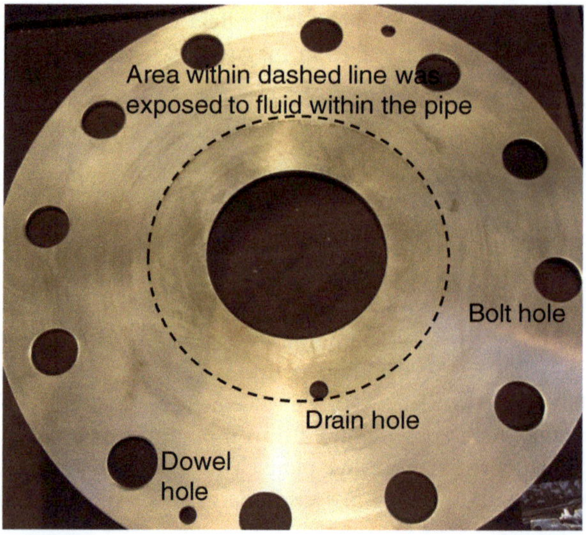

Fig. 10.1 Annotated picture of an orifice plate with a drain hole

ISO/TR 15377:2007 (ISO 2007), the only reference document, is based on a very simple theoretical model: the measured orifice diameter, d, is replaced by a corrected orifice diameter, d_c, to allow for the additional flowrate due to the orifice area represented by the drain hole of diameter d_h as shown in the following equation:

$$d_c = d\left(1 + 0.55\left(\frac{d_h}{d}\right)^2\right) \tag{10.1}$$

NOTE 1 The corrected diameter gives rise to a corrected diameter ratio.
NOTE 2 Eq. 10.1 is based on the assumption that the value for $C\varepsilon(1 - \beta^4)^{-0.5}$ for flow through the drain hole is 10 % greater than the value for flow through the orifice.

When Eq. 10.1 is used to calculate the flowrate ISO/TR 15377:2007 states that the following additional percentage uncertainty should be added arithmetically to the discharge coefficient percentage uncertainty:

$$55\left(\frac{d_h}{d}\right)^2. \tag{10.2}$$

Since Eqs. 10.1 and 10.2 are not based on experiment there was a need for experimental data to improve the understanding of the physics of flow through drain holes and then to revise the standard.

The project started with a brief literature survey and contacts with those with relevant experience. No published experimental data were found. One contributor's experience was that drain holes as large as are permissible in ISO/TR 15377 were normal to avoid blockage, another's that drain holes either larger or smaller may be used, and another's that drain holes are typically 6–10 mm in diameter, regardless of pipe size. Although drain holes are sometimes not placed on the wall all contributors agreed that they should touch the wall. Blockage with solid material is a risk with drain holes and would not generally be detected but would lead to an error whose magnitude is equal to that of the drain-hole correction.

NOTE 3 For β greater than around 0.4 (depending on drain hole size) a blocked drain hole could be detected if the differential pressure were measured using two pairs of tappings, one located on the top and the other on the side of the pipe: if the differential pressure were the same with the two pairs of tappings it should be assumed that the drain hole is blocked (see Fig. 10.8).

The correction required when there is liquid flowing through the drain hole is unknown. However, since orifice plates with drain holes should be used when the flow is nearly always dry, no correction to the total measured flow is required to take account of the small fraction of the time when liquid is flowing.

The work in this chapter is a work in progress and is described as such.

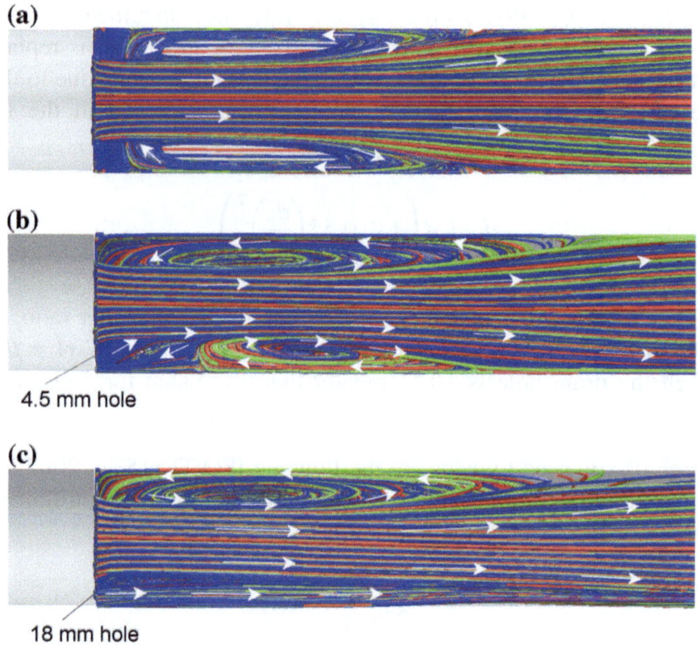

Fig. 10.2 Streamlines showing the flow pattern downstream of a $\beta = 0.6$ orifice plate in a 6″ pipe, **a** no drain hole, **b** 4.5 mm drain hole, **c** 18 mm drain hole

The significant disturbance that a drain hole makes to the flow pattern is well illustrated by computational work included in Reader-Harris et al. (2008b): see Fig. 10.2.

The collection of new data on orifice plates with drain holes is described in Sects. 10.2 and 10.3. The analysis is given in Sect. 10.4, leading to a proposed new equation for d', the corrected orifice diameter taking account of the drain hole: this is Eq. 10.26 and gives errors less than 0.25 % in magnitude, whereas Eq. 10.1 from ISO/TR 15377:2007, leads to flowrate errors up to nearly 2 % in magnitude. The conclusions of this chapter are in Sect. 10.5.

10.2 Experimental Work: Initial Data

In order to determine the effect of drain holes on orifice plates tests were carried out over a range of values of diameter ratio, β, drain hole diameter, d_h, pipe diameter, D, plate thickness, E, pipe Reynolds number, Re_D, and of tapping locations (Reader-Harris et al. 2008a, b). Throughout this chapter β remains the ratio of orifice diameter, d, to pipe diameter. From physical considerations it is reasonable to expect that the key parameters are d_h/d, β, E/d_h, and L'_2, where $L'_2 = l'_2/D$ and l'_2 is the distance between the downstream face of the orifice plate and the downstream pressure tapping.

To achieve an appropriate pattern of test data an existing 4″ orifice run with flange and corner tappings was used with new plates with $\beta = 0.4$, 0.6 and 0.75. These three plates all had $E = 3$ mm. 3 mm is near the middle of the permissible range of values in ISO 5167-2:2003 (ISO 2003). In the US $E = 3.175$ mm $\left(\frac{1}{8}''\right)$ is normal for this pipe size. It seemed reasonable to suppose that thicker plates (relative to d_h) might make a difference, as the flow through a thick plate might reattach to the wall that forms the top of the drain hole whereas the flow through a thin one might remain separated; so a plate with $\beta = 0.6$ and $E = 5$ mm was manufactured. Where $E/D = 0.05$ the plate was designated as thick.

The maximum permissible value of d_h/d in ISO/TR 15377:2007 is 0.1; so $d_h = 0.1d$ was tested together with $d_h = 0.07d$ and, where possible, a higher value of d_h too. The pattern of tests is given in Table 10.1. The drain holes (like the orifices) were inserted using spark erosion (sometimes called Electrical Discharge Machining). The measured hole diameters were so close to the requested values that it is acceptable to assume that d_h/d is exactly equal to the value specified in Table 10.1 (the error in the drain-hole diameter never changed the total open area of a plate by more than 0.011 %). All the plates with all the drain holes were tested in

Table 10.1 Summary of initial tests

Pipe size	E (mm)	β	Fluid	d_h/d	d_h (mm)	E/d_h
4″	3	0.400	Water and gas	0	0	n/a
			Water	0.07	2.863	1.05
			Water and gas	0.1	4.091	0.73
			Water	0.2	8.181	0.37
4″	3	0.600	Water and gas	0	0	n/a
			Water	0.07	4.296	0.70
			Water and gas	0.1	6.137	0.49
			Water	0.167	10.249	0.29
4″	5	0.600	Water and gas	0	0	n/a
			Water	0.07	4.296	1.16
			Water and gas	0.1	6.137	0.81
			Water	0.167	10.249	0.49
4″	3	0.750	Water and gas	0	0	n/a
			Water	0.07	5.370	0.56
			Water and gas	0.1	7.671	0.39
8″	6	0.420	Water	0	0	n/a
			Water	0.1	8.507	0.71
8″	6	0.603	Water	0	0	n/a
			Water	0.1	12.212	0.49

water, and, to examine whether there was an effect of Reynolds number, baselines and data with $d_h = 0.1d$ were obtained in gas (nitrogen) at nominal pressures of 20 barg and 60 barg.

To determine the effect of pipe diameter an 8″ orifice run was constructed and tested in water with two orifice plates, $\beta = 0.42$ and 0.6. The internal diameter upstream of the orifice plate was 202.56 mm (all three diameter measurements in the plane closest to the orifice plate were equal to 202.56 mm); the internal diameter downstream of the plate was 202.65 mm. The pattern of tests is again given in Table 10.1.

For the water tests the orifice plates were installed in the NEL gravimetric system at least 70*D* downstream of a perforated-plate flow conditioner, itself preceded by around 10*D* of straight pipe. There was at least 20*D* of straight pipe downstream of the orifice plates. At least 15*D* of the straight pipe upstream of the orifice plate and at least 4*D* of that downstream were machined. At flanges within 40*D* of the orifice plate on the upstream side dowels and 'O' rings were used.

For the gas tests the orifice plates were installed in the NEL high-pressure recirculating loop. Following a concentric reduction from the primary line size of 6″ there was 49*D* of 4″ pipe upstream of the plate. Downstream of the plate 15*D* of 4″ pipe was used; the gas temperature was measured 13.7*D* downstream of the plate. A metering package consisting of four 4″ turbine meters was used as the reference. This was installed downstream of the orifice-plate test section.

Two pairs of tappings were used for each test. An important issue is the location of the tappings relative to the drain hole. In the 4″ data the A tappings were flange tappings at 115° from the drain hole; the B tappings were corner tappings at 155° from the drain hole. In the 8″ data both sets of tappings were flange tappings, the A tappings at 180° from the drain hole, the B tappings at 90° from the drain hole.

The discharge coefficient, *C*, was obtained from the following equation:

$$q_m = \frac{C\varepsilon}{\sqrt{1-\beta^4}} \frac{\pi d^2}{4} \sqrt{2\rho_1 \Delta p} \tag{1.15}$$

where q_m is the mass flowrate of fluid, Δp is the differential pressure and ρ_1 the density at the upstream pressure tapping. ε is the expansibility factor as given by Eq. 6.13, the equation in ISO 5167-2:2003. When a drain hole was included *S*, the percentage shift in *C* from that obtained with the same plate without a drain hole, was calculated. Note that the orifice diameter, *d*, was used for both calculations of discharge coefficient (not a corrected orifice diameter for the plate with a drain hole). Moreover, in both calculations $\beta = d/D$ and the calculated value of ε was based on β.

The data were analysed as a function of $(10^6/Re_D)^{0.5}$, and are presented over the range for which they are most linear (at low Reynolds number they ceased to be very linear because of the increased uncertainty in the measured differential pressure). The water data for $\beta = 0.4$ are presented in Figs. 10.3 and 10.4; data for $d_h/d = 0.2$ were not stable and are not considered in Figs. 10.3 and 10.4. In Figs. 10.5 and 10.6 the water and gas data for $\beta = 0.4$ are presented; only those gas data for which the uncertainty in expansibility factor is less than 0.067 % are

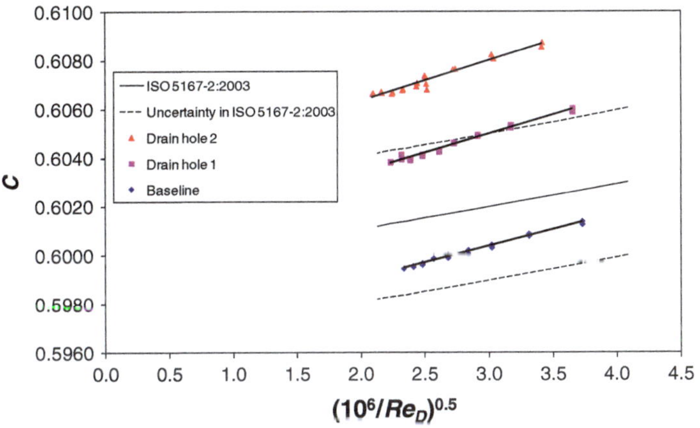

Fig. 10.3 Orifice plate in water: 4″ pipe, $\beta = 0.4$, flange tappings at 115° from the drain hole (A tappings). Drain hole 1: $d_h/d = 0.07$; drain hole 2: $d_h/d = 0.1$

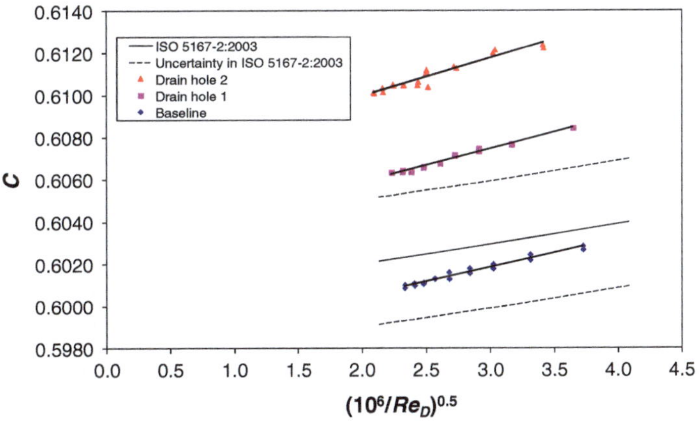

Fig. 10.4 Orifice plate in water: 4″ pipe, $\beta = 0.4$, corner tappings at 155° from the drain hole (B tappings). Drain hole 1: $d_h/d = 0.07$; drain hole 2: $d_h/d = 0.1$

included. Shifts in discharge coefficient for water as in Figs. 10.3 and 10.4 are given in Table 10.2. Shifts were evaluated by fitting lines to the data sets and then calculating the difference between the lines at a point near where their uncertainty is minimum. See Reader-Harris et al. (2008a). The lines are close to parallel to each other, and close to parallel to the Reader-Harris/Gallagher (1998) Equation as found in ISO 5167-2:2003; the Equation with its uncertainty band is plotted in Figs. 10.3, 10.4, 10.5 and 10.6. Shifts are considered as simple increases in discharge coefficient without changing either d or β.

Shifts in discharge coefficient for water for $d_h/d = 0.1$ for the data in Figs. 10.3 and 10.4, for gas for each pressure for the data in Figs. 10.5 and 10.6, and for all the

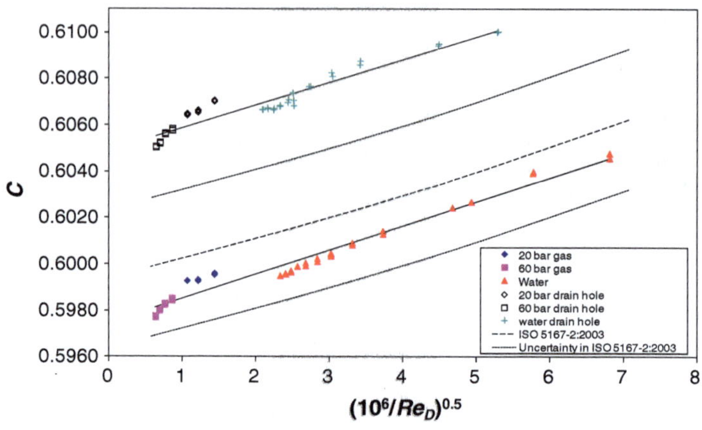

Fig. 10.5 Orifice plate in water and gas: 4″ pipe, $\beta = 0.4$, flange tappings at 115° from the drain hole (A tappings). Drain hole $d_h/d = 0.1$

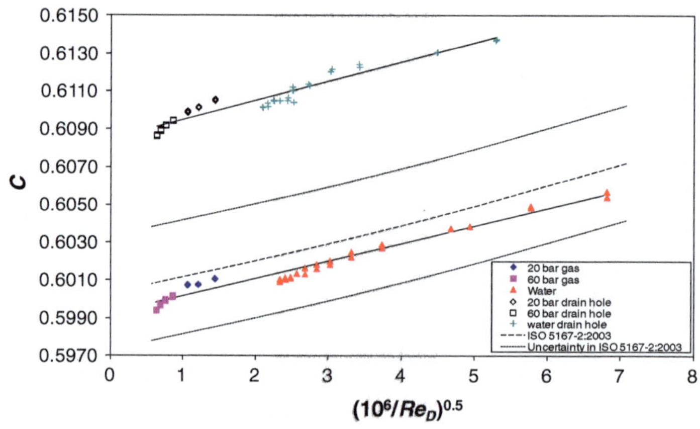

Fig. 10.6 Orifice plate in water and gas: 4″ pipe, $\beta = 0.4$, corner tappings at 155° from the drain hole (B tappings). Drain hole $d_h/d = 0.1$

data in Figs. 10.5 and 10.6 are given in Table 10.3 for comparison with one another. Equivalent data for $\beta = 0.6$, $\beta = 0.6$ (thick plate) and $\beta = 0.75$ were obtained and are included in Tables 10.2 and 10.3. Graphs of all the data are given in Reader-Harris et al. (2008a). For the A tappings for $\beta = 0.75$ in water the estimates of shift were obtained by fitting only over the top of the Reynolds number range. Water data for $\beta = 0.42$ and 0.6 obtained in 8″ pipe were obtained similarly and are included in Table 10.2.

Table 10.3 shows that the shift in discharge coefficient has very little dependence on Reynolds number.

Table 10.2 Percentage shifts in discharge coefficient in water (without correcting the orifice diameter)

Pipe size	*E*/*D*	β	d_h/d	Percentage shift in discharge coefficient	
				Tapping pair A	Tapping pair B
4″	0.03	0.4	0.07	0.751	0.913
4″	0.03	0.4	0.1	1.249	1.612
4″	0.03	0.6	0.07	0.840	1.528
4″	0.03	0.6	0.1	1.456	2.578
4″	0.03	0.6	0.167	3.487	5.163
4″	0.05	0.6	0.07	0.892	1.583
4″	0.05	0.6	0.1	1.575	2.778
4″	0.05	0.6	0.167	3.662	5.429
4″	0.03	0.75	0.07	1.590	2.330
4″	0.03	0.75	0.1	2.266	3.508
8″	0.03	0.42	0.1	1.512	1.258
8″	0.03	0.603	0.1	2.234	1.306

Table 10.3 Percentage shifts in discharge coefficient for those drain holes (all d_h/d = 0.1) for which there are water and gas data (without correcting the orifice diameter)

Pipe size	*E*/*D*	β	Tapping pair	Percentage shift in discharge coefficient			
				Water	Gas (20 barg)	Gas (60 barg)	Water and gas
4″	0.03	0.4	A	1.249	1.225	1.221	1.215
4″	0.03	0.4	B	1.612	1.566	1.547	1.573
4″	0.03	0.6	A	1.456	1.447	1.517	1.439
4″	0.03	0.6	B	2.578	2.551	2.581	2.567
4″	0.05	0.6	A	1.575	1.574	1.565	1.585
4″	0.05	0.6	B	2.778	2.742	2.719	2.752
4″	0.03	0.75	A	2.266	2.143	2.251	2.242
4″	0.03	0.75	B	3.508	3.310	3.427	3.426

An initial analysis of all the data was carried out and presented in Reader-Harris et al. (2008a). It showed that more data were required, especially for small β, for smaller *E*/*D* and for a wider range of pipe sizes.

10.3 Experimental Work: Additional Data

Additional data were then collected: in 8″ pipe β = 0.2, 0.6 and 0.75; in 2″ pipe β = 0.49 and 0.6 (see Reader-Harris and Addison 2013). The 8″ β = 0.6 plate had *E*/*D* = 0.02, since data with *E*/*D* = 0.03 (in 4″ and 8″ pipe) and 0.05 (in 4″ pipe) had

Table 10.4 Percentage shifts in discharge coefficient in water (without correcting the orifice diameter)

Pipe size	*E*/*D*	*β*	d_h/d	Percentage shift in discharge coefficient		
				Tappings at 90° from drain hole (on side)	Tappings at 180° from drain hole (on top)	
8″	0.03	0.2	0.045	0.273	0.282	
8″	0.03	0.2	0.07	0.626	0.649	
8″	0.03	0.2	0.1	1.263	1.291	
8″	0.02	0.6	0.045	0.306	0.54	
8″	0.02	0.6	0.07	0.63	1.189	
8″	0.02	0.6	0.1	1.257	2.196	
8″	0.03	0.75	0.045	0.762	1.034	
8″	0.03	0.75	0.07	1.341	2.037	
8″	0.03	0.75	0.1	2.033	3.171	
2″	0.06	0.489	0.039	0.225		*
2″	0.06	0.489	0.104	1.369		*
2″	0.06	0.6	0.032		0.279	*
2″	0.06	0.6	0.1		2.12	*

*Paid for by CNR (Canadian Natural Resources Ltd.) and published by Spearman (2012)

already been taken. The new data used flange tappings, were taken in water at NEL, and are shown in Table 10.4. The 2″ data were published by Spearman (2012). The orifice plates in 8″ pipe were calibrated using reference meters: the orifice plates were 70*D* downstream of a perforated-plate flow conditioner, itself preceded by around 10*D* of straight pipe. There was at least 20*D* of straight pipe downstream of the orifice plates. 9*D* of the straight pipe upstream of the orifice plate and at least 8*D* of that downstream were machined. At flanges within 29*D* of the orifice plate on the upstream side dowels and 'O' rings were used. The drain hole diameter, d_h, was chosen so that $d_h/d \leq 0.1$. As before, the percentage shift in discharge coefficient is the change in discharge coefficient from that obtained with the same plate without a drain hole: the orifice diameter, *d*, is used for both calculations of discharge coefficient (not a corrected orifice diameter for the drain-hole plate).

The most surprising feature of the data is that the results have such a strong dependence on the circumferential location of the pressure tappings. Accordingly in the 8″ run tappings at 150°, 120° and 60° from the drain hole were added and more data obtained: these data are shown in Table 10.5. In practice, tappings at 60° would not be used. For the points in Table 10.5 the baseline discharge coefficient was taken from the baseline values with the same plate but with different pressure tappings (also flange tappings), since the new pressure tappings were added after the drain holes were inserted in the orifice plates. Each of these orifice plates was then rotated and data obtained with tappings at 30° and 0° from the drain hole. The data at 0° from the drain hole were strongly affected by the flow past the tappings and are not shown, but those obtained with tappings at 30° from the drain hole are shown in Table 10.5 (the flow through the drain hole was shown to be unaffected by

Table 10.5 Percentage shifts in discharge coefficient in water (without correcting the orifice diameter)

Pipe size	E/D	β	d_h/d	Percentage shift in discharge coefficient			
				Tappings at 150° from drain hole	Tappings at 120° from drain hole	Tappings at 60° from drain hole	Tappings at 30° from drain hole
8″	0.03	0.42	0.1	1.446	1.354	1.209	Not taken
8″	0.02	0.6	0.1	2.046	1.708	0.911	0.885
8″	0.03	0.75	0.1	3.085	2.817	0.807	−0.410

the presence of the tappings at 0° from the drain hole by the fact that the data at 60° and 90° from the drain hole were retaken and found to be within 0.05 % of the values obtained without tappings at 0° from the drain hole).

The errors in measured flowrate using Eq. 10.1 (the equation in 5.1.2 of ISO/TR 15377:2007) and all the data in Tables 10.2, 10.4 and 10.5 (except those with tappings at 30° from the drain hole and those with d_h/d = 0.167) are shown in Fig. 10.7, together with the uncertainty given by Eq. 10.2 (from 5.1.2 of ISO/TR 15377:2007). It is clear that there is almost always an under-measurement (unless the tappings are at less than 90° from the drain hole) and that the under-measurement is often larger than the claimed uncertainty. The shift in discharge coefficient near the top of the pipe is much larger for large β than would have been expected given the size of the drain holes.

The data in Table 10.5 together with the values with the same plates using tappings at 90° from the drain hole (on the side) and at 180° from the drain hole (on top) are plotted in Fig. 10.8. Additional data with plates with drain holes for which

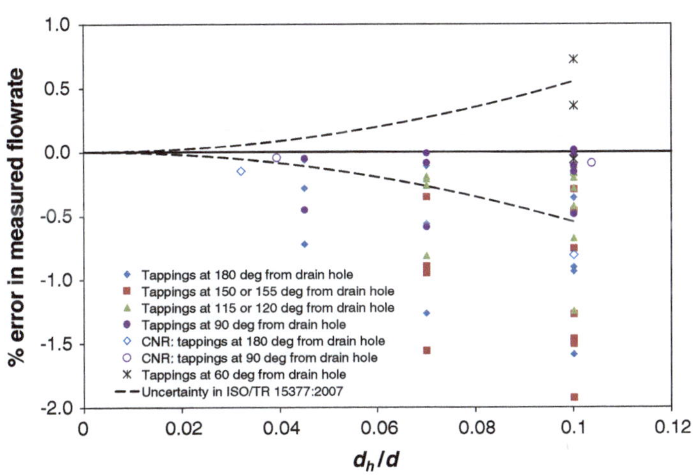

Fig. 10.7 Errors in measured flowrate using Eq. 10.1 (the equation in 5.1.2 of ISO/TR 15377:2007)

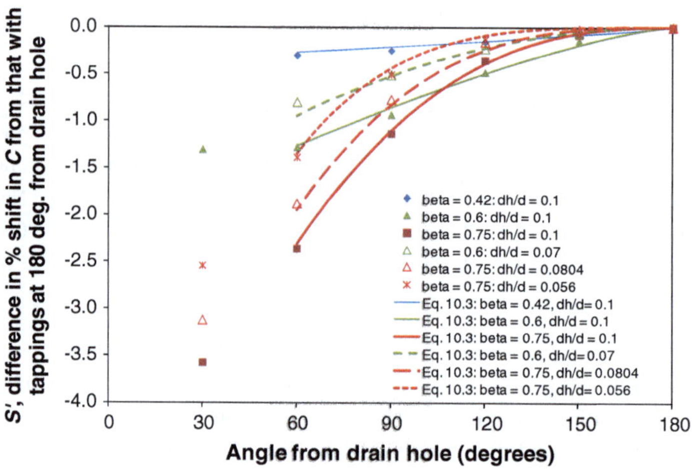

Fig. 10.8 Difference between percentage shift in C and percentage shift in C obtained with tappings at 180° from the drain hole: 8″ pipe with flange tappings

no baseline was available are also included in Fig. 10.8: baseline discharge coefficients taken from different plates made from the same drawings by the same manufacturer were used: their effect on the data in Fig. 10.8 is very small.

When the data in Fig. 10.8 from tappings at 60° or more from the drain hole were fitted, S', the percentage shift in discharge coefficient less that obtained with tappings at 180° from the drain hole, was given by

$$S' = -26.8\beta^{4.9}\left(1 - \frac{\theta}{180}\right)^{-0.95+7.5\beta^{4.9}+0.168\frac{d}{d_h}}. \tag{10.3}$$

where θ is the angle from the drain hole to the pressure tapping.

10.4 Analysis

10.4.1 Bernoulli's Theorem

Applying Bernoulli's Theorem (see Sect. 1.2.1) and adding the flows through the orifice and the drain hole gives approximately

$$q_m = \frac{\frac{\pi}{4}d^2 C(Re_D, \beta'')\varepsilon(\beta'')\sqrt{2\rho_1 \Delta p_{av}}}{\sqrt{1-\beta''^4}} + \frac{\frac{\pi}{4}d_h^2 C_h \varepsilon(\beta'')\sqrt{2\rho_1 \Delta p_{btm}}}{\sqrt{1-\beta''^4}} \tag{10.4}$$

where

β'' is the effective diameter ratio,
q_m is the total mass flowrate,
d_h is the drain hole diameter,
C is the discharge coefficient for the orifice (without a drain hole) with flange (or corner etc. as provided) tappings,
C_h is the discharge coefficient for the drain hole with the differential pressure obtained near the bottom of the pipe (0° from the drain hole),
ε is the expansibility factor,
Δp_{av} is the average differential pressure around the pipe circumference at the flange (or corner etc. as provided) location,
Δp_{btm} is the differential pressure near the bottom of the pipe (0° from the drain hole), and
ρ_1 is the density at an upstream pressure tapping.

The flow through the orifice as a fraction of the total flow is approximately

$$\frac{Cd^2}{Cd^2 + C_h d_h^2}.$$

So the flow through the orifice can be calculated on the basis that the effective diameter ratio for the orifice is given by

$$\beta'' = \sqrt{\frac{\frac{\pi}{4}d^2}{\frac{\pi}{4}D^2 \frac{Cd^2}{Cd^2+C_h d_h^2}}} = \beta\sqrt{1 + \frac{C_h d_h^2}{Cd^2}}. \tag{10.5}$$

It is assumed that C and ε are unaffected by the presence of the drain hole, except for the effect of change in diameter ratio. It might be that where there is a drain hole the discharge coefficient for the flow through the orifice is very different from $C(Re_D, \beta'')$ owing to asymmetry, for example. However, where the average differential pressure can be calculated Eq. 10.4 is in fact remarkably accurate. Where C_h is taken from Eq. 10.24 (to be derived later), the errors using Eq. 10.4 are given in Table 10.6: they are 0.17 % in magnitude at most.

NOTE If data were taken with tappings from 30° to 180° from the drain hole the differential pressure at 0° from the drain hole was calculated from those at 30° and 60° from the drain hole assuming that the differential pressure Δp is a quadratic in θ and that $\frac{\partial \Delta p}{\partial \theta} = 0$ at 0° from the drain hole; if data were taken with tappings from 60° to 180° from the drain hole the differential pressures at 0° and 30° from the drain hole were calculated from those at 60° and 90° from the drain hole assuming that Δp is a quadratic in θ and that $\frac{\partial \Delta p}{\partial \theta} = 0$ at 0° from the drain hole.

Table 10.6 Error in calculated flowrate using Eq. 10.4 and the required angle from the drain hole for the tappings so that Eq. 10.6 gives the true flowrate

Pipe size	E/D	β	d_h/d	Percentage error in calculated flowrate using Eq. 10.4	θ^*, angle required for no error in flowrate using Eq. 10.6
8″	0.03	0.42	0.1	−0.03	91.1°
8″	0.02	0.6	0.1	−0.13	89.6°
8″	0.03	0.75	0.1	−0.02	76.5°
8″	0.03	0.6	0.07	0.05	94.1°
8″	0.03	0.75	0.0804	0.17	78.6°
8″	0.03	0.75	0.056	0.08	73.2°
8″	0.03	0.6	0.1	Not available	85.1°

10.4.2 Pressure Tapping Location for Flow Measurement Without Error

It is also possible to obtain values of θ^*, the angular location for the pressure tappings at which the flowrate would be measured without error using Eq. 10.6, where Eq. 10.6 is as follows:

$$\begin{aligned} q_m &= \frac{\frac{\pi}{4} d^2 C(Re_D, \beta'')\varepsilon(\beta'')\sqrt{2\rho_1 \Delta p^*}}{\sqrt{1-\beta''^4}} + \frac{\frac{\pi}{4} d_h^2 C_h \varepsilon(\beta'')\sqrt{2\rho_1 \Delta p^*}}{\sqrt{1-\beta''^4}} \\ &= \frac{\frac{\pi}{4} d^2 C(Re_D, \beta'')\varepsilon(\beta'')\sqrt{2\rho_1 \Delta p^*}}{\sqrt{1-\beta''^4}} \left(1 + \frac{d_h^2 C_h}{d^2 C}\right) \end{aligned} \tag{10.6}$$

where

Δp^* is the differential pressure at the flange (or corner etc. as provided) location with tappings at an angle θ^* from the drainhole.

$\frac{C_h}{C}$ is expressed in this way because it only appears in the smaller term in Eq. 10.6.

These locations are given in Table 10.6. It is worth noting that where the angular location of the tappings has most effect on the shift in C (i.e. large β) there appears to be no effect of d_h/d on θ^*.

Following the practice in ISO/TR 15377 the calculated flow is to be given by using a diameter d' such that

$$q_m = \frac{\frac{\pi}{4} d'^2 C(Re_D, \beta')\varepsilon(\beta')\sqrt{2\rho_1 \Delta p_{meas}}}{\sqrt{1-\beta'^4}} \tag{10.7}$$

where

$$\beta' = \frac{d'}{D}, \quad \text{and} \tag{10.8}$$

Δp_{meas} is the measured differential pressure at a particular flange (or corner etc. as provided) location.

Since the second term in Eq. 10.4 is much smaller than the first, in the second term Δp_{btm} can be approximated by Δp_{av}. Then Eq. 10.4 becomes

$$q_m = \frac{\frac{\pi}{4} d^2 C(Re_D, \beta'')\varepsilon(\beta'')\sqrt{2\rho_1 \Delta p_{av}}}{\sqrt{1-\beta''^4}}\left(1 + \frac{d_h^2 C_h}{d^2 C}\right) \tag{10.9}$$

The aim of this work is to provide a formula for d'. Equating Eqs. 10.9 and 10.7 gives

$$\frac{d'^2}{d^2} = \frac{C(Re_D, \beta'')\varepsilon(\beta'')}{C(Re_D, \beta')\varepsilon(\beta')}\sqrt{\frac{1-\beta'^4}{1-\beta''^4}}\left(1 + \frac{d_h^2 C_h}{d^2 C}\right)\sqrt{\frac{\Delta p_{av}}{\Delta p_{meas}}} \tag{10.10}$$

Since percentage change in differential pressure is simply related to percentage change in discharge coefficient, from Eq. 10.3 it might be reasonable to suppose that

$$\Delta p_{meas} = \Delta p_{top} + a\left(\beta, L_2', \frac{d_h}{d}\right)\Delta p_{av}\left(1 - \frac{\theta}{180}\right)^n \tag{10.11}$$

where

Δp_{top} is the differential pressure at the flange (or corner etc. as provided) location at the top of the pipe (180° from the drain hole).

From Eq. 10.3 it is reasonable to suppose that n is a function of β and $\frac{d_h}{d}$ and possibly L_2', since Eq. 10.3 is based on data for only one pipe size. The available data only require $n\left(\beta, \frac{d_h}{d}\right)$.

Equation 10.11 cannot be used for all θ, because as θ tends to 0 $\frac{\partial \Delta p_{meas}}{\partial \theta}$ should tend to 0 (but it does not in Eq. 10.11); moreover, if a and n are estimated from Eq. 10.3, as $\frac{d_h}{d}$ tends to 0 Δp_{meas} from Eq. 10.11 becomes discontinuous and Δp_{btm} does not equal Δp_{top} (but they should be equal in the limit where there is no drain hole).

However, it is only necessary to assume that Eq. 10.11 holds for $\theta \geq 60°$. Moreover, comparison of Eqs. 10.9 and 10.6 shows that $\Delta p^* = \Delta p_{av}$ and therefore that θ^* is the value of θ at which the average differential pressure is measured. Then

$$\Delta p_{av} = \Delta p_{top} + a\left(\beta, L_2', \frac{d_h}{d}\right)\Delta p_{av}\left(1 - \frac{\theta^*}{180}\right)^n \qquad (10.12)$$

Eliminating Δp_{top} between Eqs. 10.11 and 10.12 gives

$$\Delta p_{meas} = \Delta p_{av}\left\{1 + a\left(1 - \frac{\theta}{180}\right)^n - a\left(1 - \frac{\theta^*}{180}\right)^n\right\} \qquad (10.13)$$

10.4.3 An Equation for the Corrected Diameter

Substituting from Eq. 10.13 into Eq. 10.10 gives

$$\frac{d'^2}{d^2} = \frac{C(Re_D, \beta'')\varepsilon(\beta'')}{C(Re_D, \beta')\varepsilon(\beta')}\sqrt{\frac{1-\beta'^4}{1-\beta''^4}}\frac{1 + \frac{d_h^2 C_h}{d^2 C}}{\sqrt{1 + a\left(1 - \frac{\theta}{180}\right)^n - a\left(1 - \frac{\theta^*}{180}\right)^n}} \qquad (10.14)$$

It remains to determine a, n and θ^*. From Eq. 10.3 it is to be expected that a should be proportional to $\beta^{m'}$, where m' is a constant approximately equal to 4.9. It is worth testing the possibility that a is proportional to $\left(\frac{d_h}{d}\right)^{k'}$. As $\frac{l_2'}{d_h}\left(= \frac{L_2' d}{\beta d_h}\right)$ increases a should decrease, since the pressure variation round the pipe decreases with distance from the orifice plate. Accordingly a might be expressed as

$$a = a'\beta^{m'}\left(\frac{d_h}{d}\right)^{k'}\exp\left(-a''\frac{L_2' d}{\beta d_h}\right). \qquad (10.15)$$

where a', k' and a'' are constants.

Similarly from Eq. 10.3 it is reasonable to expect that

$$n = n' + n''\beta^{m''} + n'''\frac{d}{d_h}. \qquad (10.16)$$

where n', n'', m'' and n''' are constants.

From Table 10.6 it is reasonable to assume that

$$\theta^* = f' - f''\beta^{f'''}. \qquad (10.17)$$

where f', f'' and f''' are constants.

For an orifice whose axis is the pipe axis the effect of gradually changing the ratio of the orifice (bore) thickness e to the orifice diameter d is gradually to change the orifice from a thin orifice, in which it is as if e were as close as possible to 0 given that a square edge is required on the orifice, to a thick orifice. For a thick orifice with

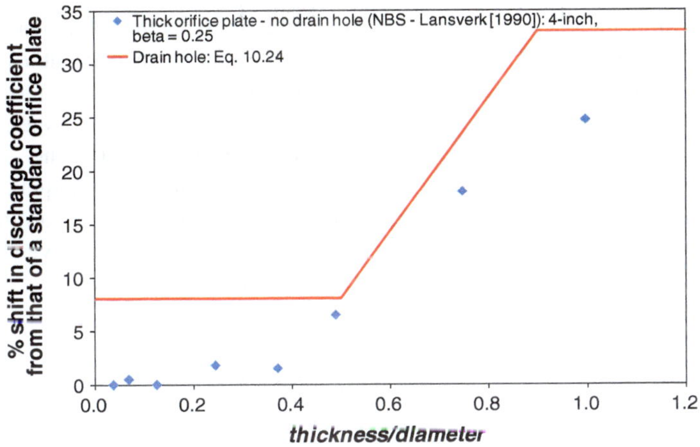

Fig. 10.9 Shift in discharge coefficient from its value in a standard orifice (for which the ratio of its (bore) thickness to its diameter is close to 0): for the NBS data the x-axis is *e*/*d*; for the drain hole it is E/d_h

e/*d* a little greater than 1 the discharge coefficient is approximately 0.8, because the flow has now reattached to the orifice bore. The discharge coefficient changes slowly where *e*/*d* is small but more rapidly around *e*/*d* = 0.7. This is well exhibited in a set of data from NBS (now NIST) (see Lansverk 1990) given in Fig. 10.9.

From Fig. 10.9 it is reasonable to suppose that $\frac{C_h}{C}$ is a function of $\frac{E}{d_h}\left(=\frac{dE}{\beta d_h D}\right)$ and that for appropriate constants C', C'', r' and r''

$$\frac{C_h}{C} = \begin{cases} C' & \text{if} \quad E/d_h \le r' \\ \dfrac{C'(r'' - E/d_h) + C''(E/d_h - r')}{r'' - r'} & \text{if} \quad r' < E/d_h < r'' \\ C'' & \text{if} \quad r'' \le E/d_h \end{cases} \tag{10.18}$$

Since the discharge-coefficient data had been calculated as

$$q_{m,\text{true}} = \frac{\frac{\pi}{4} d^2 C(Re_D, \beta)\varepsilon(\beta)\sqrt{2\rho_1 \Delta p_{meas}}}{\sqrt{1-\beta^4}}\left(1 + \frac{S}{100}\right) \tag{10.19}$$

where *S* is the percentage increase in discharge coefficient, from Eqs. 10.7 and 10.19 the percentage error in the measured flow when using Eq. 10.7 is

$$100\left(\frac{d'^2 C(Re_D, \beta')\varepsilon(\beta')}{d^2 C(Re_D, \beta)\varepsilon(\beta)(1 + 0.01S)}\sqrt{\frac{1-\beta^4}{1-\beta'^4}} - 1\right) \tag{10.20}$$

To determine the coefficients in Eqs. 10.15 and 10.16 a value of a'' was assumed, what the data points in Fig. 10.8 for $\theta \geq 60°$ would have been at $L_2' = 0$ was calculated, and then a', m', k', n', m'', n'' and n''' were calculated (given that Fig. 10.8 is for changes in discharge coefficient, Eq. 10.11 for changes in differential pressure). As might have been expected from Eq. 10.3 no significant improvement was obtained with non-zero k' or with m' and m'' unequal.

Then the data on percentage errors in measured flowrate were examined: there are many solutions with almost equal r.m.s. error. One solution is given below, in which the value of a'' has been used to refit the data in Fig. 10.8.

The coefficients are as follows:

$$a = 0.66\beta^{4.6} \exp\left(-0.15\frac{L_2' d}{\beta d_h}\right) \tag{10.21}$$

$$n = -0.45 + 7.3\beta^{4.6} + 0.117\frac{d}{d_h} \tag{10.22}$$

$$\theta^* = 92 - 62\beta^{4.6} \tag{10.23}$$

$$\frac{C_h}{C} = \begin{cases} 1.08 & \text{if} \quad E/d_h \leq 0.5 \\ 0.7675 + 0.625E/d_h & \text{if} \quad 0.5 < E/d_h < 0.9 \\ 1.33 & \text{if} \quad 0.9 \leq E/d_h \end{cases} \tag{10.24}$$

Equation 10.24 does not follow the NBS data in Fig. 10.9 exactly, but that is not surprising, since a drain hole is different from an orifice plate in that even where the plate thickness is small the fluid in a drain hole presumably remains attached to the pipe wall.

Equation 10.23 gives quite good agreement with the data in Table 10.6: see Fig. 10.10. Equation 10.23 is fitted to a wider range of data than those plotted in the figure.

Figure 10.11 shows the points included in Fig. 10.8 compared with Eq. 10.11 using the coefficients from Eqs. 10.21 and 10.22.

The errors in flowrate using Eqs. 10.14 and 10.21–10.24 for all the data in Tables 10.2, 10.4 and 10.5 (except those at 30° from the drain hole) are given in Fig. 10.12. The data marked 'baseline from a similar plate' are data taken with plates with drain holes for which no baseline was available: these are plates from which data are included in Fig. 10.8; baseline discharge coefficients taken from different plates made from the same drawings by the same manufacturer were used. Iteration was required to calculate d' using Eq. 10.14. The standard deviation of the data for $\frac{d_h}{d} \leq 0.1$ is 0.104 %. Data with $\frac{d_h}{d} > 0.1$ are shown, since they tend to confirm the form of the equation. However, the equation should not be used for $\frac{d_h}{d} > 0.1$.

The uncertainty that will be included in a revision of ISO/TR 15377 will depend, among other things, on the requirements on the upstream and downstream pipes.

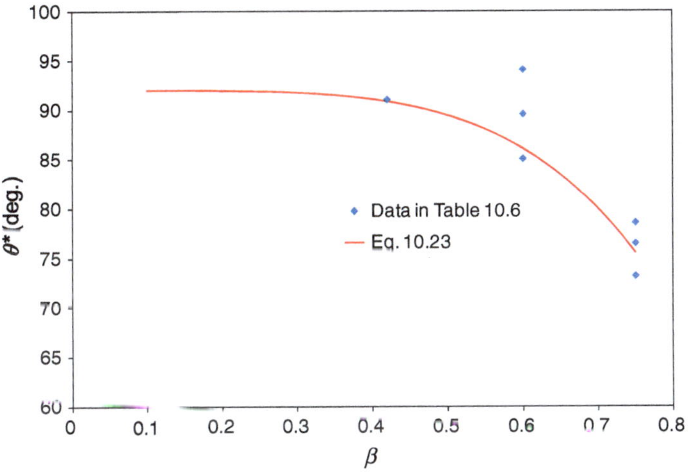

Fig. 10.10 Angle from drain hole at which a tapping should be placed to give a calculated flowrate without angular correction

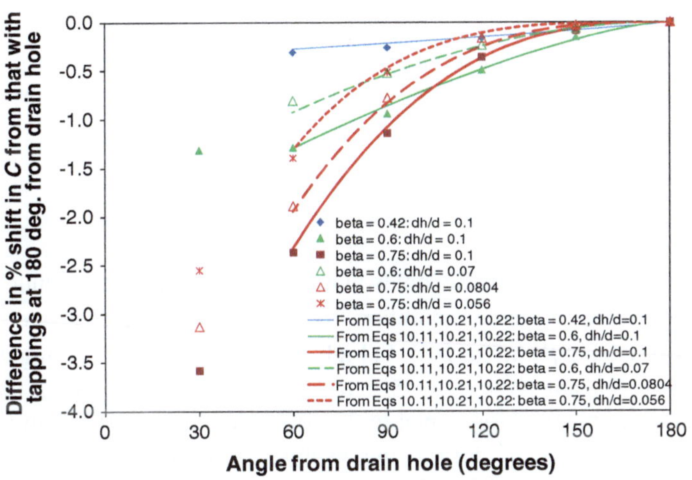

Fig. 10.11 Difference between percentage shift in *C* and percentage shift in C obtained with tappings at 180° from the drain hole: 8″ pipe with flange tappings

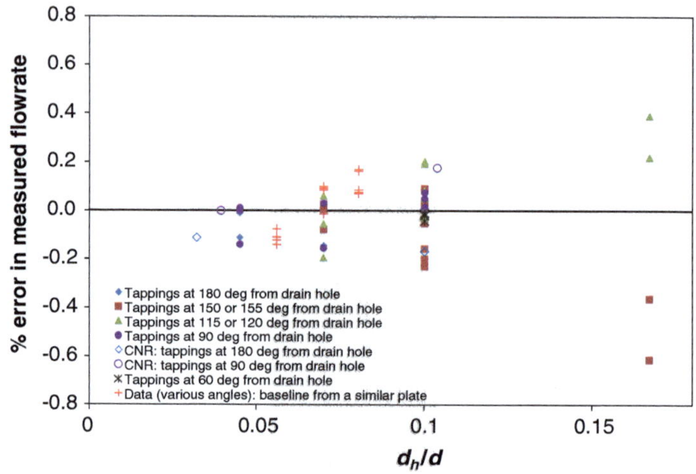

Fig. 10.12 Errors in measured flowrate using Eqs. 10.14 and 10.21–10.24

10.4.4 Practical Equations for the Corrected Diameter

In practice it is very desirable to have a fixed value for d' for an orifice plate with a drain hole, not a function of flowrate; to achieve this it would be necessary to use

$$\frac{d'^2}{d^2} = \frac{C(Re'_D, \beta'')}{C(Re'_D, \beta')}\sqrt{\frac{1-\beta'^4}{1-\beta''^4}}\frac{1+\frac{d_h^2 C_h}{d^2 C}}{\sqrt{1+a\left(1-\frac{\theta}{180}\right)^n - a\left(1-\frac{\theta^*}{180}\right)^n}} \tag{10.25}$$

where Re'_D is a fixed value of Reynolds number typical of the flow being measured but not the actual Reynolds number, Re_D, and a, n, θ' and $\frac{C_h}{C}$ are given in Eqs. 10.21–10.24 and β'' in Eq. 10.5. In high-pressure gas flows Re'_D might be taken as, say, 4×10^6. Then if $10^6 < Re_D < 5 \times 10^7$ the error in flowrate due to the use of a fixed Reynolds number of Re'_D equal to 4×10^6 is less than 0.013 % in magnitude for the values of β' and β'' used in this analysis. From Eq. 6.13 (Eq. (5) of ISO 5167-2:2003) if $\Delta p/p_1 < 0.02$ and the isentropic exponent $\kappa > 1.25$ the error in flowrate due to omitting the expansibility-factor ratio term is less than 0.014 % in magnitude (if $\Delta p/p_1 < 0.05$ and $\kappa > 1.25$ the error in flowrate due to omitting the expansibility-factor ratio term is less than 0.036 % in magnitude).

The number of iterations to convergence using Eq. 10.25 can be reduced by rearranging the equation as follows:

$$\frac{d'}{d} = \frac{1}{\left\{ \left(1-\beta''^4\right)\left(\frac{C(Re'_D,\beta')}{C(Re'_D,\beta'')}\right)^2 \frac{\left(1+a\left(1-\frac{\theta}{180}\right)^n - a\left(1-\frac{\theta^*}{180}\right)^n\right)}{\left(1+\frac{d_h^2 C_h}{d^2 C}\right)^2} + \beta^4 \right\}^{0.25}} \tag{10.26}$$

Moreover, if $\beta \leq 0.63$ or $\beta \leq 0.7$ and $\theta = 90°$ the ratio of discharge coefficients $\frac{C(Re'_D,\beta')}{C(Re'_D,\beta'')}$ can be taken as equal to 1 with an error in flowrate of 0.016 % or less in magnitude: in this case there is no need to iterate: Eq. 10.26 becomes

$$\frac{d'}{d} = \frac{1}{\left\{ \left(1-\beta''^4\right) \frac{\left(1+a\left(1-\frac{\theta}{180}\right)^n - a\left(1-\frac{\theta^*}{180}\right)^n\right)}{\left(1+\frac{d_h^2 C_h}{d^2 C}\right)^2} + \beta^4 \right\}^{0.25}} \quad \text{for } \beta \leq 0.63 \text{ or } \beta \leq 0.7 \text{ and } \theta = 90° \tag{10.27}$$

Equation 10.27 gives good results for the whole database, but there is insufficient evidence to use it for all situations.

10.5 Conclusions

Recent data show that the existing drain-hole equation in ISO/TR 15377:2007 is unsatisfactory. The data on drain holes have a surprisingly strong dependence on the circumferential location of the pressure tappings, although very little dependence on Reynolds number. A new analysis is presented based on the assumption that, although the differential pressure is disturbed by the presence of the drain hole, the discharge coefficient of the orifice (distinct from the drain hole) is unaffected by the drain hole's presence provided that the mean differential pressure is used. On this basis Eq. 10.26 has been produced for d', the corrected orifice diameter taking account of the drain hole. More data would be good, but it is very desirable to amend ISO/TR 15377:2007, given that its equation leads to flowrate errors up to nearly 2 % in magnitude, whereas the new equation gives errors less than 0.25 % in magnitude.

References

ISO 5167-2:2003 Measurement of fluid flow by means of pressure differential devices inserted in circular cross-section conduits running full—Part 2: orifice plates. International Organization for Standardization, Geneva

ISO/TR 11583:2012 Measurement of wet gas flow by means of pressure differential devices inserted in circular cross-section conduits. International Organization for Standardization, Geneva

ISO/TR 15377:2007 Measurement of fluid flow by means of pressure differential devices—guidelines for the specification of orifice plates, nozzles and venturi tubes beyond the scope of ISO 5167. International Organization for Standardization, Geneva

Lansverk NB (1990) Effects of abnormal conditions on accuracy of orifice measurement. ISHM (sent by Loy Upp)

Reader-Harris MJ, Addison D (2013) Orifice plates with drain holes. In: Proceedings FLOMEKO 16, Paris

Reader-Harris MJ, Hodges D, Rushworth R (2008a) The effect of drain holes in orifice plates on the discharge coefficient. In: Proceedings 26th International North Sea Flow Meas Workshop, St Andrews, Fife

Reader-Harris MJ, Hodges D, Barton N, Graham E (2008b) Improvements in the measurement of gas flows with entrained liquids using orifice plates. Report No 2008/265 on Project No. FFRE45. TUV NEL, East Kilbride, Glasgow

Spearman EP (2012) Operational measurement experiences in North Sea applications. In: Proceedings 30th International North Sea Flow Meas Workshop, St Andrews, Fife

Chapter 11
Wet Gas

Abstract This chapter describes the use of Venturi tubes and orifice plates to measure wet gas. A wet-gas flow is one in which there is a significant amount of free liquid carried with the gas, but the volumetric flowrate of gas is more than about 95 % of the total volumetric flowrate (or alternatively but not equivalently the Lockhart-Martinelli parameter $X \leq 0.3$). If the flowrate is not corrected the flowmeter will over-read the gas passing through it. Various correlations have been used to provide the correction: those included in ISO/TR 11583:2012 are first described, and then material on the derivation of the equations is provided. The de Leeuw Equation for Venturi tubes is also given. The controversy about correlations for Venturi tubes is fierce at present. In a wet-gas flow a measurement of liquid flowrate is generally required from outside the differential-pressure flowmeter: however, in some cases the pressure loss ratio can provide sufficient information: how and when it can be used are described.

11.1 Introduction

Differential-pressure meters are commonly selected for the measurement of wet-gas flows. A wet-gas flow is one in which there is a significant amount of free liquid carried with the gas: generally the volumetric flowrate of liquid is less than about 5 % of the total volumetric flowrate. Wet-gas flows are commonly found in wet-steam applications and where a natural gas contains some liquid hydrocarbons and/or water.

The presence of the liquid causes an increase in the measured differential pressure and results in the differential-pressure meter over-reading the actual amount of gas passing through it, if the flow is assumed to consist of gas alone. This over-reading is usually 'corrected' using available correlations derived from experimental data to determine the actual gas mass flowrate. The flowrate of the liquid, which can include a combination of water and hydrocarbons, may be determined by an external means such as using test separator data, tracer experiments or sampling etc. Information on

M. Reader-Harris, *Orifice Plates and Venturi Tubes*,
Experimental Fluid Mechanics, DOI 10.1007/978-3-319-16880-7_11

the liquid flowrate is necessary to enable the correlations to be used; in some cases the pressure loss ratio can provide sufficient information: how and when it can be used is described.

The correlations currently available for correcting the over-reading of differential-pressure meters have been derived from a limited set of data and may only be suitable to cover restricted ranges of differential-pressure meter parameters, for example, diameter ratio. Use of correlations outside the conditions used to define them can result in large errors in the calculation of the gas mass flowrate.

Venturi tubes are commonly chosen to measure wet gas: reasons for this choice include their physical robustness to withstand erosion and the impact of liquid slugs at high velocities, and familiarity with their use. Orifice plates give very good performance in wet gas but are more liable to be damaged by slugs than Venturi tubes. The performance of both Venturi tubes and orifice plates is covered by ISO/TR 11583:2012 (ISO 2012). The history of the development of ISO/TR 11583:2012 is given in Appendix 11.A.

Although the flow pattern in wet gas is more symmetrical in vertical flow than in horizontal, it is much more difficult to measure the required differential pressure in a vertical flow (see Sect. 4.2.3): the fluid in the meter line between the two tapping points has a different density from the fluid in the impulse lines. The correction for the difference in height between the tapping points requires a calculation of the mean two-phase density in the pipe line, which is subject to some uncertainty. For this reason it was found in Hall and Reader-Harris (1999) that a horizontal orientation of a Venturi tube was preferred in multiphase flow. The correlations in this chapter and in ISO/TR 11583:2012 only cover flow in horizontal pipes.

NOTE Work on Venturi tubes of $\beta = 0.6$ in vertical wet-gas flow has been published by Britton et al. (2008) and Graham et al. (2014). It is hoped that the latter work is just the initial stage of a larger project.

Tappings in wet gas are normally placed on the top of the pipe to prevent liquid collecting in the impulse lines, and the correlations in this chapter and in ISO/TR 11583:2012 have been derived for this tapping location. Using four equispaced tappings round the circumference connected with triple-Ts would be incorrect in wet gas. Flow conditioners should not be used in wet gas.

The method described in this chapter is appropriate where only the gas mass flowrate is to be determined: the liquid flowrate is not determined.

Computational Fluid Dynamics (CFD) work will be found in Reader-Harris et al. (2007): although further work is still required on CFD in wet gas, even in 2007 it showed that the liquid surface tension has a significant effect on the over-reading, and it led to the experimental work described in Appendix 11.B which shows that the effect of liquid viscosity is quite small; the rationale for that work is described in Sect. 11.B.1.

ISO/TC 193 is about to publish ISO/TR 12748: the recently balloted version is ISO/DTR 12748 (ISO 2014). This will contain much useful material on measuring wet natural gas. In particular it includes de Leeuw's equation for Venturi tubes (given here as Eq. 11.6) and Steven's Equation for orifice plates (given here as Eq. 11.11).

Section 11.2 gives the equations that are used for calculations in wet-gas flow; Sects. 11.3 and 11.4 give derivations.

11.2 Fundamental Equations

11.2.1 General

In a wet-gas flow a mixture of gas and liquid passes through the differential-pressure meter. To calculate the gas mass flowrate, $q_{m,\text{gas}}$, an additional term, the over-reading, is added to the basic equation (Eq. 1.15), so that $q_{m,\text{gas}}$ is given by

$$q_{m,\text{gas}} = \frac{C}{\sqrt{1-\beta^4}} \varepsilon \frac{\pi d^2}{4} \frac{\sqrt{2\rho_{1,\text{gas}} \Delta p}}{\phi} \tag{11.1}$$

where C is a discharge coefficient, ε is determined from the appropriate part of ISO 5167:2003, $\rho_{1,\text{gas}}$ is the upstream gas density and ϕ is the over-reading. In evaluating ε in ISO/TR 11583:2012 the actual values of p_1 and p_2 measured in wet gas are used. C and ϕ are described below. Until recently C was determined from the appropriate part of ISO 5167 or by calibration in dry gas and ϕ was determined from calibration in wet gas or from an equation, e.g. that of de Leeuw (1994, 1997). However, for a Venturi tube using ISO/TR 11583:2012 C and ϕ are both obtained from wet-gas equations.

The over-reading, ϕ, depends on

- the primary device
- the gas-liquid density ratio, $\rho_{1,\text{gas}}/\rho_{\text{liquid}}$, where ρ_{liquid} is the density of the liquid
- the Lockhart-Martinelli Parameter, X, as defined in Eq. 11.2:

$$X = \left(\frac{q_{m,\text{liquid}}}{q_{m,\text{gas}}}\right)\sqrt{\frac{\rho_{1,\text{gas}}}{\rho_{\text{liquid}}}} \tag{11.2}$$

  where $q_{m,\text{liquid}}$ is the liquid mass flowrate
- and the gas densiometric Froude number, Fr_{gas} (Eq. 11.3a):

$$Fr_{\text{gas}} = \frac{4q_{m,\text{gas}}}{\rho_{1,\text{gas}}\pi D^2\sqrt{gD}}\sqrt{\frac{\rho_{1,\text{gas}}}{\rho_{\text{liquid}} - \rho_{1,\text{gas}}}} \tag{11.3a}$$

where g is the acceleration due to gravity.

X is a measure of the wetness of the gas. It is equal to $\frac{Fr_{\text{liquid}}}{Fr_{\text{gas}}}$ where Fr_{liquid} is the liquid densiometric Froude number:

$$Fr_{\text{liquid}} = \frac{4q_{m,\text{liquid}}}{\rho_{\text{liquid}}\pi D^2\sqrt{gD}}\sqrt{\frac{\rho_{\text{liquid}}}{\rho_{\text{liquid}} - \rho_{1,\text{gas}}}} \tag{11.3b}$$

One common definition of wet-gas (as distinct from multiphase) flow is that $X \leq 0.3$.

11.2.2 Laboratory Test Work

In test work C/ϕ is determined from Eq. 11.1.
In dry-gas flow $\phi = 1$, and C is determined.
The value of C in dry-gas flow is then used to calculate ϕ for the wet-gas points.

11.2.3 Models for Field Use

11.2.3.1 General

Early work was undertaken by Murdock (1961). Usual practice today is that in models for field use ϕ is given by Chisholm's model (Chisholm 1967, 1977):

$$\phi = \sqrt{1 + C_{\text{Ch}}X + X^2} \tag{11.4}$$

where C_{Ch} is given by the following equation:

$$C_{\text{Ch}} = \left(\frac{\rho_{\text{liquid}}}{\rho_{1,\text{gas}}}\right)^n + \left(\frac{\rho_{1,\text{gas}}}{\rho_{\text{liquid}}}\right)^n \tag{11.5}$$

n depends on the differential-pressure meter and on the wet-gas correlation. In homogeneous flow $n = 0.5$.

Chisholm developed the model using steam/water experiments. In Chisholm's model for orifice plates $n = 0.25$: this gives quite good results, but Steven's Equation for n is better (see Sect. 11.2.3.3). However, $n = 0.25$ does not give good results for Venturi tubes. Murdock's equation for orifice plates has ϕ equal to $1 + 1.26X$, but has been superseded, first by Chisholm's model (with $n = 0.25$) and then by Steven's Equation.

11.2.3.2 Venturi Tube

General

Venturi meters are usually chosen for measuring wet natural gas, particularly in subsea applications where an orifice plate could be damaged but replacement would not be possible. Chisholm and Murdock correlations were used, but there remained significant bias.

Experimental work was carried out, leading to the de Leeuw Equation (de Leeuw 1994, 1997). The correlation produced a significant reduction in error, and it has been applied successfully in many applications. However, it is based on a single diameter ratio, $\beta = 0.4$.

A further development was a correlation from Reader-Harris (Reader-Harris and Graham 2009) based on a larger database. This correlation is included in ISO/TR 11583:2012 and is described here as the ISO/TR 11583:2012 Correlation.

The debate between correlations for Venturi tubes is fierce, and it is expected that further improvements will be made. Section 11.3 describes the development of the correlation in ISO/TR 11583:2012 and compares its performance with that of the de Leeuw Equation.

de Leeuw Equation

For Eq. 11.1 C is obtained from a dry-gas calibration of the Venturi tube.

Equations 11.4 and 11.5 are used; in Eq. 11.5 n is as follows:

$$n = \begin{cases} 0.41 & 0.5 \leq Fr_{\text{gas}} < 1.5 \\ 0.606\left(1 - e^{-0.746 Fr_{\text{gas}}}\right) & Fr_{\text{gas}} \geq 1.5 \end{cases} \tag{11.6}$$

de Leeuw produced his wet-gas Venturi-tube equation based on a Venturi tube of $\beta = 0.4$. According to ISO/DTR 12748:2014 with the liquid being a light liquid hydrocarbon with $\beta = 0.4$ and $D = 97.18$ mm over limited ranges of other parameters 'the gas flow rate is reported to be predicted within ±2 % at 95 % confidence level'. It is quite common for data sets other than the original data set to show deviations larger than this: see, e.g., Fig. 11.9.

ISO/TR 11583:2012 Correlation

For Eq. 11.1 C is given by Eq. 11.7:

$$C = 1 - 0.0463\,\mathrm{e}^{-0.05 Fr_{\text{gas},th}} \min\left(1, \sqrt{\frac{X}{0.016}}\right) \tag{11.7}$$

where the throat Froude number

$$Fr_{\text{gas},\,th} = \frac{Fr_{\text{gas}}}{\beta^{2.5}}. \tag{11.8}$$

Equations 11.4 and 11.5 are used; in Eq. 11.5

$$n = \max\left(0.583 - 0.18\beta^2 - 0.578\,\mathrm{e}^{-0.8Fr_{\text{gas}}/H},\ 0.392 - 0.18\beta^2\right) \tag{11.9}$$

H depends on the liquid and is equal to 1 for hydrocarbon liquid, 1.35 for water at ambient temperature, and 0.79 for liquid water in a wet-steam flow. It is primarily a function of the surface tension of the liquid (the effect of the viscosity of the liquid is shown in Appendix 11.B).

Limits of use:

- $0.4 \le \beta \le 0.75$
- $0 < X \le 0.3$
- $Fr_{\text{gas},th} > 3$
- $\frac{\rho_{1,\text{gas}}}{\rho_{\text{liquid}}} > 0.02$
- $D \ge 50$ mm

The first, fourth and fifth limits are the limits of the data in the database (in the case of the fourth limit there is a small amount of data in the database below the limit). The second limit is the normal range for wet gas. Operating below the third limit was shown to give increased errors, whereas as $Fr_{\text{gas},th}$ increases the errors generally reduce (as they reduce as Fr_{gas} increases in Fig. 11.6). There was no evidence of a diameter effect that might cause increased uncertainty for large D. As $\rho_{1,\text{gas}}/\rho_{\text{liquid}} \rightarrow 1$, from Eqs. 11.4 and 11.5 $\phi \rightarrow 1 + X$ and from Eqs. 11.3a, 11.7 and 11.8 $C \rightarrow 1$, both as expected; so the equation has the correct performance in this limit, and no maximum value of $\rho_{1,\text{gas}}/\rho_{\text{liquid}}$ has been imposed.

The uncertainty in C/ϕ is $\begin{cases} 3\,\% & \text{for} \quad X \le 0.15 \\ 2.5\,\% & \text{for} \quad 0.15 < X \le 0.3 \end{cases}$ (see Fig. 11.7)

11.2.3.3 Orifice Plate

For Eq. 11.1 C is obtained from a dry-gas calibration of the orifice plate or more usually from a dry-gas discharge-coefficient equation. In ISO/TR 11583:2012 C is given by the Reader-Harris/Gallagher (1998) Equation [Eq. 5.22; 5.3.2.1 of ISO 5167-2:2003 (ISO 2003a)]; the Reynolds number used is that which would be obtained if only the gas were flowing.

Equations 11.4 and 11.5 are used; Steven (Steven et al. 2008) derived an equation for n for use in Eq. 11.5, and this has been adopted by ISO/TR 11583:2012:

$$n = \begin{cases} 0.214 & 0.2 \le Fr_{\text{gas}} \le 1.5 \\ \left(\frac{1}{\sqrt{2}} - \frac{0.3}{\sqrt{Fr_{\text{gas}}}}\right)^2 & Fr_{\text{gas}} > 1.5 \end{cases} \quad (11.10)$$

Limits of use:

- $0.24 \le \beta \le 0.73$
- $0 < X \le 0.3$
- $Fr_{\text{gas}} \ge 0.2$
- $\frac{\rho_{1,\text{gas}}}{\rho_{\text{liquid}}} > 0.014$
- $D \ge 50$ mm

The uncertainty in C/ϕ is

$$\begin{cases} 2\,\% \text{ when the liquid is a light hydrocarbon liquid or water in a wet-steam flow} \\ 3\,\% \text{ when the liquid is water at ambient temperature} \end{cases}$$

The equation in ISO/TR 11583:2012 is in fact a special case (the case where the liquid is a light hydrocarbon liquid) of Eq. 11.11, which Steven obtained using data taken with natural gas and water and with natural gas, water and light hydrocarbon liquid in addition to the data used to obtain Eq. 11.10 (Steven et al. 2011).

$$n = \begin{cases} \left(\frac{1}{\sqrt{2}} - \frac{0.4-0.1e^{-w}}{\sqrt{1.5+0.2w}}\right)^2 & Fr_{\text{gas}} \le 1.5 + 0.2w \\ \left(\frac{1}{\sqrt{2}} - \frac{0.4-0.1e^{-w}}{\sqrt{Fr_{\text{gas}}}}\right)^2 & Fr_{\text{gas}} > 1.5 + 0.2w \end{cases} \quad (11.11)$$

where w is the water liquid mass ratio, i.e. the ratio of the water mass flowrate to the total liquid mass flowrate. The uncertainty in C/ϕ is 2 %. In due course ISO/TR 11583 should be revised to include Eq. 11.11.

11.2.4 Methods to Obtain the Lockhart-Martinelli Parameter, X (Eq. 11.2)

11.2.4.1 General

To perform the flowrate computation X is required. In the laboratory this is determined using experimental measurements of liquid flowrate. For use in the field the following methods are listed by ISO/TR 11583:2012 and described in more detail there:

(a) measuring the liquid flowrate using tracer techniques (see Clause 8 of ISO/TR 11583:2012);
(b) comparing the results from the wet-gas meter with those from gas and liquid meters downstream of a separator in series with the wet-gas meter;

(c) comparing the results with those from another wet-gas meter (see Clause 9 of ISO/TR 11583:2012);
(d) calculating from the known total mass flowrate (see Clause 10 of ISO/TR 11583:2012);
(e) using a throttling calorimeter in a wet-steam flow (see Clause 11 of ISO/TR 11583:2012);
(f) or using the third pressure tapping and applying an additional correlation.

The last method above is based on the fact that the pressure loss ratio, i.e. the ratio of the pressure loss (see Sect. 1.7) to the differential pressure, is a function of X. So, it may be possible to measure the pressure loss ratio and use it to determine X. For Venturi tubes this only works for small X: see Sect. 11.2.4.2. For orifice plates this only works for sufficiently large β; moreover, there are only limited data: the restrictions are given in Sect. 11.2.4.2.

11.2.4.2 Pressure Loss Ratio

Venturi Tube

For a limited range of X it is possible to use the pressure loss to calculate the Lockhart-Martinelli parameter. Formulae are given in ISO/TR 11583:2012 for a Venturi tube with divergent total angle in the range 7°–8° with the pressure loss, $\Delta\varpi$, measured from the upstream pressure tapping to a tapping a distance L_{down} downstream of the downstream end of the Venturi tube divergent section. $\Delta\varpi$ is the actual pressure loss: see Sect. 3.3. L_{down} is such that

$$\max(5, 20\beta - 7) \leq \frac{L_{down}}{D} \leq 9$$

In this case the following calculations are performed:

$$Y = \frac{\Delta\varpi}{\Delta p} - 0.0896 - 0.48\beta^9 \tag{11.12}$$

(Y is the increase in pressure loss ratio due to wetness) and

$$Y_{\max} = 0.61 \exp\left(-11\left(\frac{\rho_{1,\text{gas}}}{\rho_{\text{liquid}}}\right) - 0.045\left(\frac{Fr_{\text{gas}}}{H}\right)\right) \tag{11.13}$$

If $Y/Y_{\max} \geq 0.65$ it is not possible to use the pressure loss ratio to determine X. If $Y/Y_{\max} < 0.65$ X is evaluated from

$$\frac{Y}{Y_{\max}} = 1 - \exp\left(-35X^{0.75}\mathrm{e}^{-0.28Fr_{\text{gas}}/H}\right) \tag{11.14}$$

Since in this case Fr_{gas} is itself an unknown the procedure is iterative.

The uncertainty in C/ϕ is $\begin{cases} 4\,\% & \text{for} \quad \frac{Y}{Y_{max}} < 0.6 \\ 6\,\% & \text{for} \quad 0.6 \leq \frac{Y}{Y_{max}} < 0.65 \end{cases}$

Additional limits to those for the use of Eqs. 11.7 and 11.9 are that

- the divergent angle shall be between 7° and 8°
- $\frac{Y}{Y_{max}} < 0.65$
- $Fr_{gas,th} > 4$
- $\frac{Fr_{gas}}{H} \leq 5.5$
- $\frac{\rho_{1,gas}}{\rho_{liquid}} \leq 0.09$.

Orifice Plate

The pressure loss, $\Delta\varpi$, from the upstream pressure tapping to a tapping between $5D$ and $7D$ downstream of the orifice plate is measured. Then evaluate (although this is an iterative procedure, it should converge quickly)

$$Y = \frac{\Delta\varpi}{\Delta p} - \left.\frac{\Delta\varpi}{\Delta p}\right|_{dry} \tag{11.15}$$

where

$$\left.\frac{\Delta\varpi}{\Delta p}\right|_{dry}$$

is obtained using Eq. (7) of ISO 5167-2:2003 (ISO 2003a), which is Eq. 2.12.

$$X = \frac{6.41Y}{\beta^{4.9}}\left(\frac{\rho_{1,gas}}{\rho_{liquid}}\right)^{0.92}. \tag{11.16}$$

Using Eqs. 11.10 and 11.16
the uncertainty in C/ϕ is

$\begin{cases} 6\,\% \text{ when the liquid is a light hydrocarbon liquid or water in a wet-steam flow} \\ 7\,\% \text{ when the liquid is water at ambient temperature} \end{cases}$

Limits of use in addition to those in Sect. 11.2.3.3:

- $0.5 \leq \beta \leq 0.68$
- $X < 0.45\left(\frac{\rho_{1,gas}}{\rho_{liquid}}\right)^{0.46}$

- $\frac{\rho_{1,\text{gas}}}{\rho_{\text{liquid}}} \leq 0.21\beta - 0.09.$

NOTE It would be better to obtain $\left.\frac{\Delta\varpi}{\Delta p}\right|_{\text{dry}}$ from Eq. 2.14 rather than from Eq. 2.12.

11.3 Venturi Tubes

11.3.1 Over-Reading Equations

11.3.1.1 Derivation of the ISO/TR 11583:2012 Correlation

The problem with the traditional fits to data using the Chisholm model is that, as X tends towards 0, ϕ often does not tend to 0 linearly. For example data for $Fr_{\text{gas}} = 1.5$, $\beta = 0.6$, $\rho_{1,\text{gas}}/\rho_{\text{liquid}} = 0.024$ (Reader-Harris et al. 2005, 2006) are shown in Fig. 11.1. The gradient changes in the neighbourhood of the origin. One solution to this is to add an additional term to ϕ to fit the data [e.g. as in Reader-Harris et al. (2005)], but this has no obvious physical significance. However, it appears that the reason for the non-linearity is that in wet gas C is not equal to its dry-gas value. As soon as even a very small quantity of liquid is added to the gas the discharge coefficient reduces.

It is interesting to note that, whereas in dry gas a singing Venturi tube (one that emits an audible tone) is quite common, in wet gas Venturi tubes apparently do not sing; even a tiny quantity of liquid seems to change the flow in such a way as generally to prevent singing.

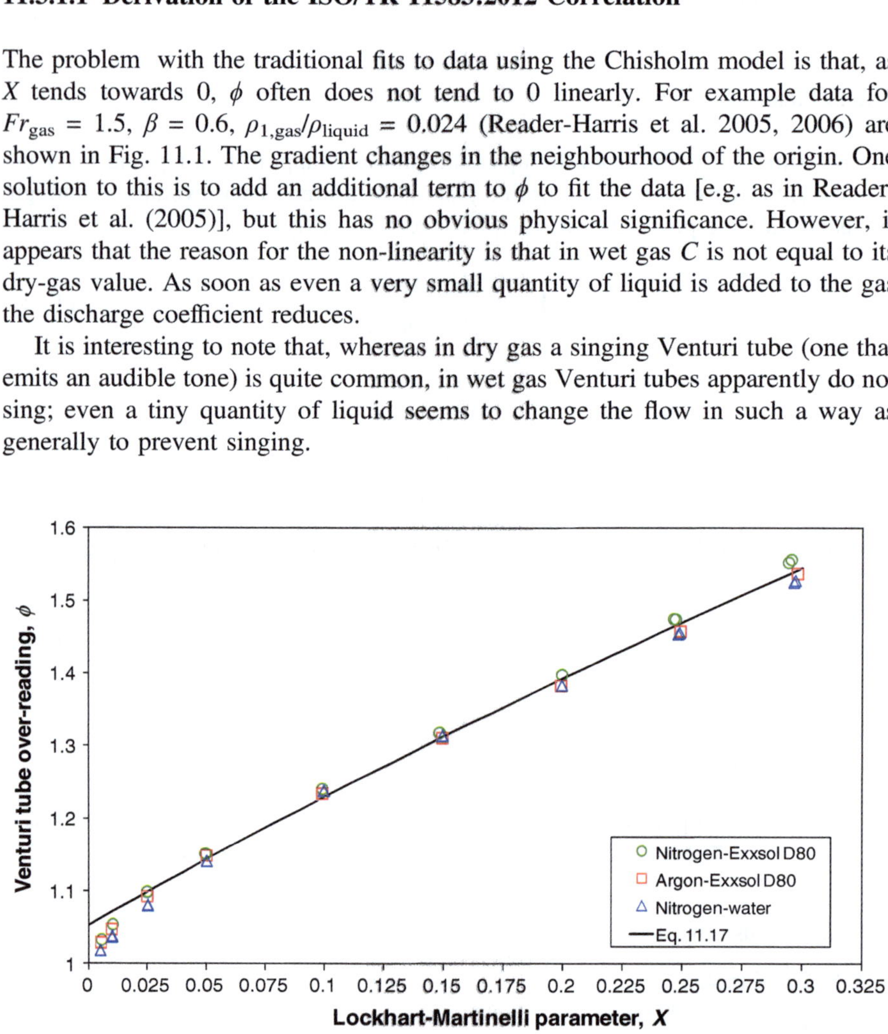

Fig. 11.1 NEL over-reading data for $Fr_{\text{gas}} = 1.5$, $\beta = 0.6$, $\rho_{1,\text{gas}}/\rho_{\text{liquid}} = 0.024$

Since in the case of the data in Fig. 11.1 the effect of fluid combinations is very small it is possible to consider all the data together. If the data for $X < 0.02$ are excluded the best fit of a Chisholm-type equation with an additional multiplicative constant is

$$\phi = 1.0532\sqrt{1 + C_{\text{Ch}}X + X^2}, \quad \text{where} \quad C_{\text{Ch}} = \left(\frac{\rho_{\text{liquid}}}{\rho_{1,\text{gas}}}\right)^{0.3162} + \left(\frac{\rho_{1,\text{gas}}}{\rho_{\text{liquid}}}\right)^{0.3162}. \quad (11.17)$$

So, for these sets of data the effective discharge coefficient in wet gas, C_{wet}, is around 5 % lower than the value of C in dry gas.

The next stage was to calculate values of C (i.e. C_{wet}) and n to fit Eqs. 11.1, 11.4 and 11.5 for $X \geq 0.02$ for the following sets of data, with each fluid combination considered separately:

- NEL 4″-β = 0.4—nitrogen/Exxsol D80 (Stewart 2002)
- NEL 4″-β = 0.6, 0.75—nitrogen/Exxsol D80, nitrogen/water, argon/Exxsol D80 (Reader-Harris et al. 2005, 2006)
- NEL 6″-β = 0.55—nitrogen/Exxsol D80 (Steven 2001)

NOTE Exxsol D80 is a light hydrocarbon oil with a kinematic viscosity of about 2.5 cSt and a density of about 775 kg/m^3 at 20 °C.

C_{wet} obviously varies from Venturi tube to Venturi tube; it is difficult to be certain on what C_{wet} depends. Possibilities are Reynolds number (or more likely throat Reynolds number), Froude number or throat Froude number. Of these throat Froude number (see Eq. 11.8) seems reasonably likely on physical grounds and a function of throat Froude number gives good agreement to the data.

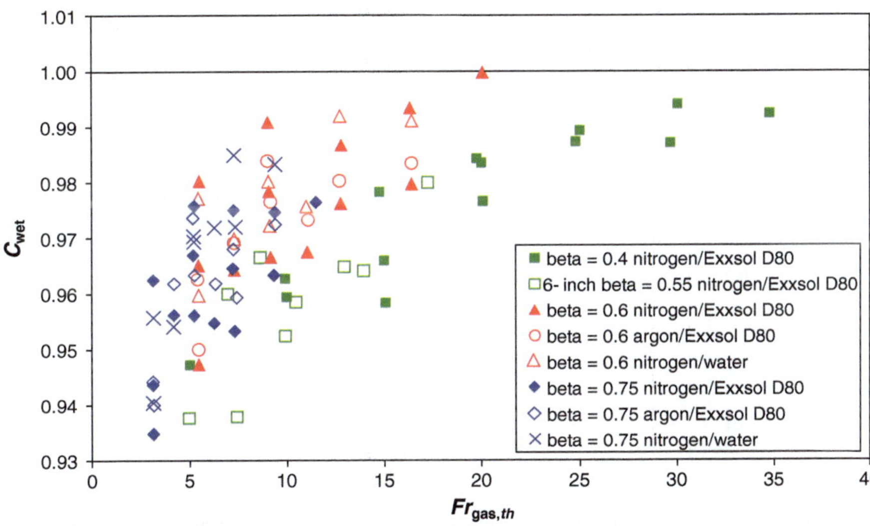

Fig. 11.2 Values of C_{wet} as a function of $Fr_{\text{gas},th}$

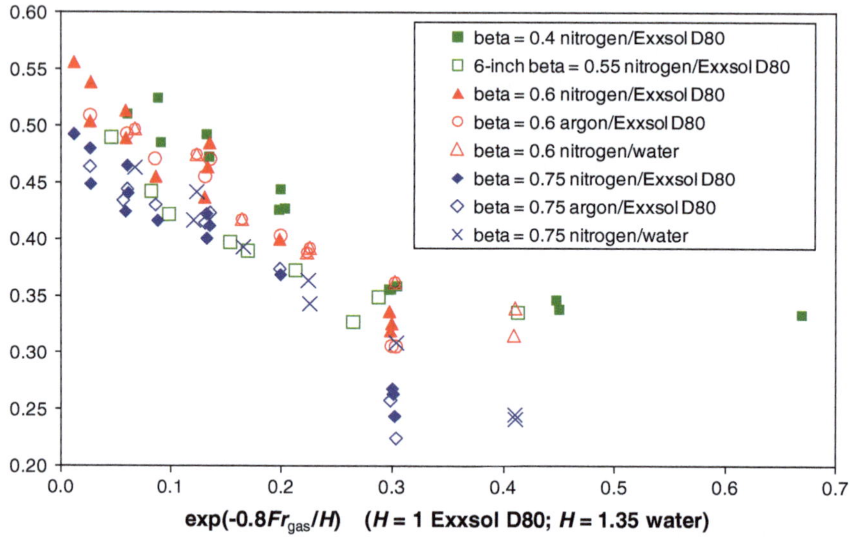

Fig. 11.3 Values of n

C_{wet} has been plotted in Fig. 11.2. Values of n have been plotted in Fig. 11.3 against a suitable function of Fr_{gas}. The effect of the choice of gas is negligible (provided that the density ratio is the same), but the effect of the choice of liquid is significant. Figure 11.4 shows a flow-régime map for a natural gas and light

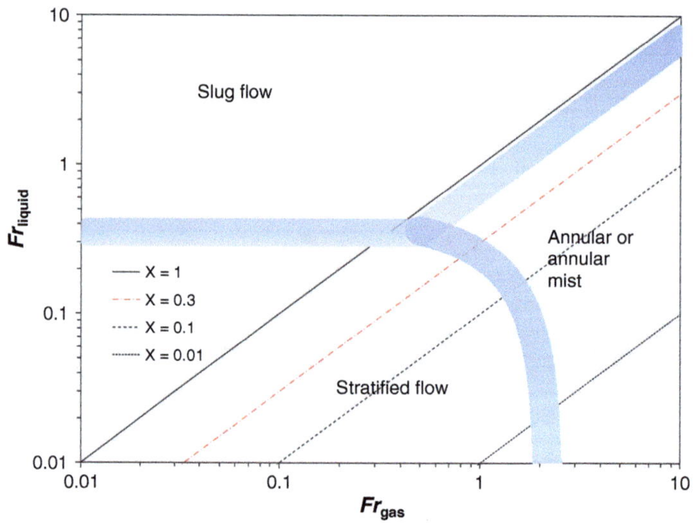

Fig. 11.4 Typical flow-régime map for a natural gas and light hydrocarbon liquid in horizontal flow

hydrocarbon liquid in horizontal flow. This is a version (with thickened borders (due to Richard Steven and Alistair Collins) between flow régimes to express the idea that the flow regime does not change suddenly as a narrow line is crossed) of the Shell flow pattern map (de Leeuw 1997), which is a well known and widely used horizontal two-phase flow pattern map, created by Shell from observations at the Trondheim wet-gas flow loop facility in Norway: the pipe was 4″, the fluids nitrogen and diesel oil, the static pressure between 15 and 90 bara. It appears from Eq. 11.9 (and the data on which the equations are based) that the flow-régime boundaries for a wet-gas flow in which the liquid is water are different from those for one in which the liquid is Exxsol D80. This is probably due to the surface tension of the liquid.

Having established appropriate dependencies for n and C_{wet} it is necessary to extend the range of data for fitting. From this point on the sets of data in Table 11.1 were used as the regression data.

The following fit was proposed:

$$n = \max\left(a - b\beta^c - de^{-fFr_{\mathrm{gas}}/H}, h - i\beta^j\right)$$
$$C_{\mathrm{wet}} = k - re^{-sFr_{\mathrm{gas},th}}$$

At very high pressure $\rho_{1,\mathrm{gas}} \rightarrow \rho_{\mathrm{liquid}}$; so $Fr_{\mathrm{gas},th} \rightarrow \infty$; and it is reasonable to suppose that $C_{\mathrm{wet}} \rightarrow C_{\mathrm{dry,typ}}$, the typical dry-gas discharge coefficient; so $k = C_{\mathrm{dry,typ}}$.

Table 11.1 Wet-gas regression data for Venturi-tube correlation

Diameter ratio	Pipe size	Gas phase	Liquid phase	Reference
0.6	2″	Natural gas	Stoddard solvent	Steven et al. (2006)
0.6	2″	Natural gas	Water	Steven et al. (2006)
0.4	4″	Nitrogen	Exxsol D80	Stewart (2003)
0.4	4″	Natural gas	Decane	Steven et al. (2007)
0.6	4″	Nitrogen	Water	Reader-Harris et al. (2005, 2006)
0.6	4″	Nitrogen	Exxsol D80	Reader-Harris et al. (2005, 2006)
0.6	4″	Argon	Exxsol D80	Reader-Harris et al. (2005, 2006)
0.6	4″	Nitrogen	Exxsol D80	–
0.6	4″	Natural gas	Exxsol D80	–[a]
0.7	4″	Steam	Very hot water	Harris and Shires (1972)
0.75	4″	Nitrogen	Water	Reader-Harris et al. (2005, 2006)
0.75	4″	Nitrogen	Exxsol D80	Reader-Harris et al. (2005, 2006)
0.75	4″	Argon	Exxsol D80	Reader-Harris et al. (2005, 2006)
0.55	6″	Nitrogen	Exxsol D80	Steven (2001)

[a]Data points with the r.m.s. of the fluctuating component of the differential pressure greater than 0.98 % of the mean differential pressure were excluded

As can be seen from Fig. 11.1 C changes when X is close to 0; so the following type of fit was used:

$$C = \begin{cases} C_{\mathrm{wet}} & X \geq X_{\mathrm{lim}} \\ C_{\mathrm{dry,\,typ}} - (C_{\mathrm{dry,\,typ}} - C_{\mathrm{wet}})\sqrt{\dfrac{X}{X_{\mathrm{lim}}}} & X < X_{\mathrm{lim}} \end{cases}$$

When the fit was undertaken various simplifications were possible: c and j were both close to 2; b and i were both close to each other; k was close to 1. A very good fit to the database was then

$$n = \max(0.583 - 0.18\beta^2 - 0.578\,\mathrm{e}^{-0.8Fr_{\mathrm{gas}}/H}, 0.392 - 0.18\beta^2) \tag{11.9}$$

$\mathrm{C} = 1 - 0.0463\,\mathrm{e}^{-0.05Fr_{\mathrm{gas},th}}$
$C_{\mathrm{dry,typ}} = 1$
$X_{\mathrm{lim}} = 0.016$

C can be written as

$$C = 1 - 0.0463\,\mathrm{e}^{-0.05Fr_{\mathrm{gas},th}}\min\left(1, \sqrt{\frac{X}{0.016}}\right) \tag{11.7}$$

H (in Eq. 11.9) depends on the liquid and is equal to 1 for hydrocarbon liquid, 1.35 for water at ambient temperature, and 0.79 for liquid water in a wet-steam flow. H has been defined as equal to 1 for hydrocarbon liquid; the other values have been obtained by fitting the data. The data for very hot water (in wet steam) are very limited. H depends chiefly on surface tension. If liquid viscosity were the key parameter then H for very hot water should be greater than H for cold water: Appendix 11.B shows that the effect of liquid viscosity is quite small. There is negligible effect of the gas properties on the over-reading (Reader-Harris et al. 2006).

It is possible to check that C is in good agreement with Eq. 11.7 by calculating the over-reading from Eqs. 11.4 and 11.5 for all the points with $0.02 < X < 0.065$ and then calculating C from Eq. 11.1. Clearly this method depends on the value of n, but by choosing small X the dependence on n is reduced. Data for $X < 0.02$ were not used as C is in transition from a wet-gas value to the dry-gas one. n was taken as in Eq. 11.9 and the values of C plotted against throat Froude number are given in Fig. 11.5.

Errors in the gas mass flowrate using Eqs. 11.7 and 11.9 are shown in Fig. 11.6 for the regression data.

The regression data in Fig. 11.6 have an r.m.s. deviation about the equation of 0.86 % (this could be called standard deviation since the mean deviation is zero, but 'r.m.s. deviation' is used throughout this chapter since in some cases the mean

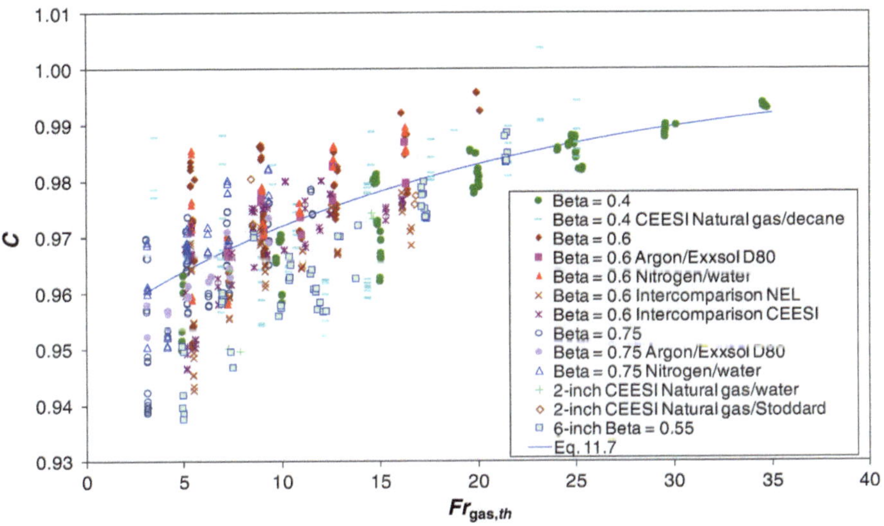

Fig. 11.5 Values of C for the regression data with $0.02 < X < 0.065$

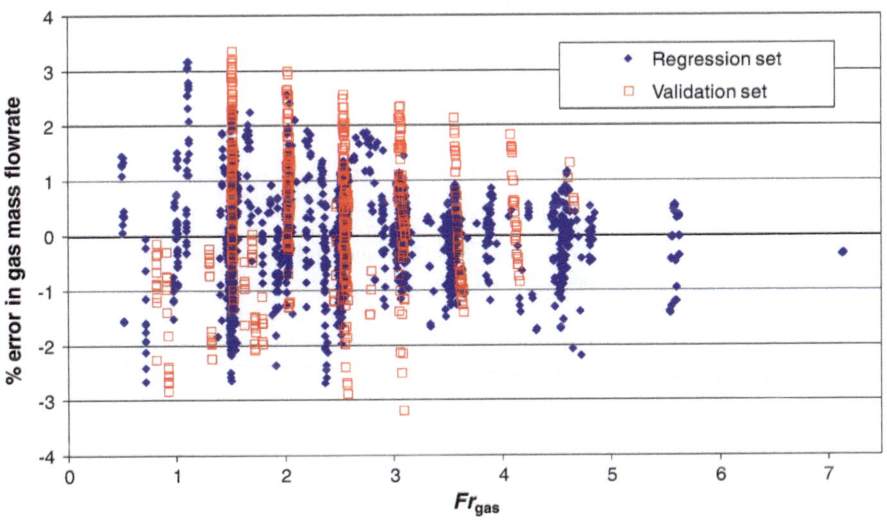

Fig. 11.6 Errors in gas mass flowrate using Eqs. 11.7 and 11.9 versus Fr_{gas}

deviation is not zero). Nevertheless it is desirable to determine the uncertainty of the equation using data to which it was not fitted.

Krohne very kindly agreed to allow their data on a large number of Venturi tubes, calibrated at NEL in nitrogen/Exxsol D80, to be used. These are given in Table 11.2.

Table 11.2 Wet-gas validation data (from Krohne) for Venturi-tube correlation

Diameter ratio	Pipe size	Gas phase	Liquid phase	Number of Venturi tubes
0.6	4″	Nitrogen	Exxsol D80	6 off
0.47	4″	Nitrogen	Exxsol D80	1 off
0.43	4″	Nitrogen	Exxsol D80	1 off
0.4	4″	Nitrogen	Exxsol D80	1 off
0.57	6″	Nitrogen	Exxsol D80	6 off
0.61	6″	Nitrogen	Exxsol D80	5 off
0.61	10″	Nitrogen	Exxsol D80	2 off

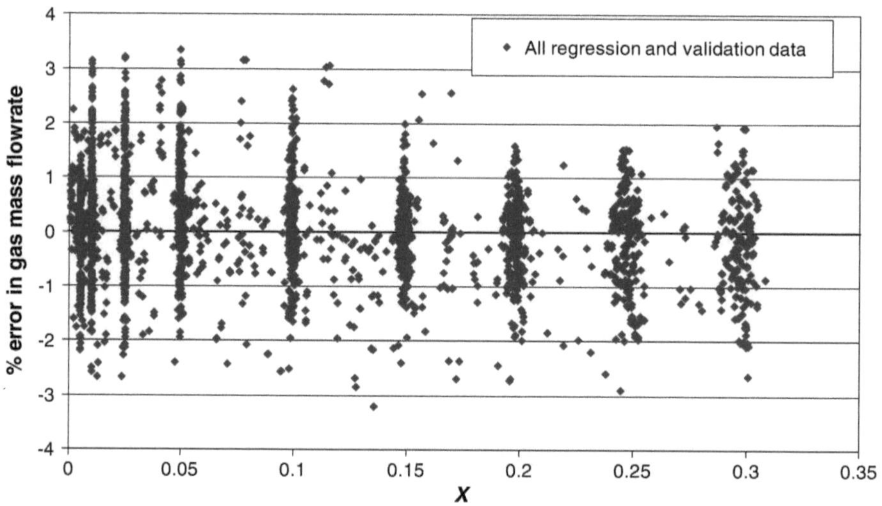

Fig. 11.7 Errors in gas mass flowrate using Eqs. 11.7 and 11.9 versus *X* for the regression data (Table 11.1) and the validation data (Table 11.2)

Data for $\beta = 0.37$ were not used as it was decided to consider $0.4 \le \beta \le 0.75$. Data for one of the 4″ and for one of the 6″ Venturi tubes were not used as each had a mean discharge coefficient in dry gas below 0.97.

Including the Krohne data in the database and using Eqs. 11.7 and 11.9 the r.m.s. deviation of the enlarged database is 0.980 %. Refitting Eqs. 11.7 and 11.9 would only reduce the r.m.s. deviation of the enlarged database to 0.969 %; so the Krohne data are just used for validation. Their scatter is also shown in Fig. 11.6. For the validation data the r.m.s. deviation of the data about Eqs. 11.7 and 11.9 is 1.34 %. It is perhaps more informative to plot the errors against *X* as in Fig. 11.7.

Equations 11.7 and 11.9 can be used in accordance with the limits of use already stated in Sect. 11.2.3.2

$$\text{with an uncertainty in } C/\phi \text{ of } \begin{cases} 3\,\% & \text{for} \quad X \le 0.15 \\ 2.5\,\% & \text{for} \quad 0.15 < X \le 0.3 \end{cases}$$

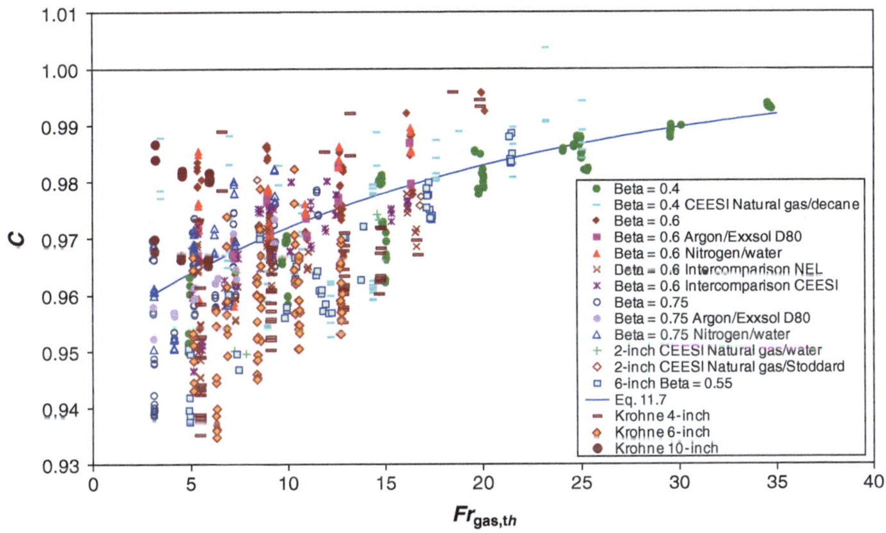

Fig. 11.8 Values of C for the regression and validation data with $0.02 < X < 0.065$

The surprising reduction in uncertainty for $X > 0.15$ from that for $X \leq 0.15$ is discussed in Sect. 11.3.1.3.

Values of C calculated as for Fig. 11.5 for all the regression and validation data with $0.02 < X < 0.065$ are given in Fig. 11.8.

11.3.1.2 Comparison with the de Leeuw Equation

It is possible to compare the errors for the ISO/TR 11583:2012 Correlation with those for the de Leeuw Equation on the same database. The errors using the de Leeuw Equation are shown in Fig. 11.9. Its performance depends on β.

The de Leeuw Equation for wet-gas flow needs a measured dry-gas discharge coefficient, the ISO/TR 11583:2012 Correlation does not (although a dry-gas calibration is useful to show that a Venturi tube has been well manufactured and to exclude poorly manufactured Venturi tubes from use).

11.3.1.3 Possible Improvement of the ISO/TR 11583:2012 Correlation

Where the measured dry-gas discharge coefficient, C_{dry}, is known it is tempting but incorrect to multiply the right hand side of Eq. 11.7 by C_{dry} [this is done by de Leeuw et al. (2011), but see Reader-Harris (2012)]. This is because where the discharge coefficient in single-phase flow differs significantly from 1 the errors in measured gas mass flowrate often reduce in magnitude as X increases. By way of illustration the errors for two Venturi tubes are given in Fig. 11.10. These are rather

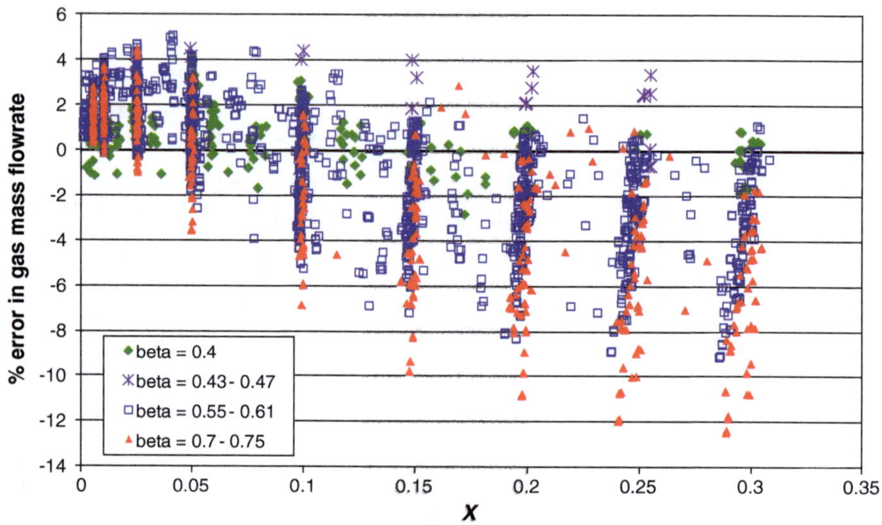

Fig. 11.9 Errors in gas mass flowrate using the de Leeuw Equation (Eq. 11.6) versus X for the regression data (Table 11.1) and the validation data (Table 11.2)

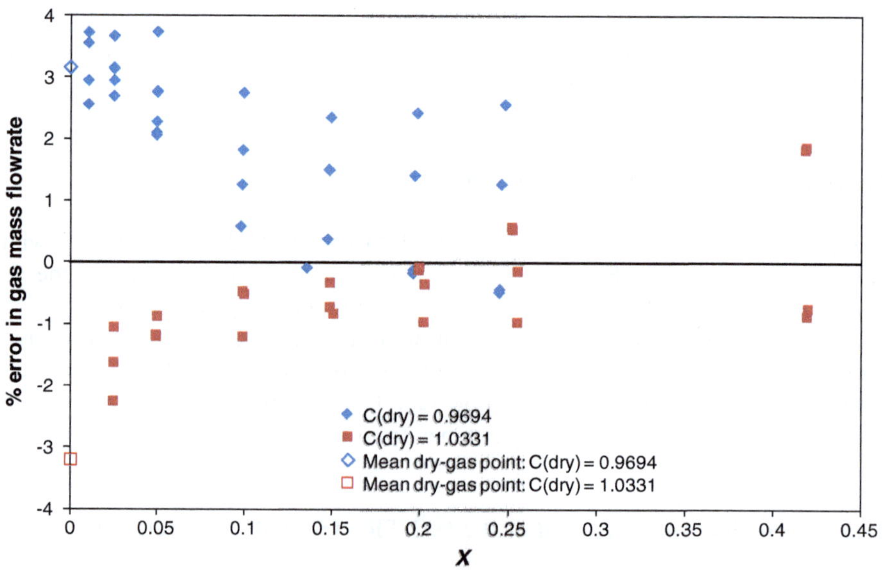

Fig. 11.10 Errors for two Venturi tubes with dry-gas discharge coefficients about 3 % different from 1.00: correction according to ISO/TR 11583:2012

extreme examples in terms of dry-gas discharge coefficient: most of the wet-gas errors are smaller than the dry-gas ones. It is possible that the big variation in dry-gas discharge coefficients is suppressed by wet gas.

NOTE: It may be asked whether it is reasonable to have an uncertainty in wet gas approximately equal to that in dry gas.

Figure 7 of de Leeuw et al. (2011) shows Venturi tubes in dry gas: all of their discharge coefficients are well within 1.00 ± 3 %, which is a necessary condition for ISO/TR 11583:2012 to be true as X tends to 0. Then there is a tendency for those Venturi tubes with a dry-gas discharge coefficient far from 1 to have their measurement errors reduce in magnitude as X increases (see below); so the uncertainty reduces slightly for $X > 0.15$.

It is also possible to consider in a more systematic way how the dry-gas discharge coefficient affects the wet-gas results. It is clear how it affects them as X tends to 0. To consider the issue more broadly, suppose the measured gas mass flowrate in ISO/TR 11583:2012 is multiplied by

$$1 + (C_{\mathrm{dry}} - 1)\exp(-aX) \tag{11.18}$$

Then if $a = 0$ the correction method in de Leeuw et al. (2011) is (at least to a very good approximation) obtained. Then over the regression and validation database the r.m.s. deviation is 1.068 % instead of 0.980 %, which was obtained using ISO/TR 11583:2012 without any extra term. So the fact that the errors in Fig. 5 of de Leeuw et al. (2011) are greater than those in Fig. 4 of de Leeuw et al. (2011) is not surprising (they have been increased by de Leeuw et al.'s correction). The minimum r.m.s. deviation using Eq. 11.18 is obtained with $a = 11.25$ and is 0.88693 %. It is also possible to consider if it would be better to use

$$1 + (C_{\mathrm{dry}} - 1)((1 - b)\exp(-aX) + b) \tag{11.19}$$

The best fit is obtained with $a = 12.8$ and $b = 0.05$ and is 0.88682 %: the inclusion of b is not a significant improvement; the inclusion of a is an improvement, but not consistent with the intention in ISO/TR 11583:2012 to have an equation for uncalibrated Venturi tubes. Errors for the data in Fig. 11.10 using the ISO/TR 11583:2012 Correlation with an additional term as in Eq. 11.18 with $a = 11.25$ are shown in Fig. 11.11.

Errors for regression and validation data using the ISO/TR 11583:2012 Correlation with an additional term as in Eq. 11.18 with $a = 11.25$ are shown in Fig. 11.12.

So an improved wet-gas correlation for Venturi tubes calibrated in dry gas is possible, but the change to the correlation is not the one that might have been expected.

It is interesting that over the range $0 < X < 0.016$ the discharge coefficient changes from dry to wet, whereas over the range $0 < X < 0.1$ the effect of C_{dry} largely disappears. It is possible that as X increases from 0 to 0.016 there is the rapid

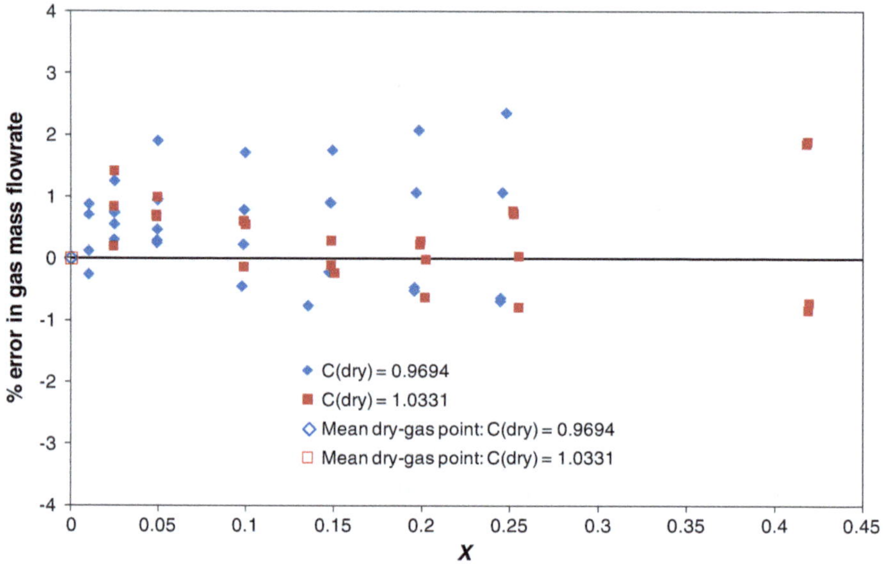

Fig. 11.11 Errors for the two Venturi tubes in Fig. 11.10 using the ISO/TR 11583:2012 correlation with an additional term as in Eq. 11.18 with $a = 11.25$

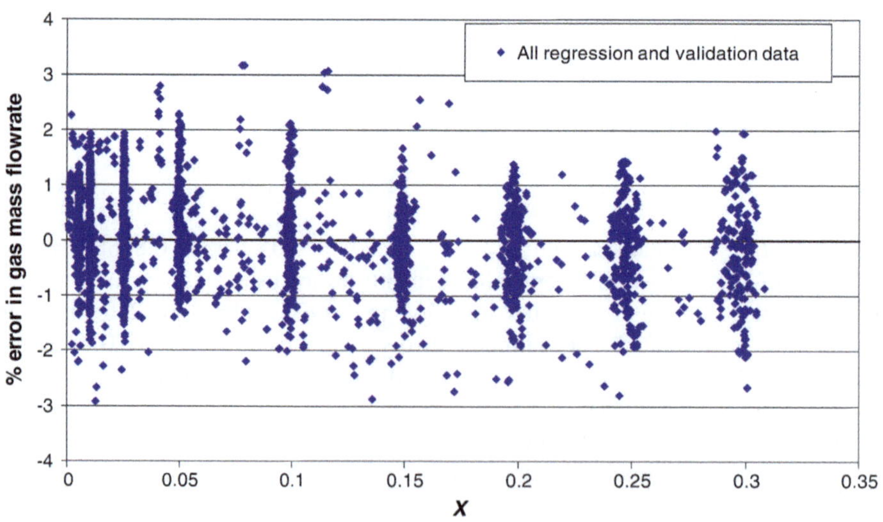

Fig. 11.12 Errors for all the regression and validation data using the ISO/TR 11583:2012 correlation with an additional term as in Eq. 11.18 with $a = 11.25$

formation of a layer of liquid on the Venturi throat, reducing the discharge coefficient, but that the layer is thinner at the top of the pipe (where the pressure tapping is) so that X has to be much larger than 0.016 before the effect of C_{dry} disappears.

11.3.2 Using Pressure-Loss Measurements

The pressure loss is a function of the wetness of the gas. Whereas in dry gas the pressure loss is generally in the range of 5–30 % of the differential pressure (5–15 % for a divergent angle of 7°) it can be much greater in wet gas. Under certain circumstances the ratio of the pressure loss to the differential pressure can be used to determine X and thus determine $q_{m,gas}$ without a separate measure of the liquid flowrate. All the subsequent work on Venturi tubes is based on those with a divergent angle of 7.5°. The downstream tapping was placed around 6D downstream of the downstream end of the Venturi tube (the location of the downstream tapping relative to the downstream end of the divergent of the Venturi tube that is required to use the equations described here is given in Sect. 11.2.4.2). The following sets were used:

- NEL 4″-β = 0.4—nitrogen/Exxsol D80
- NEL 4″-β = 0.6, 0.75—nitrogen/Exxsol D80, nitrogen/water, argon/Exxsol D80
- NEL/CEESI 4″ Intercomparison-β = 0.6—nitrogen/Exxsol D80

On the basis of these the ratio of the pressure loss, $\Delta\varpi$, to the differential pressure was determined in dry gas to be:

$$\left.\frac{\Delta\varpi}{\Delta p}\right|_{dry} = 0.0896 + 0.48\beta^9 \tag{11.20}$$

The increase in pressure loss ratio due to wetness was defined as:

$$Y = \frac{\Delta\varpi}{\Delta p} - \left.\frac{\Delta\varpi}{\Delta p}\right|_{dry} \tag{11.21}$$

So

$$Y = \frac{\Delta\varpi}{\Delta p} - 0.0896 - 0.48\beta^9 \tag{11.12}$$

Then the maximum value of the increase in pressure loss ratio due to wetness was determined:

$$Y_{max} = \left.\left(\frac{\Delta\varpi}{\Delta p} - \left.\frac{\Delta\varpi}{\Delta p}\right|_{dry}\right)\right|_{X\approx 0.3} \tag{11.22}$$

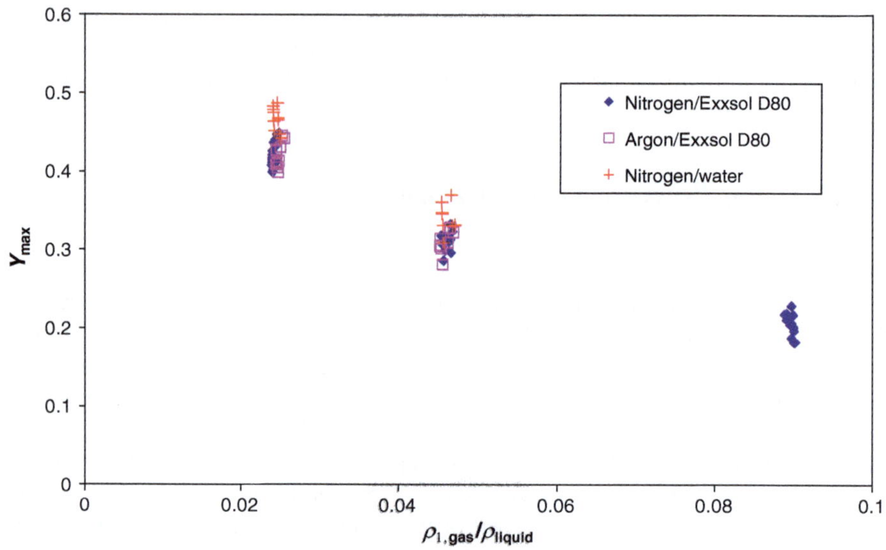

Fig. 11.13 Y_{max} versus $\rho_{1,gas}/\rho_{liquid}$ for nominal values of Fr_{gas} in the range 1.5–4.5

This was first investigated by plotting Y_{max} against $\rho_{1,gas}/\rho_{liquid}$. This is shown in Fig. 11.13.

It is clear that as expected there is an effect of the choice of liquid but no effect of the choice of gas. It is assumed that the same dependence on liquid as appeared in Eq. 11.9 is appropriate here. Figure 11.14 shows the data plotted against a suitable function of Fr_{gas}/H.

It is then clear that the dependence on $\rho_{1,gas}/\rho_{liquid}$ can be represented by a simple function that becomes extremely small as $\rho_{1,gas}/\rho_{liquid}$ tends to 1.

The following form for Y_{max} is then assumed:

$$Y_{max} = a \exp\left(-b\left(\frac{\rho_{1,gas}}{\rho_{liquid}}\right) - c\left(\frac{Fr_{gas}}{H}\right)\right)$$

On fitting the database

$$Y_{max} = 0.61 \exp\left(-11\left(\frac{\rho_{1,gas}}{\rho_{liquid}}\right) - 0.045\left(\frac{Fr_{gas}}{H}\right)\right) \tag{11.13}$$

Y/Y_{max} is then evaluated for each point and plotted in Fig. 11.15.

This confirms that obtaining the value of Y_{max} at $X = 0.3$ is reasonable. Even if it were not it would not invalidate the final formula.

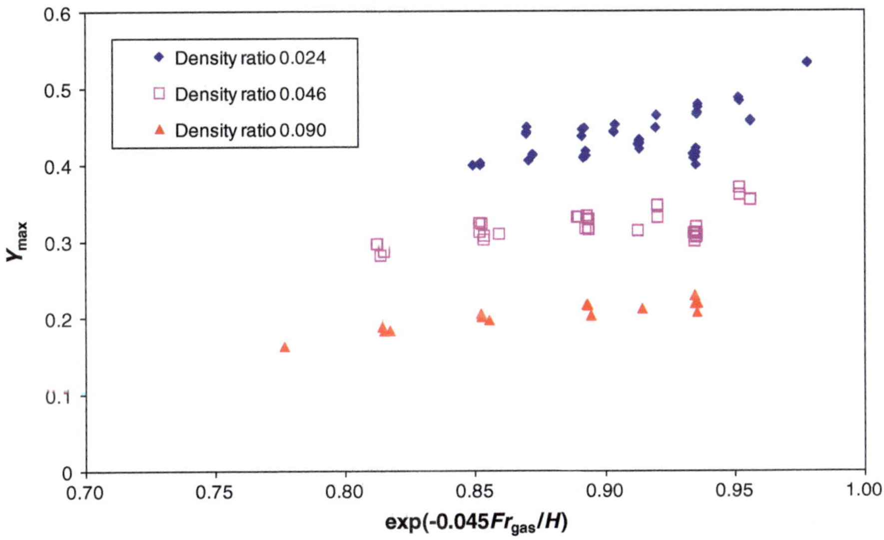

Fig. 11.14 Y_{max} as a function of Fr_{gas}/H

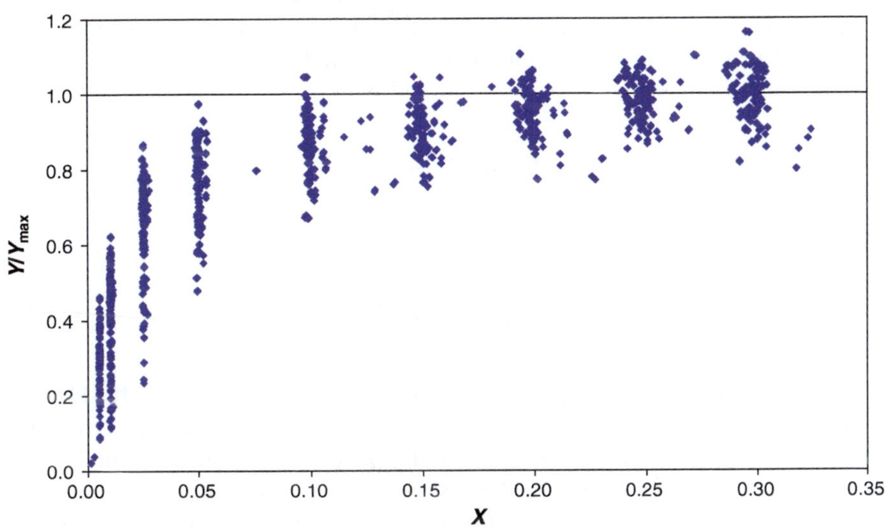

Fig. 11.15 Y/Y_{max} versus X

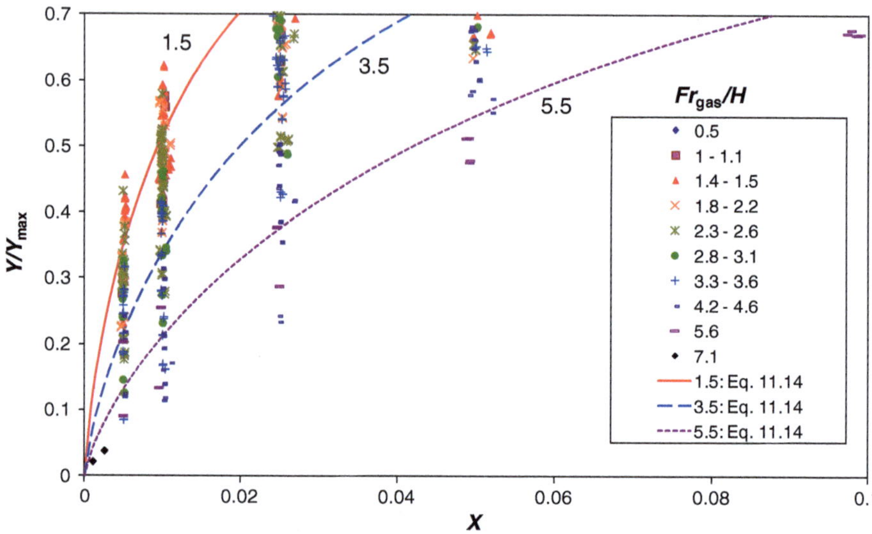

Fig. 11.16 Y/Y_{max} versus X for $Y/Y_{max} < 0.7$ and $Fr_{gas,th} > 4$

However, Fig. 11.15 is inadequate for practical use. It is better to plot the data as in Fig. 11.16 for small Y/Y_{max} divided by ranges of Fr_{gas}/H. Data obtained with $Fr_{gas,th} < 4$ increased the uncertainty and were excluded.

Fitting the following form

$$\frac{Y}{Y_{max}} = 1 - \exp\left(-bX^d e^{-cFr_{gas}/H}\right)$$

gives

$$\frac{Y}{Y_{max}} = 1 - \exp\left(-35X^{0.75} e^{-0.28Fr_{gas}/H}\right) \tag{11.14}$$

Given a value of pressure loss ratio it is possible to use Eqs. 11.12–11.14 to determine Y and Y_{max} and hence X and thus the gas mass flowrate. The errors as a function of Y/Y_{max} are shown in Fig. 11.17. The results are remarkably good, but as soon as Y/Y_{max} exceeds 0.7 the method becomes very inaccurate.

Again for use with Venturi tubes other than those from which the correlation was derived it is desirable to be conservative. A reasonable estimate for the uncertainty in C/ϕ using pressure loss and Eqs. 11.12–11.14, 11.7 and 11.9 is

$$\begin{cases} 4\,\% & \text{for} \quad \frac{Y}{Y_{max}} < 0.6 \\ 6\,\% & \text{for} \quad 0.6 \leq \frac{Y}{Y_{max}} < 0.65 \end{cases}$$

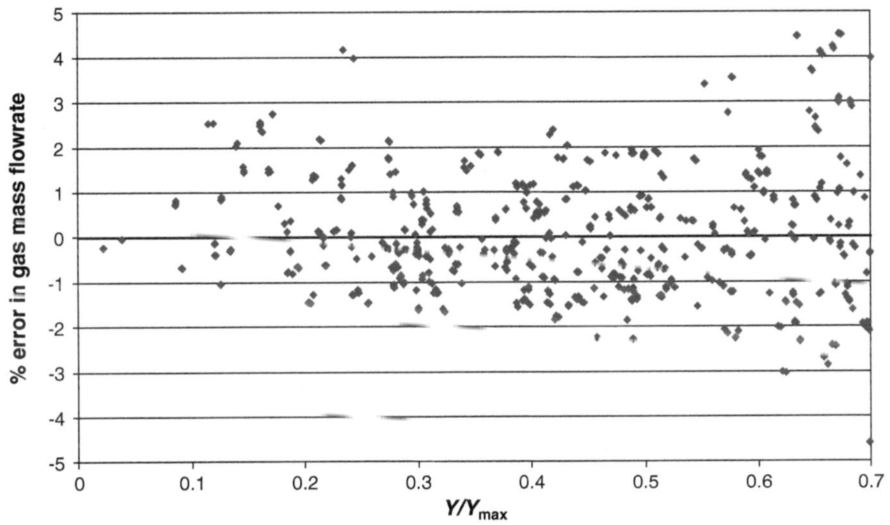

Fig. 11.17 Errors in gas mass flowrate for $Fr_{\text{gas},th} > 4$ using pressure-loss measurement and Eqs. 11.12–11.14, 11.7 and 11.9

Additional limits to those for the use of Eqs. 11.7 and 11.9 are that

- the divergent angle shall be between 7° and 8°
- $\frac{Y}{Y_{\max}} < 0.65$
- $Fr_{\text{gas},th} > 4$
- $\frac{Fr_{\text{gas}}}{H} \leq 5.5$
- $\frac{\rho_{1,\text{gas}}}{\rho_{\text{liquid}}} \leq 0.09.$

11.3.3 Mixtures of Liquids

Most data available for research purposes have only a single liquid phase of hydrocarbon or water. However, in reality natural-gas flows are usually a mixture of natural gas, liquid hydrocarbon and water. It is recognized that any correlation for a flowmeter should ideally have been tested to cover the range of conditions it is likely to encounter in the field. Unfortunately this was not possible owing to the limited amount of data that was available to derive and test correlations and the limitations of some test facilities to generate data.

The correlation in Eqs. 11.7 and 11.9 has been tested with a limited data set using a liquid mixture of hydrocarbon and water. Because of the small quantity of available data it is not claimed that Eqs. 11.7 and 11.9 can be used for mixtures of liquid with the uncertainty quoted above. Details of the data are provided in Table 11.3.

Table 11.3 Wet-gas data using a mixed liquid phase of liquid hydrocarbon and water

Diameter ratio	Pipe size	Gas phase	Liquid phase	Reference
0.6	2″	Natural gas	Stoddard solvent and water	Steven et al. (2006)

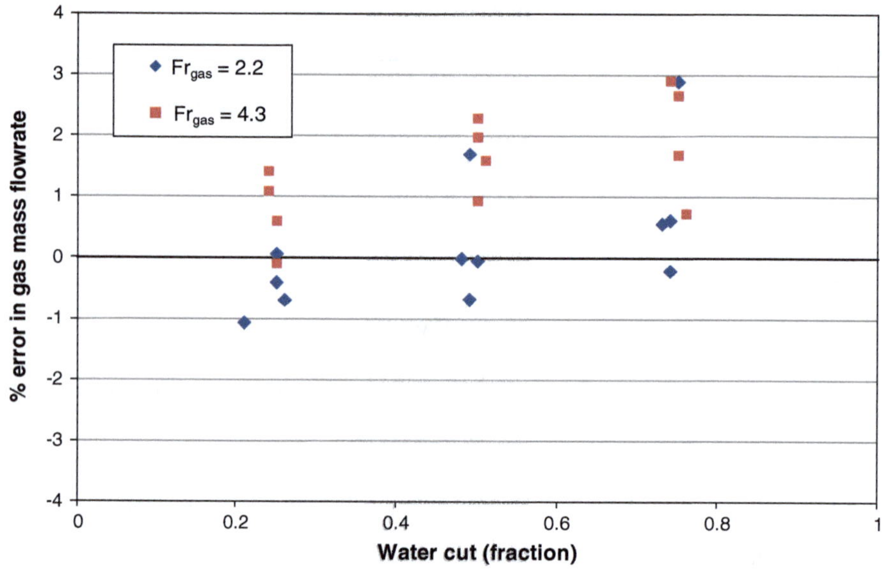

Fig. 11.18 Errors in gas mass flowrate in wet gas with a liquid mixture of Stoddard solvent and water using Eqs. 11.7 and 11.9, as a function of the water cut

The water cut of the liquid phase varied from 21 to 76 %. The value of H used for calculations was based on a linear scale between $H = 1$ for pure liquid hydrocarbon and $H = 1.35$ for pure water.

Figure 11.18 shows the percentage errors in the gas mass flowrate of wet-gas data with a liquid mixture of hydrocarbon and water. It is noticeable that there appears to be a positive error bias when using the correlation, but without the availability of a more extensive range of data it seemed unwise to fit an additional function to the data to determine H instead of interpolating linearly. The errors from using Eqs. 11.7 and 11.9 to determine the gas mass flowrate for this data with a mixed liquid phase are all within 3 %. The Lockhart-Martinelli parameter, X, is in the range 0.02–0.11. This was published by Reader-Harris and Graham (2010).

11.4 Orifice Plates

11.4.1 General

An alternative strategy is to measure wet gas using orifice plates. They are rarely the first choice for installation in wet gas, but wet gas often flows through them, and they do work well. A drain hole through the orifice plate (see Chap. 10) is not required: the equations in Sect. 11.4 have been developed for orifice plates without drain holes. Photographs and videos of wet-gas flow through orifice plates have been taken by CEESI: for the photographs see Steven et al. (2011).

The main concern with using orifice plates is that a slug of liquid might bend them. This risk can be reduced by making them as thick as is permitted by ISO 5167-2:2003, that is with the plate thickness $E = 0.05D$ (or slightly larger for a very small pipe diameter). Moreover, orifice plates are less easily bent if held between flanges rather than in fittings. The effect of using orifice plates with E/D a little larger than 0.05 is quite small (see Sect. 2.2.5.2) in terms of wet-gas measurement uncertainties, provided that the orifice (bore) thickness e meets the requirements in Sect. 2.2.5.4. However, no measurements have been made with thick orifice plates (meaning here ones with E/D a little larger than 0.05) in wet gas. Given that orifice plates generally give smaller uncertainties than Venturi tubes in wet gas they could be an attractive option, but not subsea where they could not be checked or replaced. Work on thick orifice plates in wet gas could be worthwhile.

NOTE There are some data on wet-gas flow through orifice plates with drain holes in Reader-Harris et al. (2008).

11.4.2 Derivation of the Equations in ISO/TR 11583:2012

There are modern data in the public domain on wet-gas flow through orifice plates:

- NEL 8″ $\beta = 0.42, 0.6$ (Reader-Harris et al. 2008)
- NEL 4″ JIP $\beta = 0.66$ (in Steven et al. 2007)
- CEESI 4″ $\beta = 0.34, 0.40, 0.50, 0.68$ (Steven et al. 2007).

The Chisholm model (Chisholm 1967, 1977) is commonly used for orifice plates. To do this Eqs. 11.1, 11.2, 11.4 and 11.5 are used with $n = 0.25$. The errors using the Chisholm model are given in Fig. 11.19. For $Fr_{\text{gas}} > 0.19$ the r.m.s. deviation is 0.96 %.

Rather than use C as determined in the individual projects, it is slightly better (at least for these data) to use the Reader-Harris/Gallagher (1998) Equation of ISO 5167-2:2003 (ISO 2003a) (Eq. 5.22). The errors using the Chisholm model are then shown in Fig. 11.20. For $Fr_{\text{gas}} > 0.19$ the r.m.s. deviation is 0.91 %.

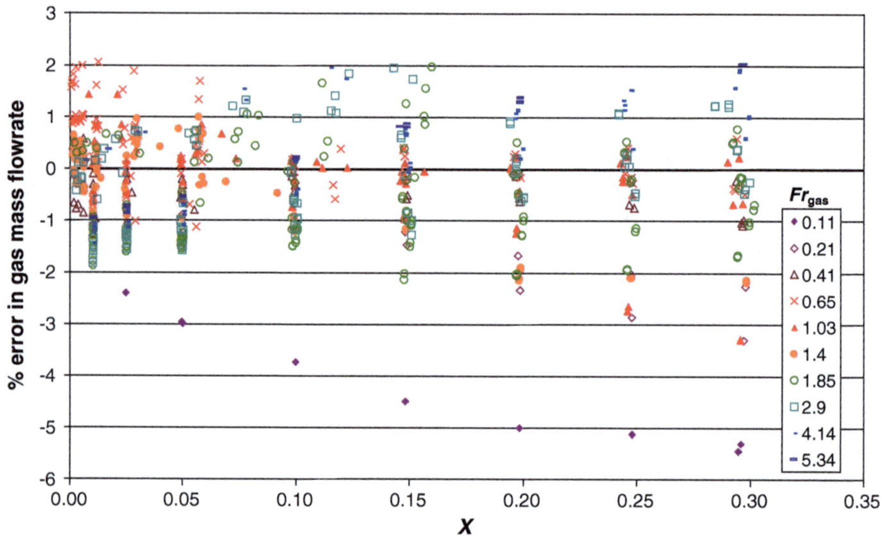

Fig. 11.19 Errors in gas mass flowrate using the Chisholm model with *C* as determined in the individual projects

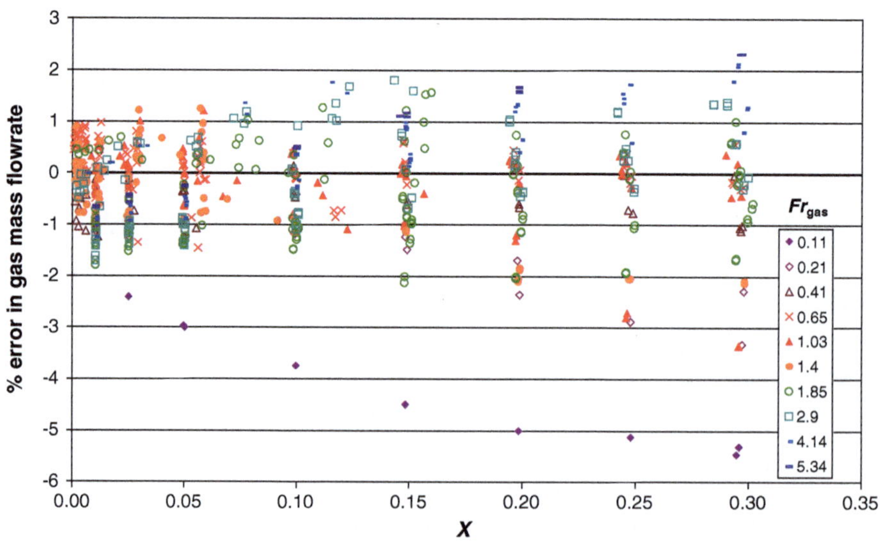

Fig. 11.20 Errors in gas mass flowrate using the Chisholm model with *C* given by the Reader-Harris/Gallagher (1998) Equation

However, there is a dependence on Fr_{gas} in the data shown in Fig. 11.20; so a better fit was obtained of the form

$$n = \max(a - b\mathrm{e}^{-fFr_{gas}}, h).$$

Using the Reader-Harris/Gallagher (1998) Equation for C and omitting data for $Fr_{gas} < 0.19$, the best fit of this type is

$$n = \max(0.4370 - 0.2612\,\mathrm{e}^{-0.134Fr_{gas}},\ 0.2302).$$

The r.m.s. deviation is 0.7834 %.

Using $a = 0.5$ gives homogeneous flow as Fr_{gas} tends to infinity. Then the best fit is

$$n = \max(0.5 - 0.321\,\mathrm{e}^{-0.1Fr_{gas}},\ 0.23). \tag{11.23}$$

Using Eq. 11.23 the errors are shown in Fig. 11.21. The r.m.s. deviation is 0.7835 %.

Similar work using a different database was undertaken by Steven (2006, 2008; Steven et al. 2008), who produced the following equation:

$$n = \begin{cases} 0.214 & 0.2 \le Fr_{gas} \le 1.5 \\ \left(\dfrac{1}{\sqrt{2}} - \dfrac{0.3}{\sqrt{Fr_{gas}}}\right)^2 & Fr_{gas} > 1.5 \end{cases} \tag{11.10}$$

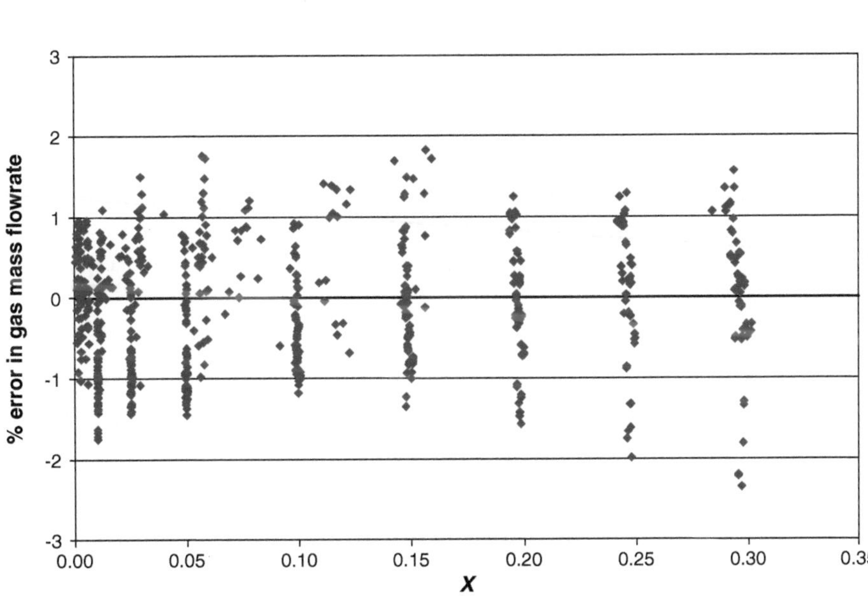

Fig. 11.21 Errors in gas mass flowrate using Eq. 11.23

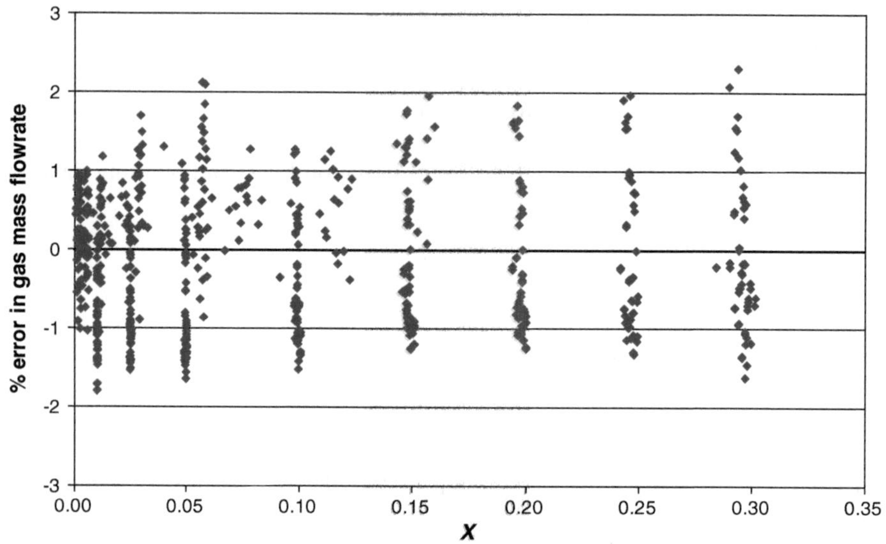

Fig. 11.22 Errors in gas mass flowrate using Eq. 11.10

Using Eq. 11.10 the errors for the data in this section are shown in Fig. 11.22. The r.m.s. deviation is 0.8840 %.

Limits of use are:

- $0.24 \leq \beta \leq 0.73$
- $0 < X \leq 0.3$
- $Fr_{\text{gas}} \geq 0.2$
- $\frac{\rho_{1,\text{gas}}}{\rho_{\text{liquid}}} > 0.014$
- $D \geq 50$ mm

Steven's Equation is included in ISO/TR 11583:2012.

NOTE Steven's Equation as in ISO/TR 11583:2012 gives very good performance on wet steam flow with errors within Table 11.4 even at 3.5 bara for a quadrant-edge orifice plate (Morita et al. 2011).

Table 11.4 The uncertainty in C/ϕ in Eq. 11.1 for an orifice plate using Eq. 11.10

	Liquid in the wet-gas flow	Relative uncertainty in C/ϕ in Eq. 11.1 (%)
X known without error	A light hydrocarbon liquid or water in a wet-steam flow	2
	Water at ambient temperature	3
X obtained from Eq. 11.16	A light hydrocarbon liquid or water in a wet-steam flow	6
	Water at ambient temperature	7

Steven et al. (2007) showed that for large β, even without allowing for Fr_{gas}, quite good estimates can be made for X from the pressure-loss ratio. From his data for $\beta = 0.5$ and 0.68 the equations in Sect. 11.2.4.2 were obtained.

The uncertainty for using these equations was calculated from the data and is given in Table 11.4, which is included in ISO/TR 11583:2012. There had been some indication that the over-reading was smaller in cold water than in a light hydrocarbon liquid.

An uncertainty of 2 % is better than can be obtained with uncalibrated Venturi tubes.

11.4.3 Subsequent Work

On adding to his database data taken with natural gas and water and with natural gas, water and light hydrocarbon liquid Steven obtained the following equation (Steven et al. 2011):

$$n = \begin{cases} \left(\frac{1}{\sqrt{2}} - \frac{0.4-0.1\,\mathrm{e}^{-w}}{\sqrt{1.5+0.2w}}\right)^2 & Fr_{gas} \leq 1.5 + 0.2w \\ \left(\frac{1}{\sqrt{2}} - \frac{0.4-0.1\,\mathrm{e}^{-w}}{\sqrt{Fr_{gas}}}\right)^2 & Fr_{gas} > 1.5 + 0.2w \end{cases} \quad (11.11)$$

where w is the water liquid mass ratio, i.e. the ratio of the water mass flowrate to the total liquid mass flowrate. The uncertainty in C/ϕ is 2 %.

Equation 11.10 is a special case of Eq. 11.11. All the wet-gas data in Steven et al. (2011), both those in which the liquid was oil and those in which it was water, are consistent with the uncertainty statements made in ISO/TR 11583:2012: the errors in the oil data are within 2 %; the water data are biased but the errors are less than 3 %. It should be possible in due course to improve ISO/TR 11583:2012 to include Eq. 11.11.

11.5 Conclusions

The equations for wet-gas measurement in ISO/TR 11583:2012 have been described, together with some material towards their derivation.

NOTE ISO/TR 11583:2012 contains two examples of how the calculations in it might be carried out, one without and one with an additional pressure tapping to measure the pressure loss.

Appendix 11.A: A Brief History of ISO/TR 11583

The project to publish ISO/TR 11583 started with a New Work Item Proposal sent out to the members of ISO/TC 30/SC 2 in September 2007. It stated that the proposed 'Technical Report will cover the measurement of wet gas flow with Venturi tubes or orifice plates. It will cover two-phase flows of gas and liquid in which the flowing fluid mixture is approximately 95 % gas by volume or more. Gases covered include air, nitrogen, natural gas and steam. Liquids covered include water, kerosene and condensate. It will be an extension to ISO 5167.' Item 6 in its outline of the proposed TR was 'Correlations for the over-reading'. This New Work Item Proposal was accepted in early 2008. To the statement 'We agree that a globally relevant International Standard on this subject is feasible and therefore agree to the addition of the proposed new work item to the program of work of the committee' there were 11 positive votes (including those of Norway and the USA) and 2 abstentions and no negative votes. One US expert was nominated in 2007 and two more in October 2008.

ISO/TR 11583 is based on guidelines prepared by technical experts for the UK Department of Trade and Industry (2003), and was developed with the input of different experts on different sections.

At around the same time ISO/TC 193 (natural gas) had identified a requirement for guidance on the measurement of wet natural gas. ISO/TC 30 proposed that TC 30 develop a part of ISO/TR 11583 addressing generic aspects of wet-gas flow (covering wet natural gas and wet steam among other wet gases), and that ISO/TC 193/SC 3 address aspects specific to natural gas production applications in a subsequent part. This structure was accepted in principle in 2008: in an e-mail to the TC 193/SC 3 members dated 11 September 2008 the TC 193/SC 3 Secretary wrote, 'This document [the then draft of ISO/TR 11583-1] serves as an initial effort to call for experts from both committees. Please circulate this draft document among your experts, and if possible, please nominate experts who could participate on the JWG on behalf of ISO/TC 193/SC 3. The intent of this draft document is to serve as a generic Part 1 to a multi-part document on wet gas.'

ISO/TC 30 worked in accordance with this arrangement, but, when the wet hydrocarbon gas part was not written, it completed ISO/TR 11583 as a stand-alone document.

The correlations for Venturi tubes that appear in ISO/TR 11583:2012 were first circulated to ISO/TC 30/SC 2/WG 15 in February 2009. They were published at the North Sea Workshop in 2009 (Reader-Harris and Graham 2009). A draft TR containing these correlations was circulated to SC 2 in August 2009. The correlation for orifice plates was taken from Steven et al. (2008).

ISO/TC 193 expressed concern about acceptance of the Venturi-tube correlation. Accordingly a meeting of experts from ISO/TC 30 and ISO/TC 193 was held at TC 193's request in November 2009. One particular concern was that the NEL data (which were the largest part of the data set used to develop the correlations) were not available for Working Group members to check the correlation. In June 2010 the data were made available to WG members subject to their agreeing to maintain confidentiality and to limit the use of the data solely to the validation process. The Venturi-tube correlation in ISO/TR 11583:2012 was in much better agreement with the database than that of de Leeuw. The ballot on ISO/TR 11583 was started in September 2010 and closed in December 2010. This passed by 6 votes to 4.

ISO/TR 11583 was at the proof stage in July 2011 when ISO/TC 193 submitted an appeal to ISO Central Secretariat and publication was stopped. Several inconclusive meetings were held: TC 30 regarded the document as generic and so within its scope and wished to publish it. TC 193 maintained the document was flawed.

Three experts involved in TC 193 published a paper at the North Sea Flow Measurement Workshop in 2011 giving arguments against its publication (de Leeuw et al. 2011).

A reply to all the points raised was sent to the authors of de Leeuw et al. (2011) in January 2012; however, no response was received. A reply was published at the North Sea Flow Measurement Workshop in 2012 (Reader-Harris 2012).

Meanwhile, ISO/TC 28 and ISO/TC 193 made a representation to the ISO Technical Management Board (TMB) requesting that oil and gas be excluded from the scope of ISO/TC 30. The TMB's resolution was that 'It reminds ISO/TC 30 that they should develop generic standards on flow measurement, and decides that further sectorial (sic) needs of the oil and gas industry should be addressed by ISO/TC 28 and ISO/TC 193 taking account of the generic standards from ISO/TC 30.' This is consistent with TC 30's understanding of its objectives. However, the TMB also required that the following words be added to ISO/TR 11583: 'This Technical Report is not intended for the oil and gas industry.' ISO/TR 11583:2012 (ISO 2012) was published including these additional words.

As the UK National Committee, BSI CPI/30/2, believed that the methods and correlations in ISO/TR 11583:2012 are generic and intended for all industries, it added the following text to the UK adoption of the ISO technical report, PD ISO/TR 11583:2012 (BSI 2012): 'The UK national committee, CPI/30/2, consider the content of this technical report to be suitable for general application in wet gas measurement using differential pressure meters.'

The main criticisms of the Venturi correlations in ISO/TR 11583 given by de Leeuw et al. (2011) and the response of Reader-Harris (2012) were as follows:

- **Comment:** the understanding of the effects of physical properties is inadequate.
- **Response:** ISO/TR 11583 is careful to state what fluids were used for the test work and not to imply that the correlations can be used for any fluids.
- **Comment:** the correlations have an incorrect behaviour in limiting cases.
- **Response:** several limits have been tested and the correlation gives good results in these cases.
- **Comment:** the correlation is only applicable to two-component flows (e.g. gas/oil) not three-component flows (e.g. gas/water/oil).
- **Response:** there are insufficient data for three-component flows in the public domain to determine whether the correlation (with additional text as to how it should be used) is adequate; so ISO/TR 11583 makes no claim that it is adequate for three-component flows (but see also Sect. 11.3.3).

- **Comment:** the term 'discharge coefficient' should not be used for anything other than the dry-gas discharge coefficient.
- **Response:** it is found that the substantial change that occurs in C/ϕ as X changes from 0 to 0.025 is most easily described as a change in C. It is, of course, desirable to distinguish clearly between dry-gas and wet-gas discharge coefficients. The TR makes clear which discharge coefficient is to be used.
- **Comment:** the most important criticism is that there are data sets which the correlation does not fit.
- **Response:** all data sets available to ISO/TC 30/SC 2/WG 15 were included and fit. The subsequent data provided in graphical form in de Leeuw et al. (2011) have been examined:

  - The Venturi tube supplied by NAM was tested both at CEESI and at SINTEF. The data taken at CEESI fit the correlation in ISO/TR 11583. Earlier data taken at SINTEF do not fit the correlation, and do not agree with the CEESI data: the explanation for this difference is not known.
  - The ISO/TR 11583 Correlation fits the vast majority of the K-Lab data (a small number of the data points are interesting).
  - Although data from a 12″ Venturi tube do not fit the ISO/TR 11583 Correlation, this Venturi tube had a dry-gas discharge coefficient of 0.936, a value well outside the expectations of ISO 5167-4:2003 (ISO 2003b). ISO/TR 11583 applies to devices that (except for the number of pressure tappings) meet the requirements of ISO 5167-4:2003.
  - The ISO/TR 11583 Correlation fits the recently published data for the "ISO-compliant" 8″ Venturi tube from CEESI.
  - In the opinion of the author of this paper the ISO/TR 11583 Correlation also fits the recently published data for the non-ISO-compliant 8″ Venturi tube from CEESI: see NOTE 2.

NOTE 1 The ISO/TR 11583:2012 Correlation fits the new data for an 8″ $\beta = 0.6$ Venturi tube in Gajan et al. (2013).
NOTE 2 CEESI took and published (Steven et al. 2012) data on wet-gas flow through two 8″ $\beta = 0.6$ Venturi tubes. An "ISO-compliant" Venturi tube was the first to be tested. In fact its upstream tapping was $1D$ from the start of the convergent, rather than $0.5D$, but the effect of this is small. The second device was a non-compliant Venturi tube (the radius of the corner upstream of the throat tapping is 14 times the maximum permitted for the compliant Venturi tube).

Both were tested using natural gas/oil mixtures and the results were consistent with ISO/TR 11583:2012.

The non-compliant meter was also tested using natural gas/water. In the paper graphs are given showing both the uncorrected over-reading data and corrected results. On the basis of the corrected results a significant fraction of the natural gas/water data for the non-compliant Venturi tube lie outside the range promised by ISO/TR 11583:2012 and it is stated that 'the results strengthen the position of ISO where they state that ISO TC30 TR11583 is not an appropriate document for use in the oil and gas industry'. However, if the uncorrected over-reading data in the same paper are correct and corrections according to ISO/TR 11583:2012 are carried out by the author of this book, then the corrected errors are within the band promised by ISO/TR 11583:2012.

There is insufficient information given to enable the data for wet gas with a mixture of liquids to be recalculated, but it appears very likely that if the corrected errors were calculated correctly (even using a simple linear fit for H in terms of water cut) the errors would lie within the range given in ISO/TR 11583:2012.

Appendix 11.B: Dependence of the Wet-Gas Correlations for Venturi Tubes on Liquid Viscosity

11.B.1 General

It was desirable to show that the wet-gas correlations for Venturi tubes depend more on surface tension than on the liquid viscosity. However, tests of the effect of changing surface tension were hard to undertake, since most oils have similar surface tension. It was, however, possible to examine the effect of liquid viscosity and to attempt to check that the effect of liquid viscosity is small over the range of liquids found in common wet-gas applications. A liquid with similar surface tension to kerosene but lower viscosity was not available for test. So three 4″ Venturi tubes with $\beta = 0.4$, 0.6 and 0.75 were tested in the NEL National Standard wet-gas facility with nitrogen and velocite (a refined hydraulic oil). The Venturi tube with $\beta = 0.6$ was also tested with nitrogen and gasoil (a blend of kerosene and hydraulic oil). The three Venturi tubes had previously been tested with nitrogen and kerosene (and indeed in the case of $\beta = 0.6$ and $\beta = 0.75$ with argon and kerosene and with nitrogen and water). The particular brand of kerosene was Exxsol D80. Kerosene, gasoil and velocite have kinematic viscosities of approximately 2.5 cSt, 6.3 cSt and 23.5 cSt, respectively, at 20 °C. Data were collected at three nominal static pressures, 15 barg, 30 barg and 60 barg; no data are presented for $\beta = 0.4$ at 60 barg.

Full details are given in Reader-Harris (2012).

Glycol has a surface tension between those of water and kerosene, but a higher viscosity than either; so tests with glycol would not have clearly shown the effect of liquid viscosity or that of surface tension.

11.B.2 Deviations from the ISO/TR 11583:2012 Correlation

The errors in gas mass flowrate using the ISO/TR 11583:2012 Correlation are given in Figs. 11.B.1, 11.B.2, 11.B.3 and 11.B.4. The data for $\beta = 0.6$ for each liquid are given first, then the data for the other values of β.

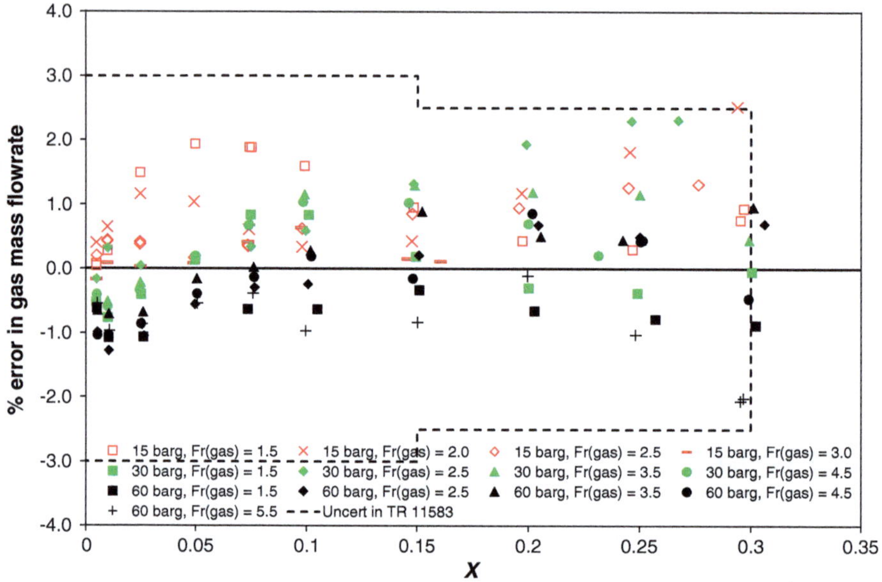

Fig. 11.B.1 Errors in gas mass flowrate using the ISO/TR 11583:2012 correlation: nitrogen and gasoil: $\beta = 0.6$

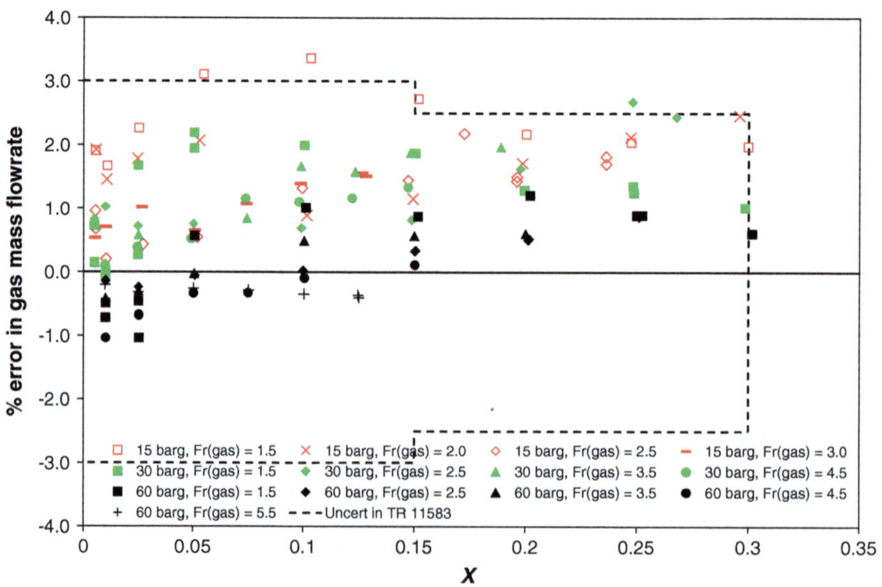

Fig. 11.B.2 Errors in gas mass flowrate using the ISO/TR 11583:2012 correlation: nitrogen and velocite: $\beta = 0.6$

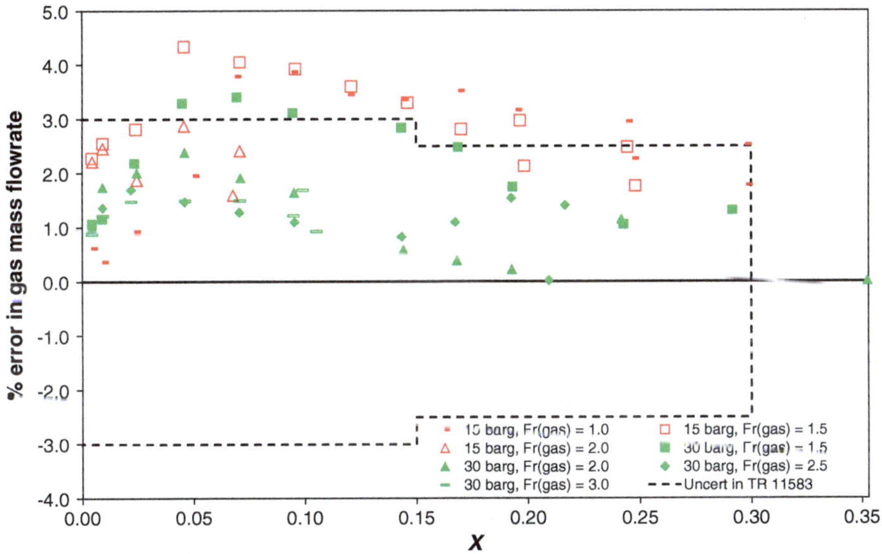

Fig. 11.B.3 Errors in gas mass flowrate using the ISO/TR 11583:2012 correlation: nitrogen and velocite: $\beta = 0.4$

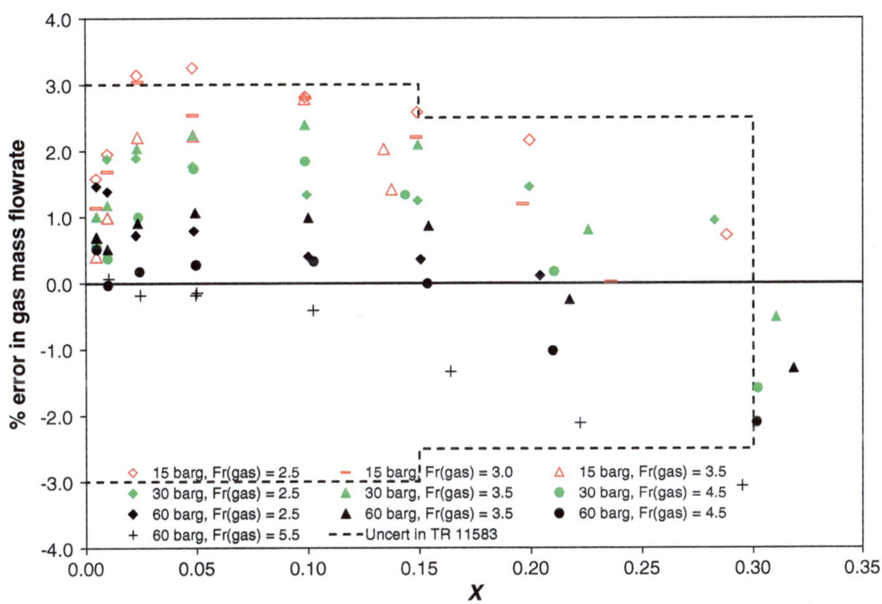

Fig. 11.B.4 Errors in gas mass flowrate using the ISO/TR 11583:2012 correlation: nitrogen and velocite: $\beta = 0.75$

ISO/TR 11583:2012 lists the fluids used to create the correlation and then adds 'It is possible that the equations do not apply to liquids significantly different from those tested, particularly to highly viscous liquids'. However, the errors using the ISO/TR 11583:2012 Correlation with liquids more viscous than those used to derive the correlation were remarkably satisfactory. For the gasoil set 99 % of the points lie within the error bands that would have been expected for less viscous liquids. Of the velocite data excluding only the data for Fr_{gas} of 1.5 or below at 15 barg 96 % of the errors lie within the error bands. Presumably for Fr_{gas} of 1.5 or below at 15 barg there is a thicker layer of velocite on the wall than that expected by the Correlation.

11.B.3 Deviations from the de Leeuw Equation

The errors in gas mass flowrate using the de Leeuw Equation are given in Figs. 11.B.5, 11.B.6, 11.B.7 and 11.B.8.

The r.m.s. deviation of the errors using the de Leeuw Equation is on average around twice that of the r.m.s. deviation using the ISO/TR 11583:2012 Correlation.

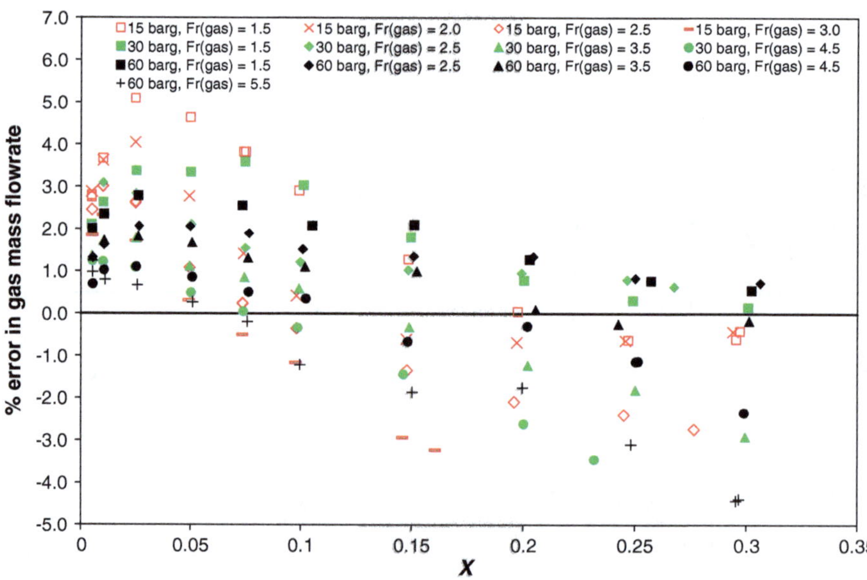

Fig. 11.B.5 Errors in gas mass flowrate using the de Leeuw Equation: nitrogen and gasoil: $\beta = 0.6$

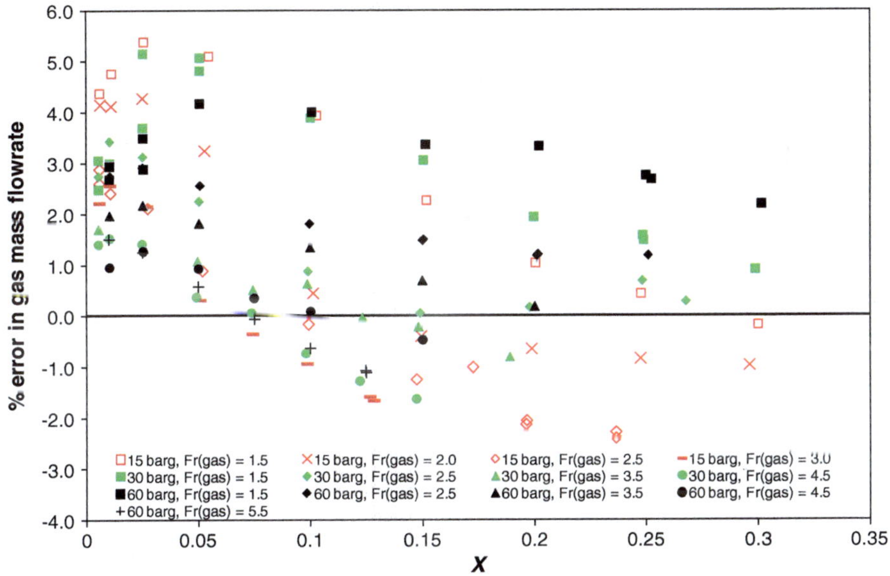

Fig. 11.B.6 Errors in gas mass flowrate using the de Leeuw Equation: nitrogen and velocite: $\beta = 0.6$

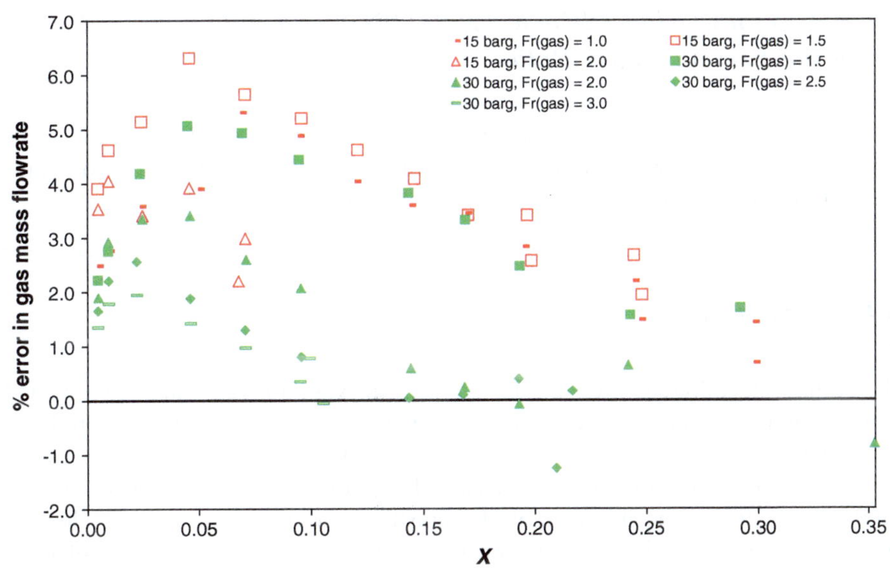

Fig. 11.B.7 Errors in gas mass flowrate using the de Leeuw Equation: nitrogen and velocite: $\beta = 0.4$

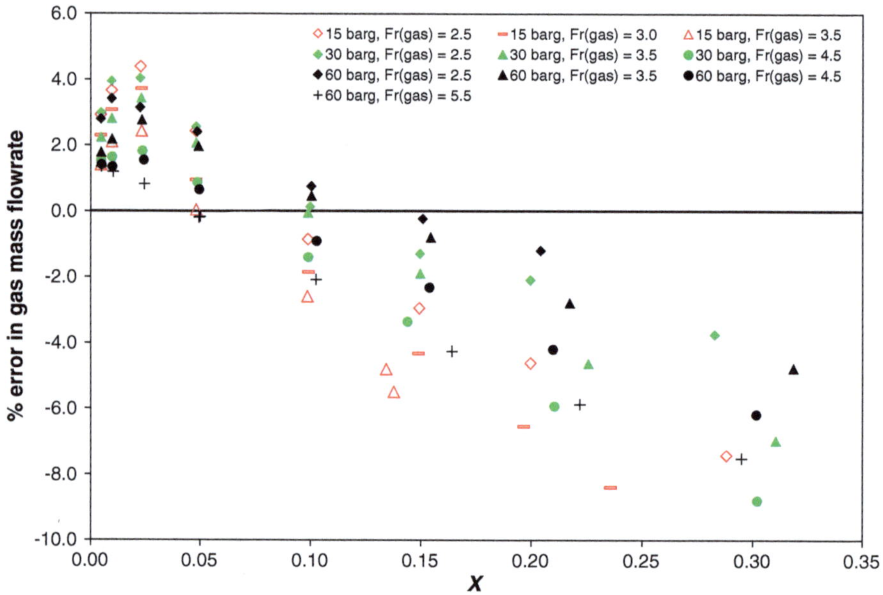

Fig. 11.B.8 Errors in gas mass flowrate using the de Leeuw Equation: nitrogen and velocite: $\beta = 0.75$

11.B.4 Errors Using ISO/TR 11583:2012 with X Determined from the Pressure Loss Ratio

Using the downstream pressure tapping and Eqs. 11.7–11.9 and 11.12–11.14 the errors in gas mass flowrate were obtained. All the points lay within the error bands in ISO/TR 11583:2012 and are presented in Reader-Harris (2012).

11.B.5 Analysis

To evaluate the difference between the error in gas mass flowrate using nitrogen and velocite and that using nitrogen and kerosene the nitrogen/kerosene data were fitted as a function of X for each value of pressure and Fr_{gas} and calculated values subtracted from the nitrogen/velocite data. The same exercise was carried out for nitrogen/gasoil data: in this case at 60 barg the difference between the values of Fr_{gas} in nitrogen/gasoil and nitrogen/kerosene at the same nominal value of Fr_{gas} was such that corrections for the effect of this difference were applied. The differences in error are given in Figs. 11.B.9, 11.B.10, 11.B.11 and 11.B.12.

Although there is a substantial scatter in Figs. 11.B.9, 11.B.10, 11.B.11 and 11.B.12, at least for lower X the error increases as Fr_{gas} reduces and is fairly

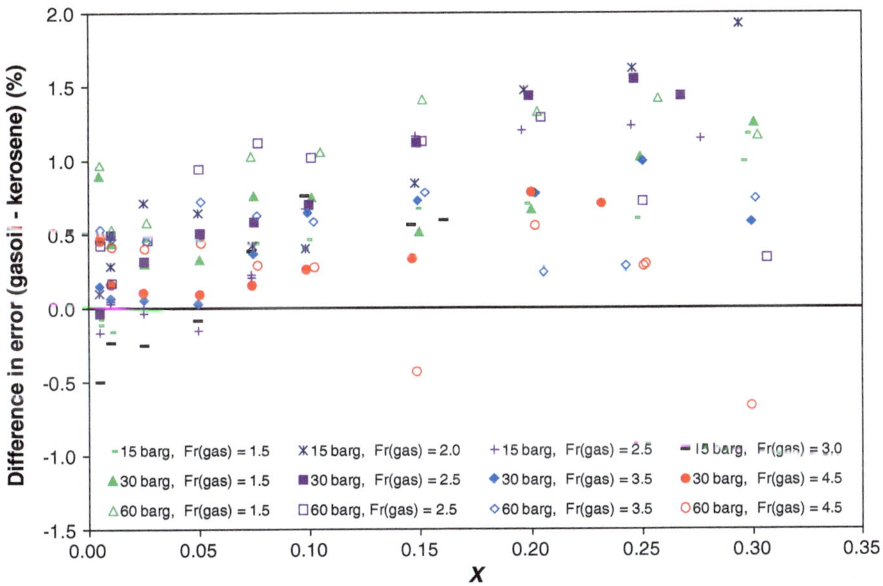

Fig. 11.B.9 Difference in the error using the ISO/TR 11583:2012 correlation when gasoil replaced kerosene: $\beta = 0.6$

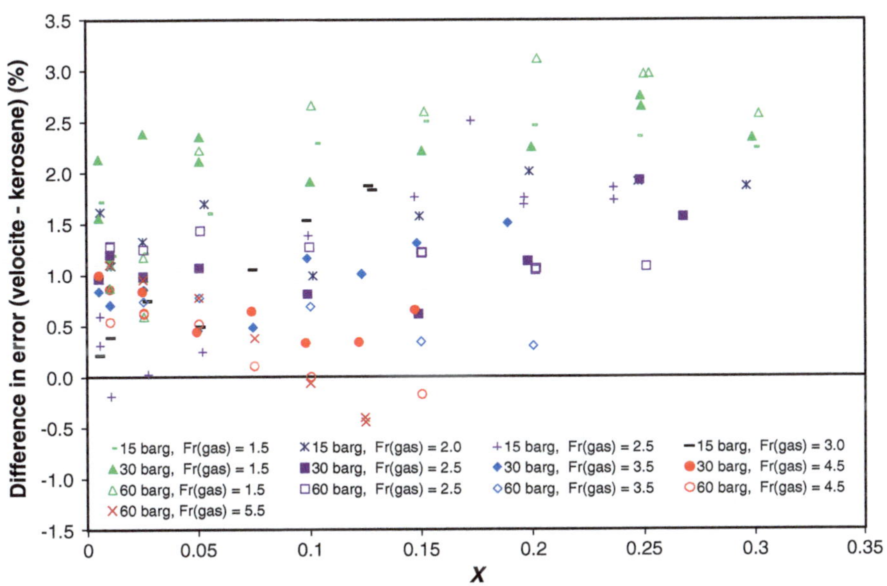

Fig. 11.B.10 Difference in the error using the ISO/TR 11583:2012 correlation when velocite replaced kerosene: $\beta = 0.6$

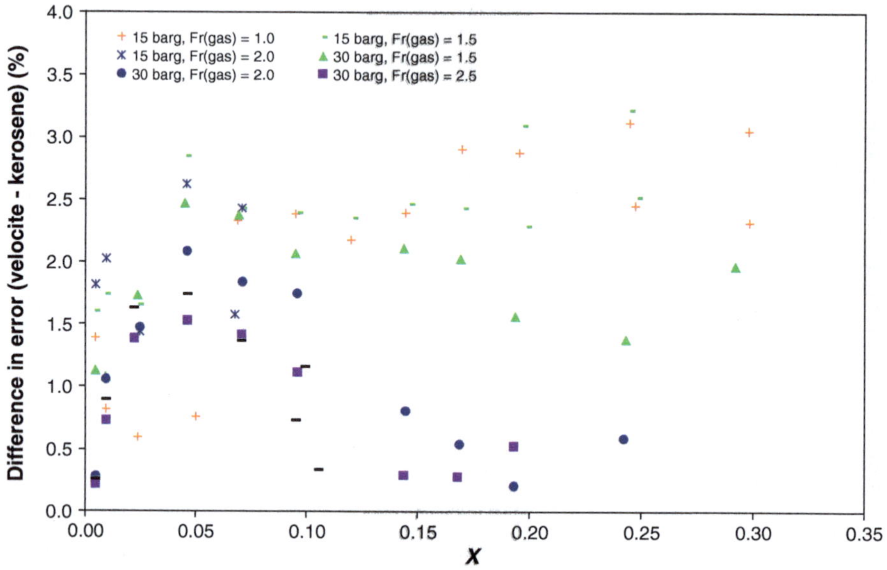

Fig. 11.B.11 Difference in the error using the ISO/TR 11583:2012 correlation when velocite replaced kerosene: $\beta = 0.4$

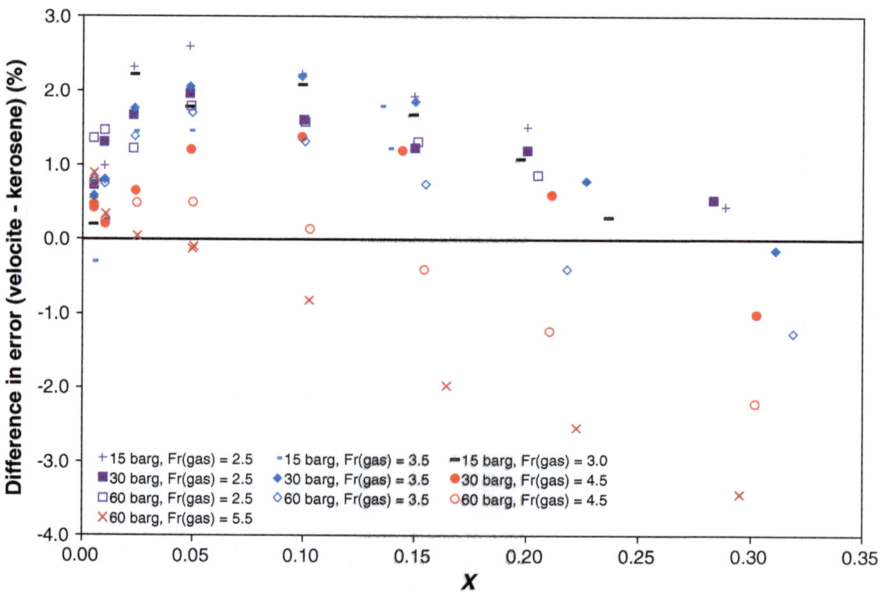

Fig. 11.B.12 Difference in the error using the ISO/TR 11583:2012 Correlation when velocite replaced kerosene: $\beta = 0.75$

constant for X from 0.025 to 0.1. It appears that at least for X up to 0.1 the main change to the flow is that there is a thicker layer of oil on the wall and so C takes a lower value than that in Eq. 11.7. This is consistent with the model in ISO/TR 11583:2012, in which there is a wet-gas discharge coefficient. To provide an approximate correction for the equation the average of the following was determined for X up to a nominal value of 0.1:

$$d_{\text{velocite-kerosene}} = \frac{e_{\text{velocite}} - e_{\text{kerosene}}}{\min\left(1, \sqrt{\frac{X}{0.016}}\right)} \tag{11.B.1}$$

where e_{velocite} is the percentage error using velocite and nitrogen and e_{kerosene} that using kerosene and nitrogen. The averages are in Fig. 11.B.13. The fitted equation is

$$d_{\text{velocite-kerosene}} = 3.21\,\mathrm{e}^{-0.33Fr_{\text{gas}}} \tag{11.B.2}$$

As $\rho_{1,\text{gas}}/\rho_{\text{liquid}} \to 1$, $Fr_{\text{gas}} \to \infty$, and the effect of changing the liquid tends to 0.

A similar calculation was carried out for the gasoil data. The data are shown in Fig. 11.B.14. The fit is

$$d_{\text{gasoil-kerosene}} = 1.03\,\mathrm{e}^{-0.33Fr_{\text{gas}}} \tag{11.B.3}$$

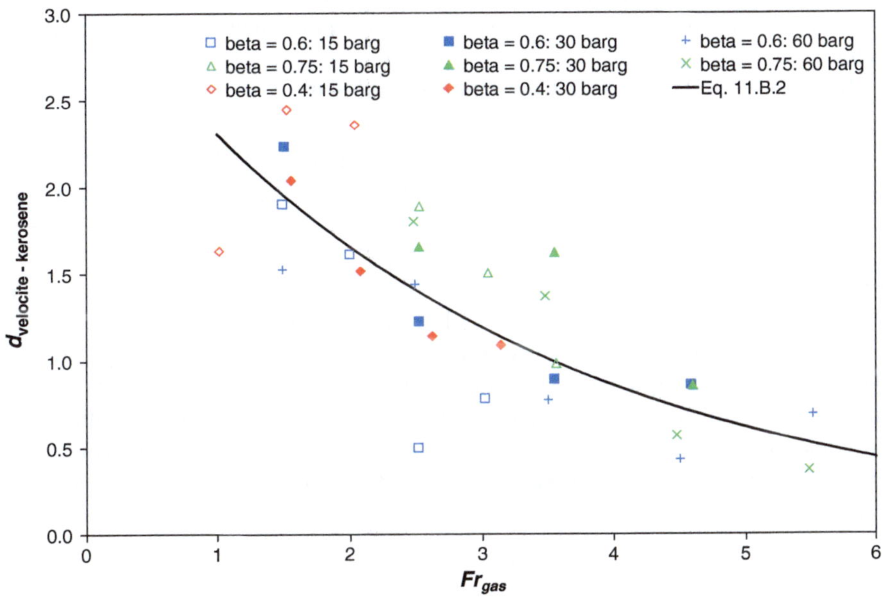

Fig. 11.B.13 Average difference in the percentage error in gas mass flowrate for X up to a nominal value of 0.1 when kerosene was replaced by velocite

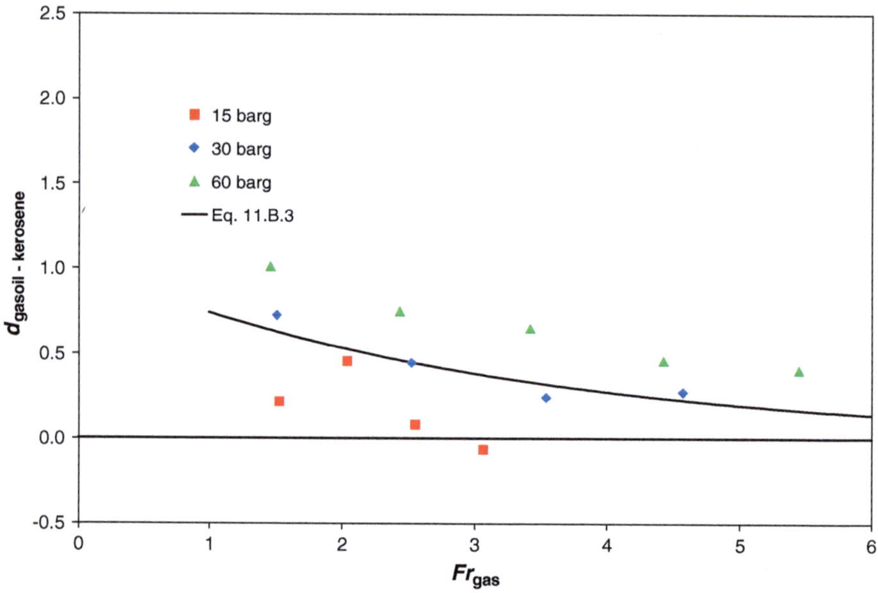

Fig. 11.B.14 Average difference in the percentage error in gas mass flowrate for X up to a nominal value of 0.1 when kerosene was replaced by gasoil

There is a dependence on β in this case, but the effect is quite small. The effect of changing from kerosene to gasoil is 31 % of the effect of changing from kerosene to velocite.

If the ISO/TR 11583:2012 Correlation is used with $H = 1$ and

$$C = \frac{1 - 0.0463\,\mathrm{e}^{-0.05 Fr_{\mathrm{gas},th}} \min\left(1, \sqrt{\frac{X}{0.016}}\right)}{1 + a\mathrm{e}^{-0.33 Fr_{\mathrm{gas}}} \min\left(1, \sqrt{\frac{X}{0.016}}\right)} \tag{11.B.4}$$

where $a = 0.0321$ for velocite and 0.0103 for gasoil, then the errors in Figs. 11.B.15, 11.B.16, 11.B.17 and 11.B.18 are obtained.

It is not intended to introduce Eq. 11.B.4 to ISO/TR 11583:2012. It was determined in order to quantify the effect of liquid viscosity. It has also shown that the main effect of liquid viscosity is to change the wet-gas discharge coefficient. If an equation for wet-gas flow with more viscous oils were required then further analysis would be required since Eq. 11.B.4 alone appears insufficient for $Fr_{\mathrm{gas}} = 4.5$ or 5.5 at high X.

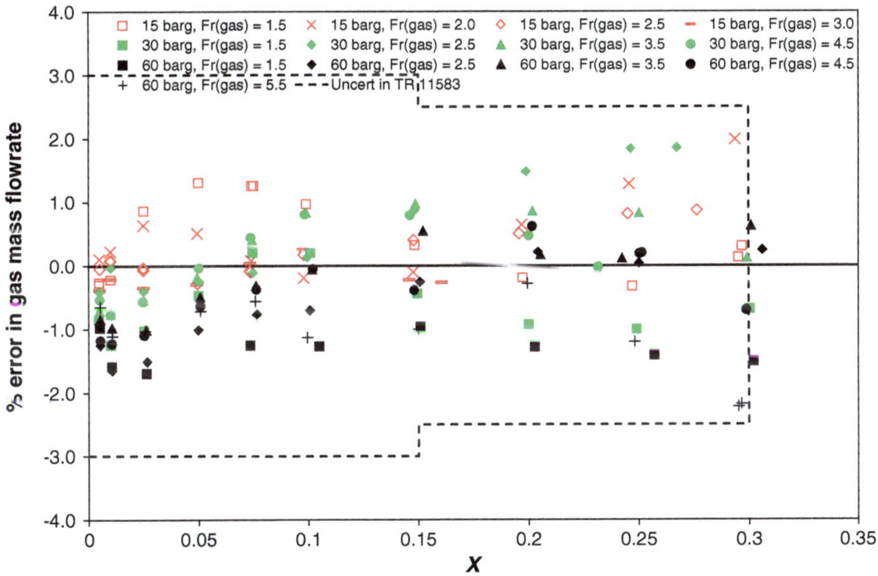

Fig. 11.B.15 Errors in gas mass flowrate using the ISO/TR 11583:2012 correlation with C as in Eq. 11.B.4: nitrogen and gasoil: $\beta = 0.6$

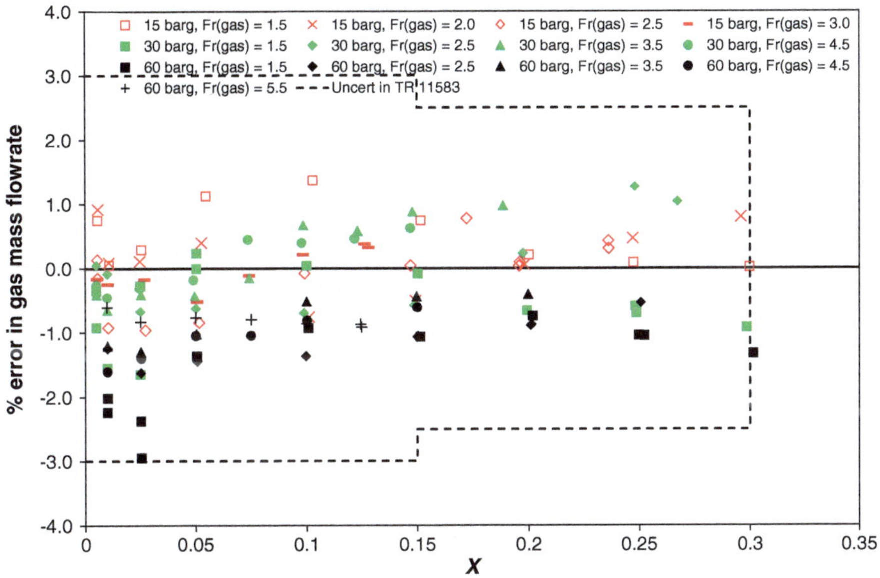

Fig. 11.B.16 Errors in gas mass flowrate using the ISO/TR 11583:2012 correlation with C as in Eq. 11.B.4: nitrogen and velocite: $\beta = 0.6$

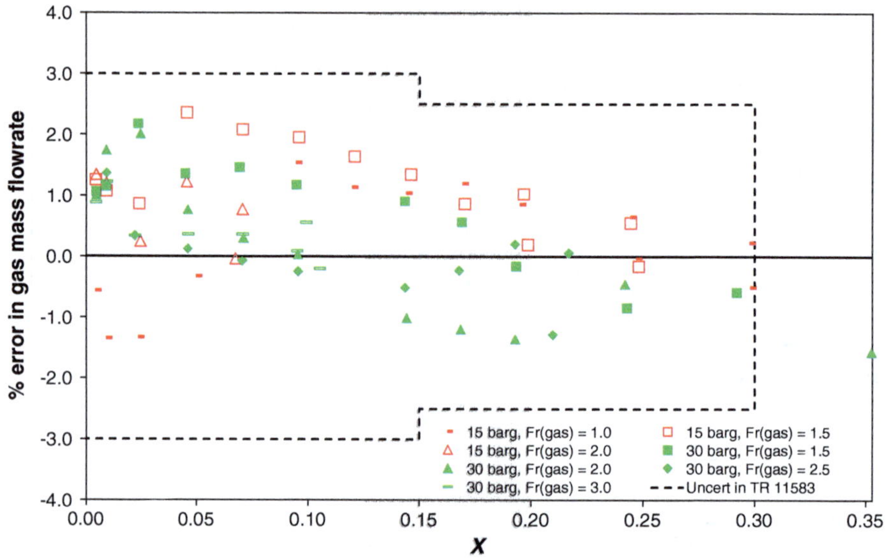

Fig. 11.B.17 Errors in gas mass flowrate using the ISO/TR 11583:2012 correlation with *C* as in Eq. 11.B.4: nitrogen and velocite: $\beta = 0.4$

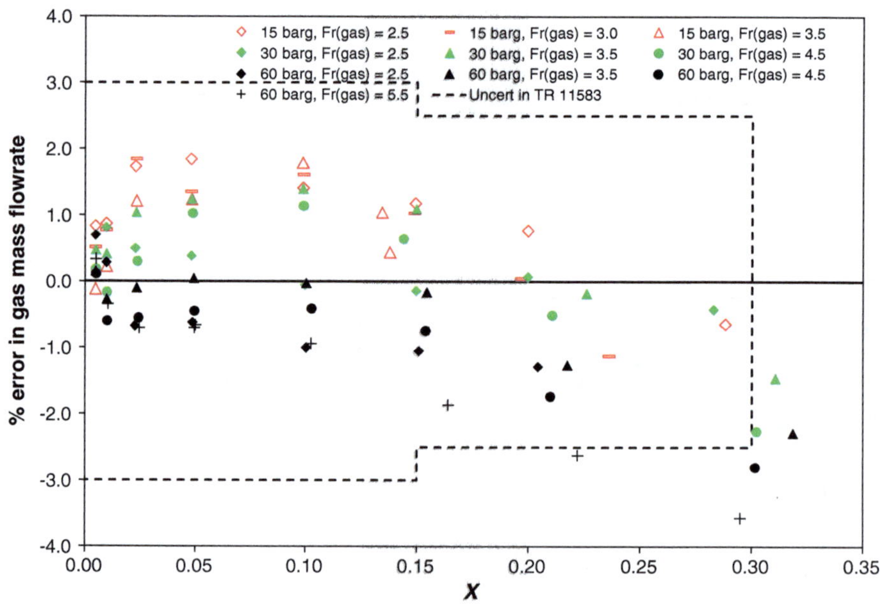

Fig. 11.B.18 Errors in gas mass flowrate using the ISO/TR 11583:2012 correlation with *C* as in Eq. 11.B.4: nitrogen and velocite: $\beta = 0.75$

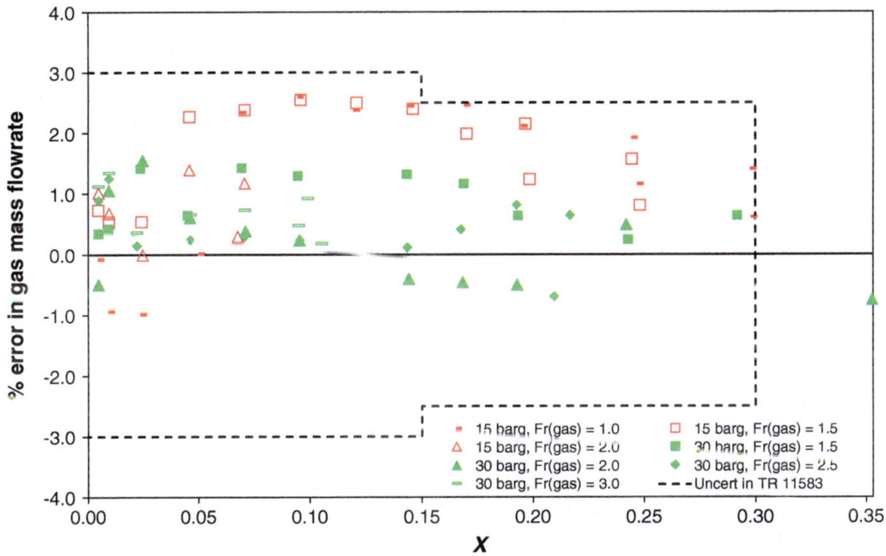

Fig. 11.B.19 Errors in gas mass flowrate using the ISO/TR 11583:2012 correlation with C as in Eq. 11.B.4: nitrogen and velocite: $\beta = 0.4$: *horizontal* tappings

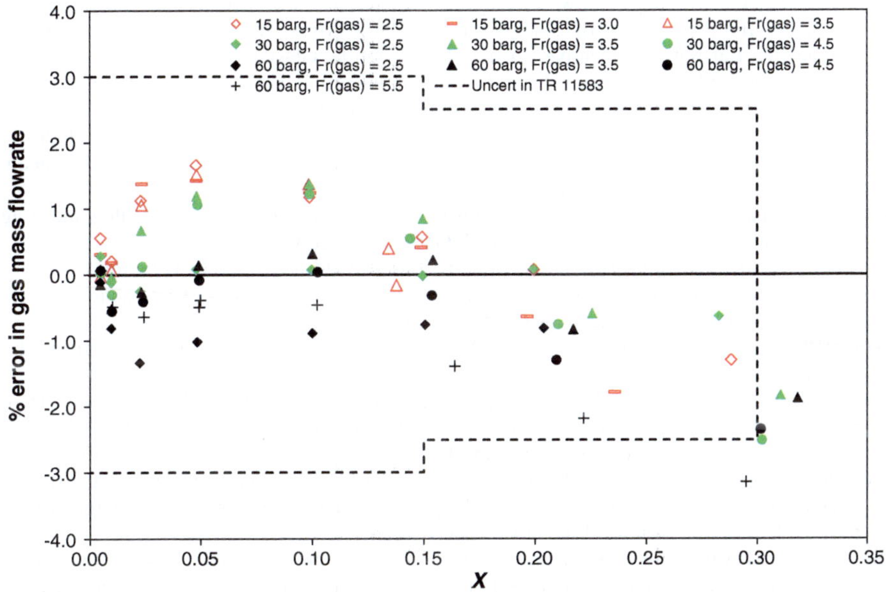

Fig. 11.B.20 Errors in gas mass flowrate using the ISO/TR 11583:2012 correlation with C as in Eq. 11.B.4: nitrogen and velocite: $\beta = 0.75$: *horizontal* tappings

11.B.6 Horizontal Tappings

Given the interest in measuring wet steam it was important to check that if horizontal tappings instead of tappings on the top of the pipe were used ISO/TR 11583:2012 would still prove satisfactory. Figures 11.B.19 and 11.B.20 (using horizontal tappings) are not very different from Figs. 11.B.17 and 11.B.18 (using tappings on the top of the pipe).

References

Britton CL, Kinney J, Steven R (2008) Liquid property and diameter effects on DP meter wet gas over-readings. In: Proceedings of 26th International North Sea Flow Meas Workshop, St Andrews, paper 7.2. NEL, East Kilbride, Glasgow

BSI (2012) Measurement of wet gas flow by means of pressure differential devices inserted in circular cross-section conduits. PD ISO/TR 11583:2012. British Standards Institution, London

Chisholm D (1967) Flow of incompressible two-phase mixtures through sharp-edged orifices. J Mech Eng Sci 9(1):72–78

Chisholm D (1977) Research note: two-phase flow through sharp-edged orifices. J Mech Eng Sci 19(3):128–130

de Leeuw H (1997) Liquid correction of Venturi meter readings in wet-gas flow. In: Proceedings of 15th North Sea Flow Meas Workshop, Norway, paper 21

de Leeuw H (1994) Wet-gas flow measurement by means of a Venturi meter and a tracer technique. In: Proceedings of 12th North Sea Flow Meas Workshop, Peebles. National Engineering Laboratory Executive Agency, East Kilbride, Glasgow

de Leeuw R, Steven R, van Maanen H (2011) Venturi meters and wet gas flow. In: Proceedings of 29th International North Sea Flow Meas Workshop, Norway

Department of Trade and Industry (2003) Module 7 of Department of Trade and Industry Licensing and Consents Unit Guidance Notes for petroleum measurement under the Petroleum (Production) Regulations, December 2003, Issue 7

Gajan P, Decaudin Q, Couput JP (2013) Analysis of high pressure tests on wet gas flow metering with a Venturi meter. In: Proceedings of FLOMEKO 2013, Paris

Graham E, Reader-Harris M, Hodges D, Hone R, Barrie A, Ramsay N (2014) Performance of a vertically installed Venturi tube in wet-gas conditions. In: Proceedings of International South East Asia Hydrocarbon Flow Meas Conference, Kuala Lumpur

Hall ARW, Reader-Harris MJ (1999) Use of Venturi meters in multiphase flow measurement. In: Proceedings of 17th North Sea Flow Meas Workshop, Norway

Harris DM, Shires GL (1972) Two-phase pressure drop in a Venturi. In: Proceedings of Two-phase flow through orifices and nozzles. Report of a Meeting at NEL, 29 Nov 1972. Paper 4. NEL Report No 549. National Engineering Laboratory, East Kilbride, Glasgow, 1973

ISO (2003a) Measurement of fluid flow by means of pressure differential devices inserted in circular cross-section conduits running full—part 2: orifice plates. international Organization for Standardization, Geneva. ISO 5167-2:2003

ISO (2003b) Measurement of fluid flow by means of pressure differential devices inserted in circular cross-section conduits running full—part 4: Venturi tubes. international Organization for Standardization, Geneva. ISO 5167-4:2003

ISO (2012) Measurement of wet gas flow by means of pressure differential devices inserted in circular cross-section conduits. international Organization for Standardization, Geneva. ISO/TR 11583:2012

ISO (2014) Natural gas—wet gas flow measurement in natural gas operations. international Organization for Standardization, Geneva. ISO/DTR 12748:2014
Morita R, Inada F, Uchiyama Y, Umezawa S, Sakai M, Amari H, Shimada H, Funaki T, Ishibashi M, Takamoto M (2011) Clarification of measurement error of orifice flow meter in wet steam flow. In: Proceedings of ASME 2011 Power conference (Power 2011) and international conference power engineering 2011 (ICOPE-11), Denver
Murdock JW (1961) Two-phase flow measurement with orifice. ASME Paper 61-WA-27
Reader-Harris MJ (2012) Wet-gas measurement: ISO/TR 11583. In: Proceedings of 30th International North Sea Flow Meas Workshop, St Andrews. NEL, East Kilbride, Glasgow
Reader-Harris MJ, Graham E (2009) An improved model for Venturi-tube over-reading in wet gas. In: Proceedings of 27th International North Sea Flow Meas Workshop, Norway
Reader-Harris MJ, Graham E (2010) Venturi tubes in wet gas—improved models for the over-reading and the pressure-loss ratio method. In: Proceedings of 9th International South East Asia Hydrocarbon Flow Meas Workshop, Kuala Lumpur
Reader-Harris MJ, Gibson J, Hodges D (2007) Venturi-tube performance in wet gas: computation and experiment. In: Proceedings of 6th International South East Asia Hydrocarbon Flow Meas Workshop, Kuala Lumpur
Reader-Harris MJ, Hodges D, Gibson J (2005) Venturi-tube performance in wet gas using different test fluids. In: Report No 2005-206, on DTI Flow Programme Project FEWG01. National Engineering Laboratory, East Kilbride, Glasgow
Reader-Harris MJ, Hodges D, Gibson J (2006) Venturi-tube performance in wet gas using different test fluids. In: Proceedings of 24th International North Sea Flow Meas Workshop, St Andrews, paper 7.1. National Engineering Laboratory, East Kilbride, Glasgow
Reader-Harris MJ, Hodges D, Barton N, Graham E (2008) Improvements in the measurement of gas flows with entrained liquids using orifice plates. Report No 2008/265 on Project No FFRE45. TUV NEL, East Kilbride, Glasgow
Steven R (2001) Wet Gas Metering. PhD Thesis, Department of Mechanical Engineering, Strathclyde University, Glasgow
Steven R (2006) Horizontally installed differential pressure meter wet gas flow performance review. In: Proceedings of 24th International North Sea Flow Meas Workshop, St Andrews. NEL, East Kilbride, Glasgow
Steven R (2008) Update on horizontally installed orifice plate meters with wet gas flows. CEESI, Colorado
Steven R, Britton C, Kinney J (2012) Horizontally installed 8″, 0.6 beta ratio Venturi meter wet natural gas flow response. In: Proceedings of 8th International Symposium Fluid Flow Meas, Colorado
Steven R, Britton C, Stewart D (2007) CEESI Wet gas JIP data release. EI/CEESI Wet gas metering seminar Aberdeen. CEESI, Colorado
Steven R, Kinney J, Britton C (2006) Liquid property and diameter effects on Venturi meters used with wet gas flows. In: Proceedings of 6th International Symposium Fluid Flow Meas, Querétaro, Mexico
Steven R, Stobie G, Hall A (2008) Further evaluation of the performance of horizontally installed orifice plate and cone differential pressure meters with wet gas flows. In: Proceedings of 7th International South East Asia Hydrocarbon Flow Meas Workshop, Kuala Lumpur
Steven R, Stobie G, Hall A, Priddy W (2011) Horizontally installed orifice plate meter response to wet gas flows. In: Proceedings of 29th International North Sea Flow Meas Workshop, Norway
Stewart DG (2002) The evaluation of dry gas meters in wet gas conditions. In: Report No 2002-100, on DTI Flow Programme Project FDMU07. National Engineering Laboratory, East Kilbride, Glasgow
Stewart DG (2003) Application of differential pressure meters to wet gas flow. In: Proceedings of 2nd International South East Asia Hydrocarbon Flow Meas Workshop, Kuala Lumpur

Chapter 12
Standards

Abstract This chapter describes the importance and value of ISO standards and lists the ISO and AGA/API standards that have been and are being used for differential-pressure flow measurement.

12.1 Introduction

It has been recognized for some time that standards are an important enabler of innovation and international technology transfer. They have an important role in national and international trade and in wealth creation. Standards contributed to about 13 % of the growth in labour productivity in the UK over the period 1948–2002 (Department of Trade and Industry 2005), and contribute £2.5 billion annually to the UK economy. Many other national economies must have benefited similarly.

By developing and adopting standards, barriers to trade are lowered, transaction costs are reduced and the operation of markets is improved through the smooth flow of goods and services. Standards enable compliance to be demonstrated, provide confidence in product performance, and help to eliminate or reduce disputes between companies and across borders. They help smaller firms by transferring technology from larger ones. Poorer countries may benefit similarly.

Flow measurement is an important area of standardization. For example UK natural gas demand is around £16 billion (€20 billion) annually, and most of this is measured (generally more than once) using the ISO standard for flow measurement using differential-pressure meters, ISO 5167. Avoidance of both inaccuracy through inadequate specification and excessive cost through over-specification is absolutely vital. Standards give consistency. Accuracy in taxation is of great importance to Government.

Standards produced largely in one sector but used in others enable technology to be transferred from one sector to another. Work done for the upstream or gas transmission sectors is valuable to refineries, power generation, aerospace and the chemical industries, and vice versa, and indeed more widely. Participation in standards-making is an opportunity not only to share knowledge but also to gain it.

M. Reader-Harris, *Orifice Plates and Venturi Tubes*,
Experimental Fluid Mechanics, DOI 10.1007/978-3-319-16880-7_12

Standardization is not only important for economic reasons. Accurate measurement of discharges, of produced water, for instance, is important to ensure a cleaner environment.

Standards are not infallible: they include engineering judgment where data are not available. Continuous improvement is always desirable.

12.2 ISO Standards

For many years industries have developed industry standards that cover their own area of work. Developing standards under the auspices of a national standards body produced a more balanced and acceptable result: this ensured a consistent approach and allowed all interested parties the opportunity to influence the standards that impinged on them. The British Standards Institution (BSI) in the UK, ANSI in the US and GOST in Russia are examples of national standards bodies. However, conflicting national standards created cross-border disputes, inconsistencies and the need to repeat tests in many countries. Creating a single, internationally acceptable, standard through ISO, the International Standards Organization, resolves the problem of inconsistent national standards. ISO standards are produced by more than 200 technical committees (TCs) and their subcommittees (SCs). The members of these committees are the national standards bodies, e.g. BSI. A TC covers a specific technical area, and its SCs are appointed to concentrate on specific areas within the remit of the TC. Under an SC (or under the TC directly) Working Groups (WGs) draft the text of the standards. The members of a WG are individual experts nominated by the national standards bodies. The national standards bodies vote on the standards.

From a search of the ISO website it appears that about 160 standards from about 40 technical committees have reference to fluid flow measurement. Two thirds of the ISO standards that include reference to fluid flow measurement are produced by four technical committees: TC 28 Petroleum products and lubricants, TC 30 Measurement of fluid flow in closed conduits, TC 113 Hydrometry, and TC 131 Fluid power systems. Many standards on natural gas properties are produced by TC 193 Natural gas.

TC 30's remit is to produce standards on flow measurement in pipes for general use. As a result standards produced by ISO/TC 30 are referred to by many other technical committees.

Outside TC 30 there are about 100 references to the standards of TC 30, the largest number being to ISO 5167. It is one of the flow-measurement standards that have very wide application. A list of the committees that made reference to TC 30 standards was given in Reader-Harris (2007). Work undertaken by TC 30 affects a wide range of industries. Users from one sector have to take account of those from other sectors in making standards.

To avoid duplication of ISO standards, the ideal arrangement between ISO/TC 30 and any other ISO TC is that a flow-measurement standard will be produced by TC 30, and standards from other TCs will refer to it, adding instructions for specific applications where required. An alternative, if necessary, is to have a standard in two parts: TC 30 produces a generic Part 1 covering all fluids, and the other TC (e.g. ISO/TC 28 or ISO/TC 193) produces a Part 2 containing any specific requirements for a particular fluid or application. This was proposed by TC 28 to TC 30; TC 30 accepted this plan and stated that Joint Working Groups should be set up to produce the two Parts; TC 30 then proposed this plan to TC 193. In practice this pattern of two-part standards has not taken place.

12.3 ISO/TC 30 Measurement of Fluid Flow in Closed Conduits

12.3.1 General

At the time of publication of this book ISO/TC 30 has three subcommittees, SCs, and one working group, WG:

- TC 30/SC 2 Pressure differential devices: this SC includes sonic nozzles
- TC 30/SC 5 Velocity and mass methods: this SC includes integration techniques, ultrasonic meters, electromagnetic meters, thermal mass meters and Coriolis meters
- TC 30/SC 7 Volume methods including water meters
- TC 30/WG 1 General methods: this WG includes uncertainties and calibration techniques

Generally the TC deals with strategy, and WGs within the three SCs and TC 30/WG 1 develop documents.

There used to be other SCs, which have been absorbed into those remaining.

For all TCs and SCs a national standards body (or an industry body on behalf of a national standards body) provides the secretariat: BSI provides the secretariat for TC 30, SC 2 and SC 7. The Swiss Association for Standardization, SNV, provides the secretariat for SC 5.

A list of standards for which TC 30/SC 2 is responsible is given in Appendix 12.A: the work of the rest of the TC is described in Reader-Harris (2008), but is not covered here.

12.3.2 ISO/TC 30/SC 2 Pressure Differential Methods

12.3.2.1 General

This committee has been very active for many years. Few standards have as significant financial consequences as ISO 5167. It and its predecessors are listed in Sect. 12.3.2.2.

NOTE 1 One reason for the continued use of earlier versions of ISO 5167 and other standards is the cost of updating flow computers and pipework. In practice because of the cost it is much rarer to change the pipework than to change the equations; if the equations are changed but the pipework is not, the errors that can be seen, for example, in Table 8.1 persist.

There are several technical reports, TRs, on steady flow that support ISO 5167:

- ISO/TR 9464 Guidelines for using ISO 5167 (see Sect. 12.3.2.3)
- ISO/TR 12767 Differential-pressure meters departing from ISO 5167 (see Sect. 12.3.2.4)
- ISO/TR 15377 Differential-pressure meters beyond the scope of ISO 5167 (see Sect. 12.3.2.5).

In each case these ISO TRs are based on previous British Standards. One consequence of the changes to ISO 5167 in 2003 was that these TRs had to be revised. Brief accounts of changes being made to standards for differential-pressure meters at that time are in Reader-Harris (2006).

There is also a TR covering pulsating flow, ISO/TR 3313 (see Sect. 12.3.2.6), and one covering differential-pressure meters in wet gas giving formulae for over-reading, ISO/TR 11583 (see Sect. 12.3.2.7).

In addition to this work on primary devices all differential-pressure meters depend on impulse lines: see Sect. 12.3.2.8 for ISO 2186.

NOTE 2 Where one standard refers to another, it is necessary to consider how revision of one of the standards would affect the other; it may be necessary that reference is made to a specific dated edition of a standard.

Future plans are covered in Sect. 12.3.2.9.

NOTE 3 One standard that is covered by SC 2 but is not a differential-pressure meter standard is ISO 9300 on sonic nozzles. The version that is current in 2014 was published in 2005. Improved quality of sonic nozzles and improved prediction of discharge coefficients may make possible in due course a revised standard with significantly improved uncertainties. Such a standard would make possible a new traceability route for some high-pressure gas facilities.

12.3.2.2 Differential-Pressure Flow Measurement Standards: ISO 5167 Etc.

There was a German flow measurement standard in the 1930s. The first British differential-pressure flow measurement standards were as follows:

- BS 1042:1943 Flow measurement
- BS 1042-1:1964 Methods for the measurement of fluid flow in pipes, Part 1 Orifice plates, nozzles and Venturi tubes

Numerous other parts and sections of BS 1042 followed, which formed the basis for many ISO standards and TRs later.

The ISO differential-pressure flow measurement standards to the present day are as follows:

- ISO/R 541-1967 Measurement of fluid flow by means of orifice plates and nozzles
- ISO/R 781-1968 Measurement of fluid flow by means of Venturi
- ISO 5167:1980 Measurement of fluid flow by means of orifice plates, nozzles and Venturi tubes inserted in circular cross-section conduits running full
  - This was the first to include the Stolz Equation (see Chap. 5).
- ISO 5167-1:1991 Measurement of fluid flow by means of pressure differential devices Part 1: Orifice plates, nozzles and Venturi tubes inserted in circular cross-section conduits running full
  - This made some minor revisions to the 1980 edition (see Stolz 1988).
  - The discharge-coefficient equation was changed to the Reader-Harris/Gallagher (1998) Equation by an amendment in 1998 (see Chap. 5):

    ISO 5167-1:1991/Amd. 1:1998 Measurement of fluid flow by means of pressure differential devices—Part 1: Orifice plates, nozzles and Venturi tubes inserted in circular cross-section conduits running full. AMENDMENT 1
- ISO 5167-1:2003 Measurement of fluid flow by means of pressure differential devices inserted in circular cross-section conduits running full—Part 1: General principles and requirements
- ISO 5167-2:2003 Measurement of fluid flow by means of pressure differential devices inserted in circular cross-section conduits running full—Part 2: Orifice plates
- ISO 5167-3:2003 Measurement of fluid flow by means of pressure differential devices inserted in circular cross-section conduits running full—Part 3: Nozzles and Venturi nozzles
- ISO 5167-4:2003 Measurement of fluid flow by means of pressure differential devices inserted in circular cross-section conduits running full—Part 4: Venturi tubes
  - These four parts are the revision of ISO 5167 that is current in 2014: it made a significant change to orifice metering in particular. One of the changes was to divide ISO 5167 into four parts as above. A summary of the changes to ISO 5167 that were made in 2003 is given in Reader-Harris (2006).

NOTE As ISO 5167 Parts 1 to 4 (2013) have been adopted as European (EN ISO) standards, they are also national standards in the 33 countries (in 2014) forming the membership of CEN, the European Committee for Standardization.

12.3.2.3 ISO/TR 9464 Guidelines for Using ISO 5167

The history of the Guidelines is as follows:

- BS 1042-1-1.4:1984 Measurement of fluid flow in closed conduits. Pressure differential devices. Guide to the use of devices specified in Sects. 1.1 and 1.2
- BS 1042-1.4:1992 Measurement of fluid flow in closed conduits. Pressure differential devices. Guide to the use of devices specified in Sects. 1.1 and 1.2
- ISO/TR 9464:1998 Guide to the use of ISO 5167-1:1991
- ISO/TR 9464:2008 Guidelines for the use of ISO 5167:2003
  - This is the revision of ISO/TR 9464 that is current in 2014: probably the most significant addition to ISO/TR 9464 in 2008 was the inclusion of an approximate formula for μ_{JT}, the Joule-Thomson coefficient.

12.3.2.4 ISO/TR 12767 Differential-Pressure Meters Departing from ISO 5167

ISO/TR 12767 covers the consequences of using a differential-pressure meter that in most respects meets the requirements of ISO 5167 but fails to do so in one respect. Its history is as follows:

- BS 1042-3:1965. Methods of measurement of fluid flow in closed conduits. Guide to the effects of departure from the methods in Part 1
- BS 1042-1-1.5:1987 Measurement of fluid flow in closed conduits. Pressure differential devices. Guide to the effect of departure from the conditions specified in Sect. 1.1
- BS 1042-1-1.5:1997 Measurement of fluid flow in closed conduits. Pressure differential devices. Guide to the effect of departure from the conditions specified in BS EN ISO 5167-1
- ISO/TR 12767:1998 Measurement of fluid flow by means of pressure- differential devices—Guidelines on the effect of departure from the specifications and operating conditions given in ISO 5167-1
- ISO/TR 12767:2007 Measurement of fluid flow by means of pressure differential devices—Guidelines on the effect of departure from the specifications and operating conditions given in ISO 5167
  - This is the revision of ISO/TR 12767 that is current in 2014: new information on the effect of grease and of short upstream lengths on orifice-plate discharge coefficients was included in 2007.

12.3.2.5 ISO/TR 15377 Differential-Pressure Meters Beyond the Scope of ISO 5167

ISO/TR 15377 covers orifice meters and nozzles with drain holes, small orifice meters with diameter less than 50 mm, meters with no upstream and/or no downstream pipeline, conical-entrance, quarter-circle and eccentric orifice plates; all these except eccentric orifice plates were covered in BS 1042-1:1964. Since then the history is as follows:

- BS 1042-1-1.2:1984 Measurement of fluid flow in closed conduits. Pressure differential devices. Specification for square-edged orifice plates and nozzles (with drain holes, in pipes below 50 mm diameter, as inlet and outlet devices) and other orifice plates and Borda inlets
- BS 1042-1-1.2:1989 Measurement of fluid flow in closed conduits. Pressure differential devices. Specification for square-edged orifice plates and nozzles (with drain holes, in pipes below 50 mm diameter, as inlet and outlet devices) and other orifice plates
- ISO/TR 15377:1998 Measurement of fluid flow by means of pressure—differential devices—Guidelines for the specification of nozzles and orifice plates beyond the scope of ISO 5167-1
- ISO/TR 15377:2007 Measurement of fluid flow by means of pressure differential devices—Guidelines for the specification of orifice plates, nozzles and Venturi tubes beyond the scope of ISO 5167
  - This is the revision of ISO/TR 15377 that is current in 2014: in 2007 there was a new section on Venturi tubes with convergent angle 10.5°. The changes that were made to ISO/TR 15377 in 2007 are described in Reader-Harris et al. (2006).

12.3.2.6 ISO/TR 3313 Pulsating Flow

ISO/TR 3313 covers pulsating flow. Its history is as follows:

- ISO/TR 3313:1974 Measurement of pulsating fluid flow in a pipe by means of orifice plates, nozzles or Venturi tubes, in particular in the case of sinusoidal or square wave intermittent periodic-type fluctuations
- ISO/TR 3313:1992 Measurement of fluid flow in closed conduits. Pressure differential devices. Method of measurement of pulsating fluid flow in a pipe, by means of orifice plates, nozzles and Venturi tubes
- ISO/TR 3313:1998 Measurement of fluid flow in closed conduits—Guidelines on the effects of flow pulsations on flow-measurement instruments
  - This also covers the effect of pulsating flow on turbine meters and vortex meters.

  - This has been withdrawn by ISO but is still current as BS ISO TR 3313:1998, and is likely to be reinstated as an ISO standard or Technical Specification (TS).

12.3.2.7 ISO/TR 11583 Wet Gas

ISO/TR 11583 covering differential-pressure meters in wet gas (giving formulae for over-reading) is a technical report supplementary to ISO 5167:

- ISO/TR 11583:2012 Measurement of wet gas flow by means of pressure differential devices inserted in circular cross-section conduits

NOTE ISO/TC 193 is about to publish ISO/TR 12748: this will contain much useful material on measuring wet natural gas (see Sect. 11.1). The recently balloted version is ISO/DTR 12748:

- ISO/DTR 12748:2014 Natural gas—wet gas flow measurement in natural gas operations

12.3.2.8 ISO 2186 Impulse Lines

The history of ISO standards on impulse lines is as follows:

- ISO 2186:1973 Fluid flow in closed conduits—Connections for pressure signal transmissions between primary and secondary elements
- ISO 2186:2007 Fluid flow in closed conduits—Connections for pressure signal transmissions between primary and secondary elements
  - This is the revision of ISO 2186 that is current in 2014: in this revision configurations that are no longer used were removed and shorter impulse-line lengths were encouraged.

12.3.2.9 Priorities for the Future as Seen in 2014

Work is being undertaken on ISO 5167-5 to cover cone meters. This will be a major extension of the scope of ISO 5167.

A part of ISO 5167 to cover wedge meters is being proposed.

Another important area is flow through orifice plates with drain holes. The existing text in ISO/TR 15377 is based on a theoretical model which experimental work described in Chap. 10 has shown not to be applicable. An amendment will be required.

Projects on the effect of contaminated orifice plates, some work paid for by the British Government under the Engineering and Flow Programme and some work sponsored by National Grid, will, in due course, make possible an improvement of the text in ISO/TR 12767.

There are small but significant differences between ISO 5167 and API MPMS 14.3 (some examples are given in Sects. 2.2.3, 2.3.2.2, 2.3.5 and 2.6 and in Chaps. 5 and 6) for flow measurement using orifice plates, and one day harmonization may prove possible.

12.3.3 The TC Itself

12.3.3.1 General

This committee sets the policy for TC 30 and its subcommittees.

It also looks at interactions between TC 30 and other ISO TCs (including TC 28 and TC 193), the ISO Technical Management Board, and important outside organizations such as the American Petroleum Institute (API) and the Energy Institute (EI).

One issue that is of great importance to this technical committee as to other technical committees is project management. The timely production of ISO standards is important to users, so that the latest wisdom is available, but also to committee members, so that their time is used efficiently. One way in which project management has been improved was by the introduction of a Chairman's Advisory Group, consisting of the Chairman and Secretary of TC 30 and of each of its subcommittees. This group has met in Geneva from time to time to check up on progress on each project and to suggest ways of improving progress on projects that are running behind schedule. By meeting in Geneva the Group has had the opportunity to meet the ISO Technical Programme Manager and Technical Editors and thus to reduce delays due to poor communication. The Chairman's Advisory Group has proved to be valuable in terms of project management and its existence was formalized by resolution of the committee in 2006. It was resolved to put it on a more official basis. To improve liaison the chairmen of TC 28 and TC 193 were invited to the last meeting. However, there has not been a meeting recently.

The future programme of work for this technical committee is important. Continuing the present work of the committee is important, but in addition to the work in existing areas new standards will be required in new areas. Especially in areas outside those traditionally covered by TC 30 there is a danger that standards will be developed by other committees that duplicate those of TC 30. To avoid one possibility of this, improved contact with the Carbon capture and storage committee, ISO/TC 265, has been achieved. An analysis of other committees outside TC 30 that have flow measurement standards was carried out, and key contacts have been made. Harmonization is important; avoiding duplication is also efficient in terms of experts' time and costs.

12.3.3.2 Priorities for the Future as Seen in 2014

There is always a need to establish a programme of work covering new topics. Hydrogen, medical gases and cryogenics may require extension of the current flow measurement standards.

12.4 AGA/API Standards

The first American standards for orifice plates were produced by the American Gas Association (AGA), formerly the Natural Gas Association: Report No 1 was issued in 1930. Report No 2 was published in 1935 and intended to be a supplement to Report No 1: Report No 2 made it possible to use orifice plates over a much wider range of conditions than had been possible under Report No 1. The range of conditions was again extended, and AGA Report No 3 was issued in 1955. Report No 3 was revised in 1969 and adopted by the American Petroleum Institute (API) in 1975 as Chapter 14.3 of the API Manual of Petroleum Measurement Standards. This standard was further revised by AGA, API and the Gas Processors Association (GPA) in 1983, and on receiving ANSI approval it became

- Orifice metering of natural gas and other related hydrocarbon fluids. ANSI/API 2530, 2nd Edition: 1985.

ANSI/API 2530-1985 was then replaced by the API Manual of Petroleum Measurement Standards, Chapter 14—Natural Gas Fluids Measurement, Section 3—Concentric, Square-Edged Orifice Meters, consisting of four parts (all jointly published with AGA as parts of AGA Report No 3: the report number remained 3 for convenience):

- Part 1—General Equations and Uncertainty Guidelines. MPMS Chapter 14, Section 3, Part 1, 3rd Edition: 1990
  - This introduced the Reader-Harris/Gallagher (RG) Equation (see Chap. 5).
- Part 2—Specification and Installation Requirements. MPMS Chapter 14, Section 3, Part 2, 3rd Edition: 1991
- Part 3—Natural Gas Applications. MPMS Chapter 14, Section 3, Part 3, 3rd Edition: 1992
- Part 4—Background, Development, Implementation Procedures and Subroutine Documentation. MPMS Chapter 14, Section 3, Part 4, 3rd Edition: 1992

The first three parts have been revised (all jointly published with AGA as parts of AGA Report No 3):

- Orifice Metering of Natural Gas and Other Related Hydrocarbon Fluids—Concentric, Square-edged Orifice Meters, Part 1—General Equations and Uncertainty Guidelines. API MPMS, Chapter 14.3.1, 4th Edition: 2012 (Errata 2013)

- This part introduced a new expansion-factor (expansibility-factor) equation (see Chap. 6).

- Natural Gas Fluids Measurement. Part 2—Specification and Installation Requirements—Concentric, Square-edged Orifice Meters. API MPMS Chapter 14.3.2, 4th Edition: 2000
  - This part introduced new required straight lengths (see Chap. 8).
- Natural Gas Fluids Measurement: Orifice Metering of Natural Gas and Other Related Hydrocarbon Fluids—Concentric, Square-Edged Orifice Meters—Part 3: Natural Gas Applications, API MPMS, Chapter 14.3.3, 4th Edition: 2013

12.5 Conclusions

ISO/TC 30 continues to have an important work programme. There will be new and improved differential-pressure meter standards which have great financial significance. Some new areas, for example hydrogen, may require an extension to existing standards. In addition to these areas new standards may in time be required in new areas, for example microfluidics. In due course there may be developments in sonic nozzles that lead to improved standards and thus to a significant improvement in gas-measurement uncertainties. Harmonization of standards with those of other organizations is an important concern as this both avoids discrepancies and leads to efficiency in their production.

Appendix 12.A: The Standards of ISO/TC 30/SC 2

Standards in 2014

ISO 2186:2007 Fluid flow in closed conduits—Connections for pressure signal transmissions between primary and secondary elements

ISO/TR 3313:1998 Measurement of fluid flow in closed conduits—Guidelines on the effects of flow pulsations on flow-measurement instruments (withdrawn by ISO but likely to be reissued)

ISO 5167-1:2003 Measurement of fluid flow by means of pressure differential devices inserted in circular cross-section conduits running full—Part 1: General principles and requirements

ISO 5167-2:2003 Measurement of fluid flow by means of pressure differential devices inserted in circular cross-section conduits running full—Part 2: Orifice plates

ISO 5167-3:2003 Measurement of fluid flow by means of pressure differential devices inserted in circular cross-section conduits running full—Part 3: Nozzles and Venturi nozzles

ISO 5167-4:2003 Measurement of fluid flow by means of pressure differential devices inserted in circular cross-section conduits running full—Part 4: Venturi tubes

ISO 9300:2005 Measurement of gas flow by means of critical flow Venturi nozzles

ISO/TR 9464:2008 Guidelines for the use of ISO 5167:2003

ISO/TR 11583:2012 Measurement of wet gas flow by means of pressure differential devices inserted in circular cross-section conduits

ISO/TR 12767:2007 Measurement of fluid flow by means of pressure differential devices—Guidelines on the effect of departure from the specifications and operating conditions given in ISO 5167

ISO/TR 15377:2007 Measurement of fluid flow by means of pressure differential devices—Guidelines for the specification of orifice plates, nozzles and Venturi tubes beyond the scope of ISO 5167

Standards under Development (October 2014)

ISO/CD 5167-5 Measurement of fluid flow by means of pressure differential devices inserted in circular cross-section conduits running full—Part 5: Cone meters

References

Department of Trade and Industry (2005) The empirical economics of standards. DTI, London. DTI Economics Paper No 12. Available as www.berr.gov.uk/files/file9655.pdf

Reader-Harris MJ (2006) ISO standards on differential-pressure flow measurement: developments. In: Proceedings 5th International South East Asia Hydrocarbon Flow Meas Workshop, Kuala Lumpur

Reader-Harris MJ (2007) ISO flow measurement standards—report on the ISO/TC 30 meeting in November 2006. Flow Meas Instrum 18(3–4):114–120

Reader-Harris MJ (2008) ISO/TC 30 standards—autumn 2008. In: Proceedings 26th International North Sea Flow Meas Workshop. St Andrews. NEL, East Kilbride, Glasgow

Reader-Harris MJ, Gibson J, Hodges D, Rushworth R (2006) The revision of ISO/TR 15377 (orifice plates, nozzles and Venturi tubes beyond the scope of ISO 5167). In: Proceedings 6th International Symposium Fluid Flow Meas, Querétaro, Mexico

Stolz J (1988) The first revision of ISO 5167. In: Proceedings 6th North Sea Flow Meas Workshop. National Engineering Laboratory, East Kilbride, Glasgow: paper 3.1

Index

M. Reader-Harris, *Orifice Plates and Venturi Tubes*,
Experimental Fluid Mechanics, DOI 10.1007/978-3-319-16880-7

MIX
Papier aus verantwortungsvollen Quellen
Paper from responsible sources
FSC® C105338

If you have any concerns about our products,
you can contact us on
ProductSafety@springernature.com

In case Publisher is established outside the EU,
the EU authorized representative is:
Springer Nature Customer Service Center GmbH
Europaplatz 3, 69115 Heidelberg, Germany

Printed by Libri Plureos GmbH
in Hamburg, Germany